·数据库技术丛书·

MySQL 9
从入门到性能优化

（视频教学版）

王英英 著

清华大学出版社
北京

内 容 简 介

MySQL是流行的关系数据库管理系统之一，由于其体积小、速度快、总体拥有成本低、开放源码等特点，一般中小型企业甚至大型互联网企业的应用开发都选择MySQL作为数据库。本书注重实战操作，详解MySQL 9数据库的操作、运维、优化和设计方法，配套示例源码、PPT课件、同步教学视频、作者微信群答疑服务，帮助读者快速掌握MySQL数据库。

本书共分20章，主要内容包括MySQL的安装与配置，数据库和数据表基本操作，数据类型和运算符，MySQL函数，查询数据，插入、更新与删除数据，索引的设计和使用，存储过程和函数，视图，MySQL触发器，数据备份与恢复，MySQL日志，MySQL权限与安全管理，SQL性能优化，MySQL服务器性能优化，MySQL性能监控，提升MySQL数据库的性能，MySQL终极优化实战，企业人事管理系统数据库设计，在线购物系统数据库设计。

本书适合MySQL初学者、MySQL应用开发人员、MySQL数据库管理员和系统运维人员阅读，同时也可以作为高等院校或高职高专院校MySQL数据库课程的教学用书。

本书封面贴有清华大学出版社防伪标签，无标签者不得销售。
版权所有，侵权必究。举报：010-62782989，beiqinquan@tup.tsinghua.edu.cn。

图书在版编目（CIP）数据

MySQL 9从入门到性能优化：视频教学版 / 王英英著. -- 北京：清华大学出版社，2024.9. --（数据库技术丛书）. -- ISBN 978-7-302-67191-6

Ⅰ. TP311.138

中国国家版本馆CIP数据核字第2024Z5B997号

责任编辑：夏毓彦
封面设计：王　翔
责任校对：闫秀华
责任印制：刘海龙

出版发行：清华大学出版社
网　　址：https://www.tup.com.cn, https://www.wqxuetang.com
地　　址：北京清华大学学研大厦A座
邮　　编：100084
社 总 机：010-83470000
邮　　购：010-62786544
投稿与读者服务：010-62776969, c-service@tup.tsinghua.edu.cn
质量反馈：010-62772015, zhiliang@tup.tsinghua.edu.cn
印 装 者：北京同文印刷有限责任公司
经　　销：全国新华书店
开　　本：190mm×260mm
印　　张：28.5
字　　数：769千字
版　　次：2024年9月第1版
印　　次：2024年9月第1次印刷
定　　价：139.00元

产品编号：105863-01

前　　言

本书是介绍MySQL数据库管理系统的一本高质量的图书。目前国内业界对MySQL数据库技术的需求非常旺盛，各大知名企业都在高薪招聘技术能力强的MySQL应用开发人员和管理人员。本书根据这样的需求，为初学者量身打造，全面而详尽地讲解MySQL的相关知识，并且通过对实例的操作与分析，引领读者快速学习和掌握MySQL数据库的操作、运维、优化和设计技术。

本书特色

内容全面：涵盖MySQL的所有实用知识点，讲解由浅入深，帮助读者循序渐进地掌握MySQL数据库管理、优化和设计技术。

图文并茂：注重操作，图文并茂，并且在案例的讲解过程中，每一个操作均有对应步骤和过程说明。这种图文结合的方式使读者在学习过程中能够直观、清晰地看到操作的过程及其效果，便于读者更快地理解和掌握相关知识。

易学易用：颠覆传统"看"书的观念，变成一本能"操作"的图书。

案例丰富：把知识点融汇于系统的案例实训当中，并且结合实战进行讲解和拓展，进而达到"知其然，并知其所以然"的效果。

提示技巧：本书对读者在学习过程中可能会遇到的疑难问题，以"提示"和"技巧"的形式进行说明，以免读者在学习的过程中走弯路。

本书配套资源下载

本书配套示例源码、PPT课件、同步教学视频、作者微信群答疑服务，请读者用自己的微信扫描下面的二维码下载。如果学习本书的过程中发现问题或疑问，可发送邮件至booksaga@163.com，邮件主题为"MySQL 9从入门到性能优化"。

读者对象

- MySQL 数据库初学者。
- MySQL 应用开发人员。
- MySQL 数据库管理员。
- 系统运维人员。
- 大数据管理人员。
- SQL 语言初学者。
- 高等院校或高职高专院校相关专业的学生。

鸣谢

本书由王英英主创，部分内容由刘增杰编写。虽然本书倾注了笔者的大量心血，但由于水平有限、时间仓促，书中难免存在疏漏之处，敬请读者谅解。

笔 者

2024年8月

目 录

第1章 MySQL的安装与配置 ……………… 1
- 1.1 什么是MySQL ………………………… 1
  - 1.1.1 客户端/服务器软件 ……………… 1
  - 1.1.2 MySQL版本 ……………………… 2
- 1.2 Windows平台下安装与配置MySQL …… 2
  - 1.2.1 安装MySQL ……………………… 2
  - 1.2.2 配置MySQL ……………………… 6
- 1.3 启动服务并登录MySQL数据库 ……… 11
  - 1.3.1 启动MySQL服务 ………………… 11
  - 1.3.2 登录MySQL数据库 ……………… 12
  - 1.3.3 配置Path变量 …………………… 13
- 1.4 MySQL常用的图形化管理工具 ……… 15
- 1.5 Linux平台下安装与配置MySQL ……… 16
  - 1.5.1 Linux操作系统下的MySQL版本介绍 …………………………… 16
  - 1.5.2 安装和配置MySQL的RPM包 ……………………………… 17

第2章 数据库和数据表的基本操作 ……… 19
- 2.1 创建数据库 …………………………… 19
- 2.2 删除数据库 …………………………… 20
- 2.3 创建数据表 …………………………… 21
  - 2.3.1 创建表的语法形式 ……………… 21
  - 2.3.2 使用主键约束 …………………… 22
  - 2.3.3 使用外键约束 …………………… 23
  - 2.3.4 使用非空约束 …………………… 25
  - 2.3.5 使用唯一性约束 ………………… 25
  - 2.3.6 使用默认约束 …………………… 26
  - 2.3.7 设置表的属性值自动增加 ……… 26
- 2.4 查看数据表结构 ……………………… 27
  - 2.4.1 查看表基本结构的语句 DESCRIBE …………………………… 27
  - 2.4.2 查看表详细结构的语句 SHOW CREATE TABLE ……… 28
- 2.5 修改数据表 …………………………… 29
  - 2.5.1 修改表名 ………………………… 29
  - 2.5.2 修改字段的数据类型 …………… 30
  - 2.5.3 修改字段名 ……………………… 31
  - 2.5.4 添加字段 ………………………… 32
  - 2.5.5 删除字段 ………………………… 34
  - 2.5.6 修改字段的排列位置 …………… 35
  - 2.5.7 删除表的外键约束 ……………… 36
- 2.6 删除数据表 …………………………… 37
  - 2.6.1 删除没有被关联的表 …………… 37
  - 2.6.2 删除被其他表关联的主表 ……… 38

第3章 数据类型和运算符 ………………… 40
- 3.1 MySQL的数据类型 …………………… 40
  - 3.1.1 整数类型 ………………………… 40
  - 3.1.2 小数类型 ………………………… 41
  - 3.1.3 日期与时间类型 ………………… 43
  - 3.1.4 文本字符串类型 ………………… 53
  - 3.1.5 二进制字符串类型 ……………… 58
- 3.2 如何选择数据类型 …………………… 60
- 3.3 运算符 ………………………………… 62
  - 3.3.1 运算符概述 ……………………… 62
  - 3.3.2 算术运算符 ……………………… 63
  - 3.3.3 比较运算符 ……………………… 64
  - 3.3.4 逻辑运算符 ……………………… 71

3.3.5 位运算符 ·············· 73
3.3.6 运算符的优先级 ·········· 76

第4章 MySQL函数 ·············· 77

4.1 MySQL函数简介 ·············· 77
4.2 数学函数 ·············· 77
    4.2.1 绝对值函数ABS(x)和返回圆周率的函数PI() ·········· 77
    4.2.2 平方根函数SQRT(x)和求余函数MOD(x,y) ·········· 78
    4.2.3 获取整数的函数CEIL(x)、CEILING(x)和FLOOR(x) ·········· 79
    4.2.4 获取随机数的函数RAND()和RAND(x) ·········· 79
    4.2.5 函数ROUND(x)、ROUND(x,y)和TRUNCATE(x,y) ·········· 80
    4.2.6 符号函数SIGN(x) ·········· 81
    4.2.7 幂运算函数POW(x,y)、POWER(x,y)和EXP(x) ·········· 81
    4.2.8 对数运算函数LOG(x)和LOG10(x) ·········· 82
    4.2.9 角度与弧度相互转换的函数RADIANS(x)和DEGREES(x) ·········· 82
    4.2.10 正弦函数SIN(x)和反正弦函数ASIN(x) ·········· 83
    4.2.11 余弦函数COS(x)和反余弦函数ACOS(x) ·········· 83
    4.2.12 正切函数、反正切函数和余切函数 ·········· 84
4.3 字符串函数 ·············· 85
    4.3.1 计算字符串的字符数的函数和计算字符串长度的函数 ·········· 85
    4.3.2 合并字符串函数CONCAT(s1,s2,...)、CONCAT_WS(x,s1,s2,...) ·········· 85
    4.3.3 替换字符串的函数INSERT(s1,x,len,s2) ·········· 86
    4.3.4 字母大小写转换函数 ·········· 87
    4.3.5 获取指定长度的字符串的函数LEFT(s,n)和RIGHT(s,n) ·········· 87
    4.3.6 填充字符串的函数LPAD(s1,len,s2)和RPAD(s1,len,s2) ·········· 88
    4.3.7 删除空格的函数LTRIM(s)、RTRIM(s)和TRIM(s) ·········· 88
    4.3.8 删除指定字符串的函数TRIM(s1 FROM s) ·········· 89
    4.3.9 重复生成字符串的函数REPEAT(s,n) ·········· 90
    4.3.10 空格函数SPACE(n)和替换函数REPLACE(s,s1,s2) ·········· 90
    4.3.11 比较字符串大小的函数STRCMP(s1,s2) ·········· 91
    4.3.12 获取子串的函数SUBSTRING(s,n,len)和MID(s,n,len) ·········· 91
    4.3.13 匹配子字符串开始位置的函数 ·········· 92
    4.3.14 字符串逆序的函数REVERSE(s) ·········· 92
    4.3.15 返回指定位置的字符串的函数 ·········· 92
    4.3.16 返回指定字符串位置的函数FIELD(s,s1,s2,...,sn) ·········· 93
    4.3.17 返回子字符串位置的函数FIND_IN_SET(s1,s2) ·········· 93
    4.3.18 选取字符串的函数MAKE_SET(x,s1,s2,...,sn) ·········· 94
4.4 日期和时间函数 ·············· 94
    4.4.1 获取当前日期的函数和获取当前时间的函数 ·········· 94
    4.4.2 获取当前日期和时间的函数 ·········· 95
    4.4.3 UNIX时间戳函数 ·········· 95
    4.4.4 返回UTC日期的函数和返回UTC时间的函数 ·········· 96
    4.4.5 获取月份的函数MONTH(date)和MONTHNAME(date) ·········· 97

| 4.4.6 | 获取星期的函数DAYNAME(d)、DAYOFWEEK(d)和WEEKDAY(d) ·················· 97
| 4.4.7 | 获取星期数的函数WEEK(d)和WEEKOFYEAR(d) ·················· 98
| 4.4.8 | 获取天数的函数DAYOFYEAR(d)和DAYOFMONTH(d) ·················· 99
| 4.4.9 | 获取年份、季度、小时、分钟和秒钟的函数 ·················· 99
| 4.4.10 | 获取日期的指定值的函数EXTRACT(type FROM date) ·················· 100
| 4.4.11 | 时间和秒数转换的函数 ·················· 101
| 4.4.12 | 计算日期和时间的函数 ·················· 101
| 4.4.13 | 将日期和时间格式化的函数 ·················· 104

4.5 条件判断函数 ·················· 107
    4.5.1 IF()函数 ·················· 107
    4.5.2 IFNULL()函数 ·················· 107
    4.5.3 CASE()函数 ·················· 108

4.6 系统信息函数 ·················· 109
    4.6.1 获取MySQL版本号、连接数和数据库名的函数 ·················· 109
    4.6.2 获取用户名的函数 ·················· 110
    4.6.3 获取字符串的字符集和排序方式的函数 ·················· 111
    4.6.4 获取最后一个自动生成的ID值的函数 ·················· 111

4.7 加密函数 ·················· 113
    4.7.1 加密函数MD5(str) ·················· 113
    4.7.2 加密函数SHA(str) ·················· 113
    4.7.3 加密函数SHA2(str, hash_length) ·················· 114

4.8 其他函数 ·················· 114
    4.8.1 格式化函数FORMAT(x,n) ·················· 114
    4.8.2 不同进制的数字进行转换的函数 ·················· 115
    4.8.3 IP地址与数字相互转换的函数 ·················· 115
    4.8.4 加锁函数和解锁函数 ·················· 116
    4.8.5 重复执行指定操作的函数 ·················· 116
    4.8.6 改变字符集的函数 ·················· 117
    4.8.7 改变数据类型的函数 ·················· 118

4.9 窗口函数 ·················· 118

第5章 查询数据 ·················· 120

5.1 基本查询语句 ·················· 120
5.2 单表查询 ·················· 122
    5.2.1 查询所有字段 ·················· 122
    5.2.2 查询指定字段 ·················· 123
    5.2.3 查询指定记录 ·················· 125
    5.2.4 带IN关键字的查询 ·················· 126
    5.2.5 带BETWEEN AND的范围查询 ·················· 127
    5.2.6 带LIKE的字符匹配查询 ·················· 128
    5.2.7 查询空值 ·················· 130
    5.2.8 带AND的多条件查询 ·················· 131
    5.2.9 带OR的多条件查询 ·················· 132
    5.2.10 查询结果不重复 ·················· 133
    5.2.11 对查询结果排序 ·················· 134
    5.2.12 分组查询 ·················· 138
    5.2.13 使用LIMIT限制查询结果的数量 ·················· 142

5.3 使用聚合函数查询 ·················· 143
    5.3.1 COUNT()函数 ·················· 144
    5.3.2 SUM()函数 ·················· 145
    5.3.3 AVG()函数 ·················· 145
    5.3.4 MAX()函数 ·················· 146
    5.3.5 MIN()函数 ·················· 147

5.4 连接查询 ·················· 148
    5.4.1 内连接查询 ·················· 148
    5.4.2 外连接查询 ·················· 151
    5.4.3 复合条件连接查询 ·················· 153

5.5 子查询 ·················· 154

5.5.1　带ANY、SOME关键字的
　　　　　子查询……………………154
　　5.5.2　带ALL关键字的子查询………155
　　5.5.3　带EXISTS关键字的子查询……155
　　5.5.4　带IN关键字的子查询…………156
　　5.5.5　带比较运算符的子查询………158
5.6　合并查询结果………………………159
5.7　为表和字段取别名…………………162
　　5.7.1　为表取别名……………………162
　　5.7.2　为字段取别名…………………163
5.8　使用正则表达式查询………………164
　　5.8.1　查询以特定字符或字符串开头
　　　　　的记录……………………………165
　　5.8.2　查询以特定字符或字符串结尾
　　　　　的记录……………………………166
　　5.8.3　用符号"."来替代字符串中的
　　　　　任意一个字符……………………166
　　5.8.4　使用"*"和"+"来匹配多个
　　　　　字符………………………………167
　　5.8.5　匹配指定字符串………………167
　　5.8.6　匹配指定字符中的任意一个…168
　　5.8.7　匹配指定字符以外的字符……169
　　5.8.8　使用{n,}或者{n,m}来指定
　　　　　字符串连续出现的次数…………170
5.9　通用表表达式………………………171

第6章　插入、更新与删除数据…………175

6.1　插入数据……………………………175
　　6.1.1　为表的所有字段插入数据……175
　　6.1.2　为表的指定字段插入数据……177
　　6.1.3　同时插入多条数据……………178
　　6.1.4　将查询结果插入表中…………180
6.2　更新数据……………………………181
6.3　删除数据……………………………183
6.4　为表增加计算列……………………185
6.5　DDL的原子化………………………186

第7章　索引的设计和使用………………188

7.1　索引简介……………………………188
　　7.1.1　索引的含义和特点……………188
　　7.1.2　索引的分类……………………189
　　7.1.3　索引的设计原则………………190
7.2　创建索引……………………………190
　　7.2.1　创建表的时候创建索引………190
　　7.2.2　在已经存在的表上创建索引…196
7.3　删除索引……………………………202
7.4　使用降序索引………………………204

第8章　存储过程和存储函数……………207

8.1　创建存储过程和存储函数…………207
　　8.1.1　创建存储过程…………………207
　　8.1.2　创建存储函数…………………209
　　8.1.3　变量的使用……………………210
　　8.1.4　定义条件和处理程序…………212
　　8.1.5　光标的使用……………………214
　　8.1.6　流程控制的使用………………216
8.2　调用存储过程和存储函数…………220
　　8.2.1　调用存储过程…………………220
　　8.2.2　调用存储函数…………………221
8.3　查看存储过程和存储函数…………221
　　8.3.1　使用SHOW STATUS语句
　　　　　查看存储过程和存储函数
　　　　　的状态……………………………222
　　8.3.2　使用SHOW CREATE语句
　　　　　查看存储过程和存储函数
　　　　　的定义……………………………222
　　8.3.3　从information_schema.Routines
　　　　　表中查看存储过程和存储函数的
　　　　　信息………………………………223
8.4　修改存储过程和存储函数…………224
8.5　删除存储过程和存储函数…………226
8.6　全局变量的持久化…………………226

第9章 视图 …………………… 228

- 9.1 视图概述 ………………………228
  - 9.1.1 视图的含义 ………………228
  - 9.1.2 视图的作用 ………………229
- 9.2 创建视图 ………………………229
  - 9.2.1 创建视图的语法形式 ……230
  - 9.2.2 在单表上创建视图 ………230
  - 9.2.3 在多表上创建视图 ………231
- 9.3 查看视图 ………………………232
  - 9.3.1 使用DESCRIBE语句查看视图基本信息 ………………232
  - 9.3.2 使用SHOW TABLE STATUS语句查看视图基本信息 ……232
  - 9.3.3 使用SHOW CREATE VIEW语句查看视图详细信息 ……234
  - 9.3.4 在views表中查看视图详细信息 ………………………234
- 9.4 修改视图 ………………………235
  - 9.4.1 使用CREATE OR REPLACE VIEW语句修改视图 ………235
  - 9.4.2 使用ALTER语句修改视图 …236
- 9.5 更新视图 ………………………237
- 9.6 删除视图 ………………………239

第10章 MySQL触发器 ………… 241

- 10.1 创建触发器 …………………241
  - 10.1.1 创建只有一个执行语句的触发器 ……………………241
  - 10.1.2 创建有多个执行语句的触发器 ……………………242
- 10.2 查看触发器 …………………244
  - 10.2.1 利用SHOW TRIGGERS语句查看触发器信息 ……………244
  - 10.2.2 在triggers表中查看触发器信息 ……………………246
- 10.3 触发器的使用 ………………247
- 10.4 删除触发器 …………………248

第11章 数据备份与恢复 ………… 249

- 11.1 数据备份 ……………………249
  - 11.1.1 使用mysqldump命令备份数据 ………………………249
  - 11.1.2 直接复制整个数据库目录 …254
  - 11.1.3 使用MySQLhotcopy工具快速备份 …………………255
- 11.2 数据恢复 ……………………255
  - 11.2.1 使用mysql命令恢复数据 …256
  - 11.2.2 直接复制到数据库目录 …256
  - 11.2.3 mysqlhotcopy快速恢复 …257
- 11.3 数据库迁移 …………………257
  - 11.3.1 相同版本的MySQL数据库之间的迁移 ………………257
  - 11.3.2 不同版本的MySQL数据库之间的迁移 ………………258
  - 11.3.3 不同数据库之间的迁移 …258
- 11.4 数据的导出和导入 …………259
  - 11.4.1 使用SELECT…INTO OUTFILE导出文本文件 …………………259
  - 11.4.2 使用mysqldump命令导出文本文件 …………………262
  - 11.4.3 使用mysql命令导出文本文件 …………………………265
  - 11.4.4 使用LOAD DATA INFILE方式导入文本文件 …………267
  - 11.4.5 使用mysqlimport命令导入文本文件 …………………269

第12章 MySQL日志 …………… 272

- 12.1 日志简介 ……………………272
- 12.2 二进制日志 …………………273
  - 12.2.1 启动和设置二进制日志 …273
  - 12.2.2 查看二进制日志 …………274
  - 12.2.3 删除二进制日志 …………276
  - 12.2.4 使用二进制日志恢复数据库 …………………………277
  - 12.2.5 暂时停止二进制日志功能 …278

12.3	错误日志	278
	12.3.1 启动和设置错误日志	278
	12.3.2 查看错误日志	279
	12.3.3 删除错误日志	280
12.4	通用查询日志	280
	12.4.1 启动通用查询日志	280
	12.4.2 查看通用查询日志	281
	12.4.3 删除通用查询日志	281
12.5	慢查询日志	282
	12.5.1 启动和设置慢查询日志	282
	12.5.2 查看慢查询日志	282
	12.5.3 删除慢查询日志	283

第13章 MySQL权限与安全管理 284

13.1	权限表	284
	13.1.1 user表	284
	13.1.2 db表	287
	13.1.3 tables_priv表和columns_priv表	288
	13.1.4 procs_priv表	289
13.2	账户管理	290
	13.2.1 登录和退出MySQL服务器	290
	13.2.2 新建普通用户	291
	13.2.3 删除普通用户	293
	13.2.4 root用户修改普通用户密码	294
13.3	权限管理	295
	13.3.1 MySQL的各种权限	296
	13.3.2 授权	298
	13.3.3 收回权限	299
	13.3.4 查看权限	300
13.4	访问控制	301
	13.4.1 连接核实阶段	301
	13.4.2 请求核实阶段	301
13.5	提升安全性	302
	13.5.1 密码到期更换策略	302

	13.5.2 安全模式	304
13.6	管理角色	305

第14章 MySQL性能优化 307

14.1	优化简介	307
14.2	优化查询	308
	14.2.1 分析查询语句	308
	14.2.2 索引对查询速度的影响	311
	14.2.3 使用索引查询	312
	14.2.4 优化子查询	315
14.3	优化数据库结构	315
	14.3.1 将字段很多的表分解成多张表	315
	14.3.2 增加中间表	317
	14.3.3 增加冗余字段	318
	14.3.4 优化插入记录的速度	318
	14.3.5 分析表、检查表和优化表	320
14.4	优化临时表性能	322
14.5	创建全局通用表空间	324

第15章 MySQL服务器性能优化 325

15.1	优化MySQL服务器	325
	15.1.1 优化服务器硬件	325
	15.1.2 优化MySQL的参数	326
15.2	影响MySQL服务器性能的重要参数	327
	15.2.1 查看性能参数的方法	327
	15.2.2 key_buffer_size的设置	331
	15.2.3 内存参数的设置	334
	15.2.4 日志和事务参数的设置	335
	15.2.5 存储和I/O相关参数的设置	337
	15.2.6 其他重要参数的设置	338
15.3	MySQL日志设置优化	339
15.4	MySQL I/O设置优化	341
15.5	MySQL并发设置优化	343
15.6	服务器语句超时处理	344

15.7	线程和临时表的优化 …………… 344
15.7.1	线程的优化 ………………… 344
15.7.2	临时表的优化 ……………… 345
15.8	增加资源组 ………………………… 346

第16章 MySQL性能监控 …………… 348

- 16.1 监控系统的基本方法 …………… 348
  - 16.1.1 ps命令 …………………… 348
  - 16.1.2 top命令 …………………… 349
  - 16.1.3 vmstat命令 ……………… 351
  - 16.1.4 mytop命令 ……………… 352
  - 16.1.5 sysstat工具 ……………… 354
- 16.2 开源监控利器Nagios实战 ……… 359
  - 16.2.1 安装Nagios之前的准备工作 ……………………… 359
  - 16.2.2 安装Nagios主程序 ……… 361
  - 16.2.3 整合Nagios到Apache服务 ……………………… 362
  - 16.2.4 安装Nagios插件包 ……… 365
  - 16.2.5 监控服务器的CPU、负载、磁盘I/O使用情况 ……… 367
  - 16.2.6 配置Nagios监控MySQL服务器 …………………… 371
- 16.3 MySQL监控利器Cacti实战 …… 373
  - 16.3.1 Cacti工具的安装 ………… 373
  - 16.3.2 Cacti监控MySQL服务器 … 378

第17章 提升MySQL数据库的性能 …… 384

- 17.1 默认字符集改为utf8mb4 ……… 384
- 17.2 自增变量的持久化 ……………… 385
- 17.3 GROUP BY不再隐式排序 ……… 387
- 17.4 统计直方图 ……………………… 389
  - 17.4.1 直方图的优点 …………… 389
  - 17.4.2 直方图的基本操作 ……… 390
- 17.5 日志分类更详细 ………………… 391
- 17.6 支持不可见索引 ………………… 392
- 17.7 支持JSON类型 ………………… 393
- 17.8 全文索引的加强 ………………… 396

- 17.9 动态修改InnoDB缓冲池的大小 …… 397
- 17.10 表空间数据加密 ……………… 399
- 17.11 跳过锁等待 …………………… 400
- 17.12 MySQL 9.0新特性1——支持将JSON输出保存到用户变量 …… 400
- 17.13 MySQL 9.0新特性2——支持准备语句 ……………………… 401
- 17.14 MySQL 9.0新特性3——支持面向AI的向量存储 …………… 402

第18章 MySQL终极优化实战 ………… 404

- 18.1 选择合适的存储引擎 …………… 404
- 18.2 通过分区表提升MySQL执行效率 …………………………… 408
  - 18.2.1 认识分区表 ……………… 408
  - 18.2.2 RANGE分区 …………… 408
  - 18.2.3 LIST分区 ……………… 410
  - 18.2.4 HASH分区 ……………… 411
  - 18.2.5 线性HASH分区 ………… 411
  - 18.2.6 KEY分区 ………………… 413
  - 18.2.7 复合分区 ………………… 413
- 18.3 优化数据表的锁 ………………… 415
  - 18.3.1 MyISAM表级锁优化建议 ……………………… 415
  - 18.3.2 InnoDB行级锁优化建议 ……………………… 416
- 18.4 优化事务控制 …………………… 416

第19章 企业人事管理系统数据库设计 ………………………… 421

- 19.1 需求分析 ………………………… 421
- 19.2 系统功能结构 …………………… 421
- 19.3 数据库设计 ……………………… 422
  - 19.3.1 数据库实体E-R图 ……… 422
  - 19.3.2 数据库表的设计 ………… 425
- 19.4 使用MySQL Workbench创建数据表 …………………………… 428
  - 19.4.1 创建数据库连接 ………… 428

	19.4.2	创建新的数据库…………429	20.2	系统主要功能……………………434
	19.4.3	创建数据表………………430	20.3	数据库与数据表设计………………435
第20章		**在线购物系统数据库设计……433**		20.3.1 数据库实体E-R图…………435
20.1	系统分析…………………………433			20.3.2 数据库和数据表分析………436
	20.1.1	系统总体设计……………433	20.4	使用MySQL Workbench数据建模…438
	20.1.2	系统界面设计……………433		20.4.1 建立E-R模型……………438
				20.4.2 导入E-R模型……………442

第 1 章
MySQL的安装与配置

MySQL支持多种平台，不同平台下的安装与配置过程也各不相同。在Windows平台下，可以使用二进制的安装软件包或免安装版的软件包进行安装。二进制的安装包提供了图形化的安装导向过程，而免安装版直接解压缩即可使用。在Linux平台下，使用命令行安装MySQL，但由于Linux是开源操作系统，有众多的分发版本，因此不同的Linux平台需要下载相应的MySQL安装包。本章将主要讲述Windows和Linux两个平台下MySQL 9的安装和配置过程。

1.1 什么是MySQL

MySQL是一个小型的关系型数据库管理系统。与大型数据库管理系统（例如Oracle、DB2、SQL Server等）相比，MySQL规模小、功能有限，但也因此具有体积小、速度快、成本低等优势。对于大部分中小型应用来说，MySQL提供的功能已经够用。这些特性使得MySQL成为世界上最受欢迎的开放源代码数据库之一。本节将介绍MySQL的特点。

1.1.1 客户端/服务器软件

主从式架构（Client-Server Model）或客户端/服务器（Client/Server）结构（简称C/S结构）是一种网络架构，在该网络架构下的软件通常可以分为客户端（Client）和服务器（Server）。

服务器是整个应用系统资源的存储与管理中心；多个客户端则各自处理相应的功能，共同实现完整的应用。在客户端/服务器结构中，客户端的请求被传送到数据库服务器；数据库服务器对请求进行处理后，将结果返回给客户端，从而减少网络数据传输量。

用户使用应用程序时，首先启动客户端，然后通过有关命令告知服务器进行连接以完成各种操作，而服务器则按照此命令提供相应服务。每一个客户端软件的实例，都可以向一个统一的服务器或应用程序服务器发出请求。

这种架构的特点就是，客户端和服务器程序不在同一台计算机上运行，它们通常归属不同的计算机。

在主从式架构中，客户端和服务器通过不同的途径应用于许多不同类型的应用程序，如网页应用。例如，在当当网站上购物时，用户的计算机和网页浏览器被看作客户端，而当当网的服务器、数据库和应用程序则被视为服务器。当用户通过浏览器向当当网发出搜索请求时，服务器会从数据库中提取相关图书信息，并生成一个网页返回给用户。

服务器通常使用高性能的计算机，并配备不同类型的数据库，如Oracle、SQL Server或MySQL等。客户端则需要安装专门的软件，如浏览器等。这种架构使得用户可以方便地访问网络服务器上的资源和服务，同时也使得服务提供者可以高效地管理和提供服务。

1.1.2 MySQL版本

针对不同用户，MySQL分为两个不同的版本：

- MySQL Community Server（社区版服务器）：该版本完全免费，但是官方不提供技术支持。
- MySQL Enterprise Server（企业版服务器）：能够以很高的性价比为企业提供数据仓库应用，支持ACID事物处理，提供完整的提交、回滚、崩溃恢复和行级锁定功能。但是该版本需付费使用，官方提供技术支持。

> 提示　MySQL Cluster主要用于架设集群服务器，需要在社区版或企业版基础上使用。

MySQL的命名机制由3个数字和1个后缀组成。这些数字和后缀共同构成了MySQL的版本号。例如，在MySQL-8.0.28-beta这个版本中：

- 第1个数字（8）是主版本号，描述了文件格式，所有版本8的发行都有相同的文件格式。
- 第2个数字（0）是发行级别，主版本号和发行级别组合在一起便构成了发行序列号。
- 第3个数字（28）是此发行系列的版本号，随每次新分发版本递增。
- 后缀（beta）表示这是一个测试版本。

这种命名机制使得MySQL的版本号具有明确的意义，方便用户了解和识别不同版本的MySQL。

1.2 Windows平台下安装与配置MySQL

要在Windows平台下安装MySQL，可以使用图行化的安装包。图形化的安装包提供了详细的安装向导，以便于用户一步一步地完成对MySQL的安装。本节将详细介绍使用图形化安装包安装MySQL的方法。

1.2.1 安装MySQL

要想在Windows中运行MySQL，需要32位或64位Windows操作系统，例如Windows 7、

Windows 10、Windows 11、Windows Server 2022等。Windows可以将MySQL服务器作为服务来运行。通常，在安装时MySQL需要具有系统的管理员权限。

Windows平台下提供了两种安装方式：MySQL二进制分发版（.msi安装文件）和免安装版（.zip压缩文件）。一般来讲，应当使用二进制分发版，因为该版本比其他的分发版使用起来要简单，不需要其他工具启动就可以运行MySQL。下面来介绍MySQL二进制分发版具体的安装步骤。

1. 下载MySQL安装文件

下载MySQL安装文件的具体操作步骤如下：

步骤01 在浏览器的地址栏中输入网址"https://dev.mysql.com/downloads/mysql/"，打开MySQL Community Server下载页面，选择Microsoft Windows平台，单击右侧的【Download】按钮，如图1.1所示。

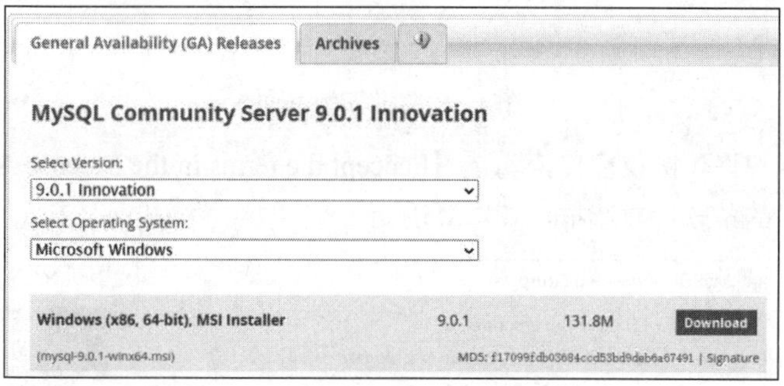

图1.1　MySQL下载页面

步骤02 在弹出的页面中，直接单击"No thanks, just start my download."链接，即可开始下载，如图1.2所示。

图1.2　开始下载页面

2. 安装MySQL

MySQL下载完成后，开始安装，具体操作步骤如下：

步骤01 双击下载的mysql-9.0.1-winx64.msi文件，如图1.3所示。

图1.3　MySQL安装文件名称

步骤02 进入欢迎安装界面，单击【Next】（下一步）按钮，如图1.4所示。

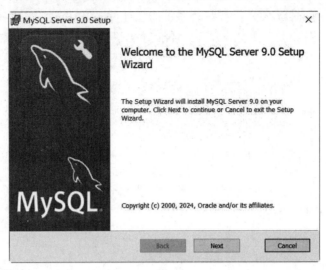

图 1.4　欢迎安装界面

步骤03 打开最终用户许可协议窗口，勾选【I accept the terms in the License Agreement】（我接受许可协议中的条款）复选框，如图1.5所示。

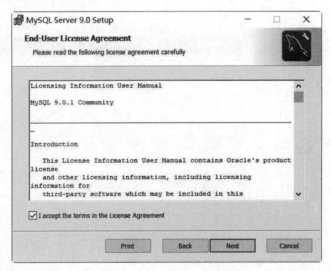

图 1.5　最终用户许可协议窗口

步骤04 单击【Next】（下一步）按钮，进入选择安装模式窗口。在其中列出了3种安装类型，分别是Typical（典型安装类型）、Custom（自定义安装类型）和Complete（完全安装）。这里单击【Custom】（自定义安装类型）按钮，如图1.6所示。

步骤05 进入自定义安装窗口，单击【Browse】（浏览）按钮，可以自定义安装路径，这里采用默认设置，如图1.7所示。

步骤06 单击【Next】（下一步）按钮，打开准备安装窗口，单击【Install】（安装）按钮，如图1.8所示。

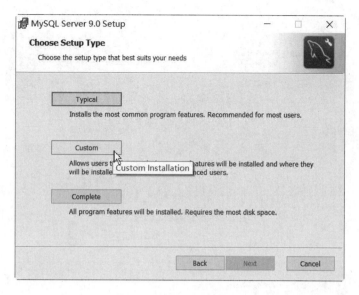

图 1.6　选择安装模式窗口

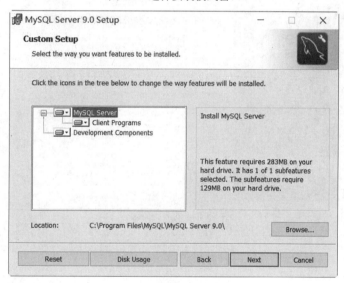

图 1.7　自定义安装窗口

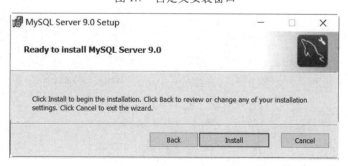

图 1.8　准备安装窗口

步骤 07　系统自动安装MySQL服务器，并显示安装进度，如图1.9所示。

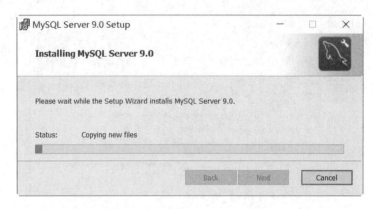

图 1.9　安装 MySQL 服务器

步骤 08　安装完成后，单击【Finish】（完成）按钮即可，如图1.10所示。

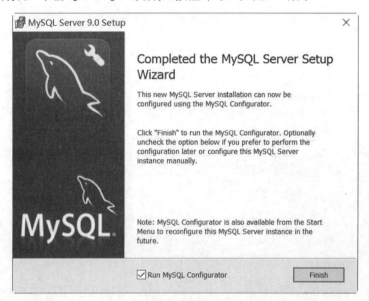

图 1.10　安装完成

1.2.2　配置 MySQL

MySQL安装完毕之后，需要对服务器进行配置，具体的配置步骤如下：

步骤 01　在1.2.1节的最后一步中，单击【Finish】（完成）按钮，将进入欢迎进入MySQL服务器配置（Welcome to the MySQL Server Configurator）窗口，如图1.11所示。

步骤 02　单击【Next】（下一步）按钮，进入设置数据保存路径（Data Directory）窗口，这里可以根据自己的需要设置数据的保存路径，如图1.12所示。

步骤 03　单击【Next】（下一步）按钮，打开MySQL服务器类型配置（Type and Networking）窗口，这里采用默认的设置即可，如图1.13所示。

步骤 04　单击【Next】（下一步）按钮，打开账户和角色（Accounts and Roles）窗口，设置MySQL root账户的密码，输入两次同样的登录密码，如图1.14所示。

图 1.11　欢迎进入 MySQL 服务器配置窗口

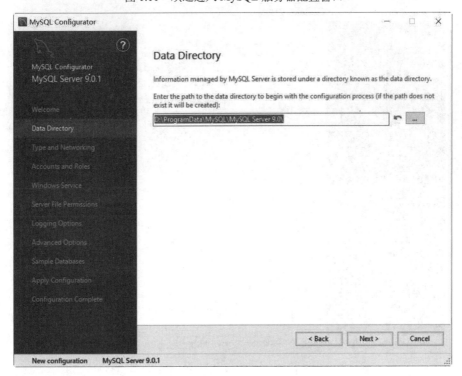

图 1.12　设置数据保存路径窗口

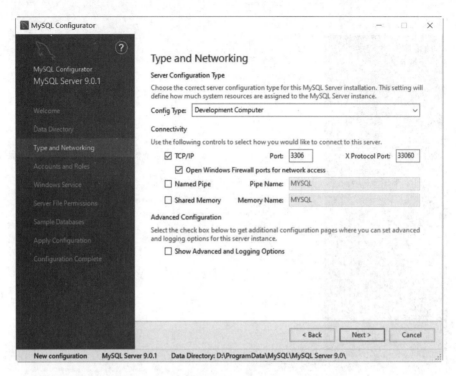

图 1.13　MySQL 服务器类型配置窗口

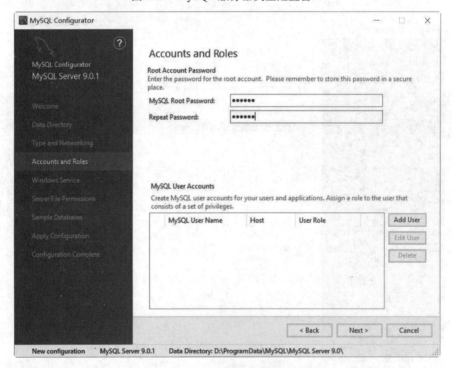

图 1.14　设置服务器的密码窗口

步骤 05　单击【Next】（下一步）按钮，打开Windows服务（Windows Service）窗口，设置服务器名称，本案例设置服务器名称为"MySQL90"，如图1.15所示。

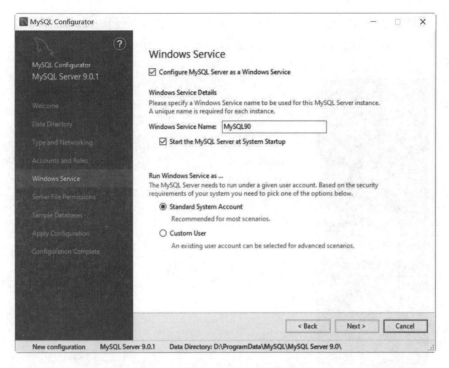

图 1.15　设置 Windows 服务的名称

步骤 06 单击【Next】（下一步）按钮，打开服务器文件许可（Server File Permissions）窗口，这里单击第三个单选按钮，如图1.16所示。

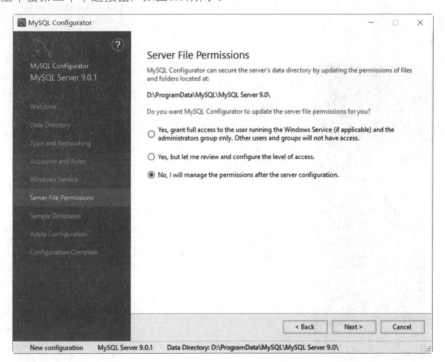

图 1.16　服务器文件许可窗口

步骤07 单击【Next】（下一步）按钮，打开创建数据库实例（Sample Databases）窗口，可以勾选创建Sakila、World样例数据库，这里没有勾选，如图1.17所示。

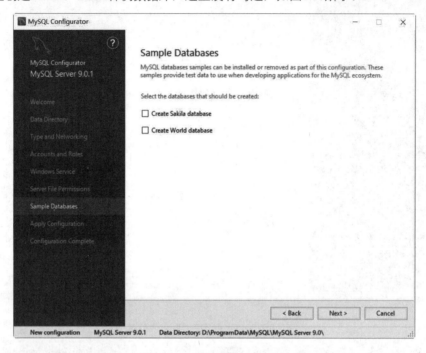

图1.17　创建数据库实例窗口

步骤08 单击【Next】（下一步）按钮，打开应用配置（Apply Configuration）窗口，单击【Execute】（执行）按钮，如图1.18所示。

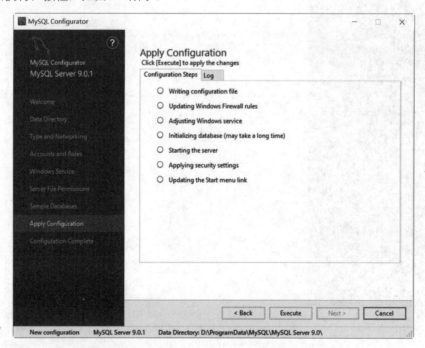

图1.18　确认配置服务器

步骤09 系统自动配置MySQL服务器。配置完成（Configuration Complete）后，单击【Finish】（完成）按钮，即可完成服务器的配置，如图1.19所示。

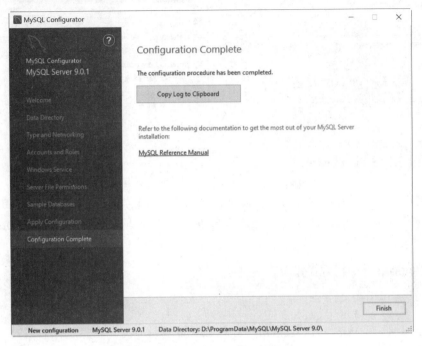

图 1.19　完成服务器的配置

1.3　启动服务并登录MySQL数据库

MySQL安装完毕之后，需要启动服务器进程，否则客户端无法连接数据库。本节将介绍启动MySQL服务器和登录MySQL的方法。

1.3.1　启动 MySQL 服务

在前面的配置过程中，已经将MySQL安装为Windows服务，当Windows启动、停止时，MySQL也自动启动、停止。不过，用户还可以使用图形服务工具来控制MySQL服务器或通过命令行使用NET命令。

具体可以通过Windows的服务管理器来查看，操作步骤如下：

步骤01 右击桌面左下角的【开始】按钮，选择【运行】菜单命令，在打开的【运行】对话框中输入"services.msc"命令，然后单击【确定】按钮，如图1.20所示。

步骤02 打开Windows的服务管理器，在其中可以看到名称为"MySQL90"的服务项，其状态为"正在运行"，表明该服务已经启动，如图1.21所示。

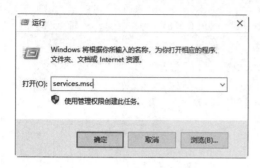

图 1.20　【运行】对话框　　　　　　　　图 1.21　服务管理器窗口

1.3.2　登录 MySQL 数据库

当MySQL服务启动完成后，便可以通过客户端来登录MySQL数据库。在Windows操作系统下，有两种登录MySQL数据库的方式：以Windows命令行方式登录和使用MySQL Command Line Client登录。

1. 以Windows命令行方式登录

具体的操作步骤如下：

步骤01　右击桌面左下角的【开始】按钮，选择【运行】菜单命令，在打开的【运行】对话框中输入"cmd"命令，单击【确定】按钮，如图1.22所示。

步骤02　打开DOS窗口（或者终端管理员窗口），输入以下命令并按Enter键确认，如图1.23所示。

```
cd C:\Program Files\MySQL\MySQL Server 9.0\bin\
```

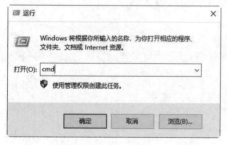

图 1.22　运行对话框　　　　　　　　图 1.23　DOS 窗口

步骤03　在DOS窗口中可以通过登录命令连接到MySQL数据库。连接MySQL的命令格式如下：

```
mysql -h hostname -u username -p
```

其中，mysql为登录命令；-h后面的参数是服务器的主机地址，在这里因为客户端和服务器在同一台机器上，所以输入"localhost"或者本机IP地址"127.0.0.1"；-u后面跟登录数据库的用户名称，在这里为root；-p后面是用户登录密码（安装MySQL时配置的密码）。

因此，输入如下命令：

```
mysql -h localhost -u root -p
```

然后按Enter键，系统会提示"Enter password"（输入密码），这里输入前面在配置向导中设置的密码，密码验证正确后，即可登录到MySQL数据库，如图1.24所示。

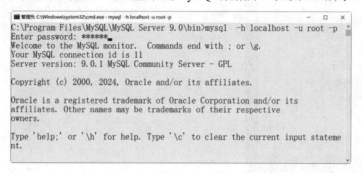

图1.24　Windows命令行登录窗口

> **提示**　当窗口中出现如图1.24所示的说明信息，命令提示符变为"mysql>"时，表明已经成功登录MySQL服务器，可以开始对数据库进行操作了。

2. 使用MySQL Command Line Client登录

依次选择【开始】|【所有程序】|【MySQL】|【MySQL 9.0 Command Line Client】菜单命令，输入正确的密码之后，就可以登录到MySQL数据库了，如图1.25所示。

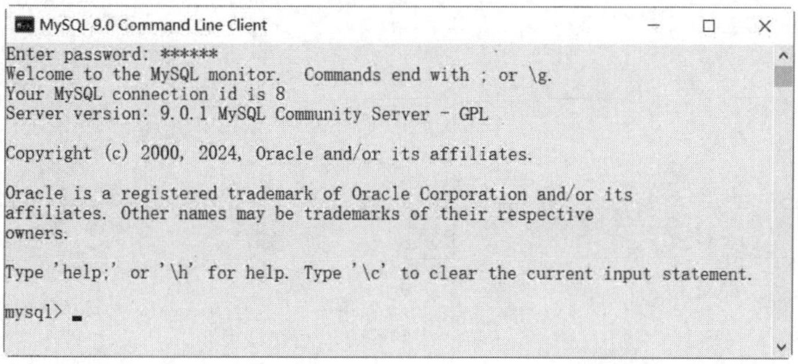

图1.25　MySQL命令行登录窗口

1.3.3　配置Path变量

前面登录MySQL服务器的时候，不能直接输入MySQL登录命令，是因为还没有把MySQL的bin目录添加到系统的环境变量里。如果每次登录都要输入"cd C:\Program Files\MySQL\MySQL Server 9.0\bin"才能使用MySQL客户端等其他命令工具，就会比较烦琐。因此，我们需要配置Path变量，以便可以在任何目录下直接输入MySQL命令来登录数据库。

下面介绍怎样手动配置PATH变量，具体步骤如下：

步骤01　在桌面上右击【此电脑】图标，在弹出的快捷菜单中选择【属性】菜单命令，如图1.26所示。

步骤02　打开【系统】窗口，单击【高级系统设置】链接，如图1.27所示。

图 1.26　【此电脑】属性菜单　　　　　　　　图 1.27　【系统】窗口

步骤 03 打开【系统属性】对话框，选择【高级】选项卡，然后单击【环境变量】按钮，如图 1.28 所示。

步骤 04 打开【环境变量】对话框，在系统变量列表中选择【Path】变量，如图 1.29 所示。

图 1.28　【系统属性】对话框　　　　　　　　图 1.29　【环境变量】对话框

步骤 05 单击【编辑】按钮，在【编辑环境变量】对话框中，将 MySQL 应用程序的 bin 目录（C:\Program Files\MySQL\MySQL Server 9.0\bin）添加到变量值中，并用分号将它与其他路径隔开，如图 1.30 所示。

步骤 06 添加完成之后，单击【确定】按钮，这样就完成了配置 Path 变量的操作，之后就可以直接输入 MySQL 命令来登录数据库了。

图 1.30 【编辑环境变量】对话框

1.4 MySQL常用的图形化管理工具

MySQL的图形化管理工具极大地方便了数据库的操作与管理。常用的图形化管理工具有MySQL Workbench和phpMyAdmin，当然，读者也可以选用自己喜欢的其他工具。

1. MySQL Workbench

MySQL Workbench 是一款专门为用户提供创建、修改、执行和优化SQL的可视化工具，并且为开发者提供了一整套可视化操作，用于创建、编辑和管理SQL查询，以及管理数据库连接。在可视化SQL编辑工作模式下，用户要创建表、删除表、修改表信息等，只需使用简单的可编辑列表就能完成。

MySQL Workbench在数据库管理方面也提供了可视化的操作。为了管理用户，可以使用Workbench来授予和收回用户权限。在数据库管理中，可以使用Workbench查看数据库的状态，其中包括数据库中开启了多少个客户端、数据库缓存的大小以及管理数据库日志等信息。

目前MySQL Workbench提供了开源和商业化两个版本，同时支持Windows和Linux系统。MySQL Workbench的下载地址是http://dev.MySQL.com/downloads/workbench/。

2. phpMyAdmin

phpMyAdmin使用PHP编写，必须安装在Web服务器中，通过Web方式控制和操作MySQL数据库。通过phpMyAdmin可以轻松地对数据库进行操作，例如建立、复制、删除数据等。使用phpMyAdmin管理数据库非常方便，并且支持中文，不足之处在于对大数据库的备份和恢复不那么方便。

phpMyAdmin的下载地址是http://www.phpmyadmin.net/。

1.5 Linux平台下安装与配置MySQL

Linux操作系统有众多的发行版，不同的平台需要安装不同的MySQL版本。MySQL主要支持的Linux版本有Ubuntu、Debian、SUSE、Red Hat、Fedora、Oracle、macOS等。本节将介绍如何在Linux平台下安装MySQL。

1.5.1 Linux 操作系统下的 MySQL 版本介绍

Linux操作系统是自由软件和开放源代码发展中最著名的例子。自它诞生以来，经过全世界计算机爱好者的共同努力，已经成为世界上使用最多的一种类UNIX操作系统，目前已经开发了超过300个发行版本，比较流行的版本有Ubuntu、Debian、SUSE、Red Hat、Fedora等（具体可以参看MySQL官方下载页面）。Linux各个平台的安装过程基本相同，读者可以针对个人的喜好，选择使用不同的安装包。

Linux操作系统MySQL安装包分为以下3类：

- RPM：RPM软件包是一种在Linux平台下的安装文件，通过安装命令可以很方便地安装与卸载。MySQL的RPM安装文件包分为两个：服务器端文件包和客户端文件包，需要分别下载和安装。
- Generic Binaries：二进制软件包，经过编译生成的二进制文件软件包。
- 源码包：源码包中的是MySQL数据库的源代码，需要用户将其编译成二进制文件之后才能安装。

下面简要介绍SUSE Linux Enterprise Server和Red Hat Enterprise Linux的MySQL安装包。

1. SUSE Linux Enterprise Server

SUSE于1992年年末创办，采用了很多Red Hat Linux的特质，于2004年1月被Novell公司收购。目前其最新版本为SUSE Linux Enterprise Server 15（2024年）。针对SUSE Linux Enterprise Server 15，MySQL官方提供MySQL 9的安装包。

读者可以在http://dev.mysql.com/downloads/mysql/页面中选择【SUSE Linux Enterprise Server】平台，下载服务器端和客户端的RPM包。

> 提示 其中，MySQL Server代表服务器端的RPM包，Client Utilities代表客户端的RPM包。官方同时提供二进制和源码的MySQL安装包。

2. Red Hat Enterprise Linux

2004年4月30日，Red Hat公司正式停止对Red Hat 9.0版本的支持，标志着Red Hat Linux的正式完结。Red Hat公司不再开发桌面版的Linux发行包，而是集中力量开发服务器版，也就是

Red Hat Enterprise Linux版。目前Red Hat Enterprise Linux 9为最新的版本，MySQL官方网站提供了针对此版本的MySQL 9安装包。

根据不同的处理器架构，Red Hat Enterprise Linux下的MySQL安装包的版本也有所不同。

读者可以在http://dev.mysql.com/downloads/mysql/页面中选择【Red Hat Enterprise Linux/ Oracle Linux】平台，根据自己版本需要下载服务器端和客户端RPM包，如图1.31所示。

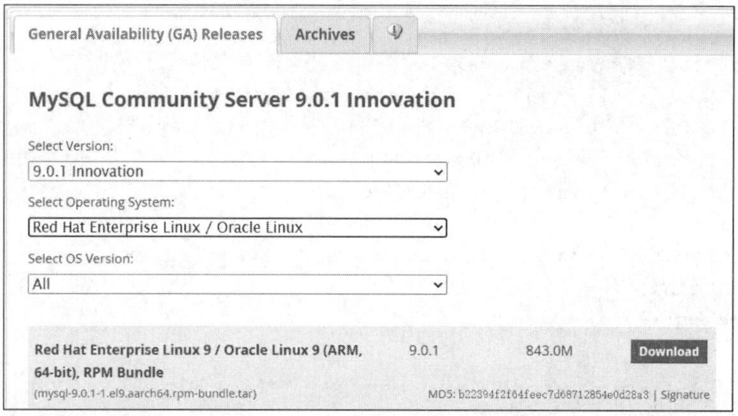

图 1.31　下载 Red Hat Enterprise Linux 的 MySQL 安装包

1.5.2　安装和配置 MySQL 的 RPM 包

MySQL推荐使用RPM包进行Linux平台下的安装，从官方下载的RPM包能够在所有支持RPM packages、glibc2.3的Linux系统下安装和使用。对于标准安装，只需要安装MySQL-server和MySQL-client。下面开始通过RPM包进行安装，具体的操作步骤如下：

步骤01 进入http://dev.mysql.com/downloads/mysql/页面，下载RPM包。在平台下拉列表中选择【Red Hat Enterprise Linux /Oracle Linux】选项。

步骤02 从RPM列表中选择要下载安装的包，单击【Download】按钮，开始下载安装包。

步骤03 下载完成后，在/usr/local/目录下创建mysql目录，命令如下：

```
mkdir mysql
```

步骤04 将MySQL的安装包传到/usr/local/mysql目录下，进行解压，命令如下：

```
tar -xvf mysql-9.0.1-1.el9.x86_64.rpm-bundle.tar
```

tar是Linux/UNIX系统上的一个打包工具，通过tar -help可以查看tar的使用帮助。

如果在操作的过程中提示权限不够，则可以通过超级管理员权限登录系统，命令如下：

```
sudo -s
```

步骤05 安装MySQL Server 9.0，命令如下：

```
rpm -ivh mysql-community-common-9.0.1-1.el9.x86_64.rpm --force --nodeps
rpm -ivh mysql-community-libs-9.0.0-1.el9.x86_64.rpm --force --nodeps
rpm -ivh mysql-community-client-9.0.1-1.el9.x86_64.rpm --force --nodeps
rpm -ivh mysql-community-server-9.0.1-1.el9.x86_64.rpm --force --nodeps
```

步骤 06 安装完成后，可以查看MySQL的版本，命令如下：

```
mysql --version
```

步骤 07 启动MySQL服务，命令如下：

```
systemctl start mysqld
```

步骤 08 安装成功之后，使用命令查看临时登录密码，然后使用临时密码登录MySQL服务器。

```
[root@localhost mysql]# cd /var/log
[root@localhost log]# grep -n password mysqld.log
6:2024-02-21T06:56:30.598513Z 6 [Note] [MY-010454] [Server] A temporary password is generated for root@localhost: ogHYgZ)D#7)u
[root@localhost log]# mysql -u root -p
Enter password:
Welcome to the MySQL monitor.  Commands end with ; or \g.
Your MySQL connection id is 9
Server version: 9.0.1

Copyright (c) 2000, 2024, Oracle and/or its affiliates.

Oracle is a registered trademark of Oracle Corporation and/or its
affiliates. Other names may be trademarks of their respective
owners.

Type 'help;' or '\h' for help. Type '\c' to clear the current input statement.
```

能够看到上面的信息就说明登录成功，接下来可以对MySQL数据库进行操作了。

步骤 09 更改root密码，命令如下：

```
mysql> alter user 'root'@'localhost' identified by 'Fyuew123456#';
```

执行完该命令，root的密码被改为Fyuew123456#，读者可以按自己的需要修改。

第 2 章
数据库和数据表的基本操作

MySQL安装好以后,首先需要创建数据库和数据表,这是使用MySQL各种功能的前提。在数据库中,数据表是最重要、最基本的操作对象,是数据存储的基本单位。数据表被定义为列的集合,数据在表中是按照行和列的格式来存储的。每一行代表一条唯一的记录,每一列代表记录中的一个域。本章将详细介绍数据库和数据表的基本操作。

2.1 创建数据库

MySQL安装完成之后,会在其data目录下自动创建几个必需的数据库。可以使用"SHOW DATABASES;"语句来查看当前存在的所有数据库,示例如下:

```
mysql> SHOW DATABASES;
+--------------------+
| Database           |
+--------------------+
| information_schema |
| mysql              |
| performance_schema |
| sys                |
+--------------------+
```

可以看到,数据库列表中包含了4个数据库,其中mysql数据库是必需的,它描述用户访问权限。

创建数据库是在系统磁盘上划分一块区域用于数据的存储和管理。如果管理员在设置权限的时候为用户创建了数据库,则用户可以直接使用该数据库,否则需要自己创建数据库。在MySQL中创建数据库的基本SQL语法格式为:

```
CREATE DATABASE database_name;
```

其中,database_name为要创建的数据库的名称,该名称不能与已经存在的数据库重名。

【例2.1】创建测试数据库test_db。SQL语句如下：

```
CREATE DATABASE test_db;
```

> **提示** MySQL中的SQL语句是不区分大小写的，因此CREATE DATABASE和create database的作用是相同的。但是，许多开发人员习惯将关键字大写，将数据列和表名小写，读者也应该养成一个良好的编程习惯，这样写出来的代码更容易阅读和维护。

数据库创建好之后，可以使用SHOW CREATE DATABASE声明查看数据库的定义。

【例2.2】查看创建好的数据库test_db的定义，SQL语句如下：

```
mysql> SHOW CREATE DATABASE test_db\G
*** 1. row ***
       Database: test_db
Create Database: CREATE DATABASE `test_db` /*!40100 DEFAULT CHARACTER SET utf8mb4 COLLATE utf8mb4_0900_ai_ci */ /*!80016 DEFAULT ENCRYPTION='N' */
```

可以看到，如果数据库已成功创建，则显示数据库的创建信息。

再次使用"SHOW DATABASES;"语句来查看当前存在的所有数据库：

```
mysql> SHOW DATABASES;
+--------------------+
| Database           |
+--------------------+
| information_schema |
| mysql              |
| performance_schema |
| sys                |
| test_db            |
+--------------------+
```

可以看到，数据库列表中多了一个刚刚创建的数据库test_db。

2.2 删除数据库

删除数据库是将已经存在的数据库从磁盘空间上清除，并且数据库中的所有数据也将一同被删除。在MySQL中删除数据库的基本语法格式为：

```
DROP DATABASE database_name;
```

其中，database_name为要删除的数据库的名称。若指定的数据库不存在，则删除出错。

【例2.3】删除测试数据库test_db，SQL语句如下：

```
DROP DATABASE test_db;
```

语句执行完毕之后，数据库test_db将被删除。再次使用SHOW CREATE DATABASE声明查看数据库的定义，结果如下：

```
mysql> SHOW CREATE DATABASE test_db\G
ERROR 1049 (42000): Unknown database 'test_db'
```

执行结果给出一条错误信息"ERROR 1049 (42000)：Unknown database 'test_db'"，即数据库test_db已不存在，说明删除成功。

2.3 创建数据表

在创建完数据库之后，接下来的工作就是创建数据表。所谓创建数据表，就是在已经创建好的数据库中建立新表。创建数据表的过程是规定数据列的属性的过程，同时也是实施数据完整性（包括实体完整性、引用完整性和域完整性等）约束的过程。本节将介绍创建数据表的语法形式以及如何添加主键约束（Primary Key Constraint）、外键约束（Foreign Key Constraint）、非空约束（Not Null Constraint）等。

2.3.1 创建表的语法形式

数据表属于数据库，因此在创建数据表之前，应该使用"USE <数据库名>"语句指定操作在哪个数据库中进行。如果没有选择数据库，就会抛出"No database selected"的错误。

创建数据表的语句为CREATE TABLE，语法格式如下：

```
CREATE  TABLE <表名>
(
字段名1,数据类型 [列级别约束条件] [默认值],
字段名2,数据类型 [列级别约束条件] [默认值],
...
[表级别约束条件]
);
```

使用CREATE TABLE创建表时，必须指定以下信息：

（1）要创建的表的名称,不区分大小写,不能使用SQL语言中的关键字,如DROP、ALTER、INSERT等。

（2）数据表中每一列（字段）的名称和数据类型，如果创建多列，各列之间就要用英文逗号隔开。

【例2.4】创建员工表tb_emp1，结构如表2.1所示。

表 2.1 tb_emp1 表结构

字段名称	数据类型	备 注
id	INT	员工编号
name	VARCHAR（25）	员工名称
deptId	INT	所在部门编号
salary	FLOAT	工资

(1)创建数据库,SQL语句如下:

```
CREATE  DATABASE test_db;
```

(2)选择创建表的数据库,SQL语句如下:

```
USE test_db;
```

(3)创建tb_emp1表,SQL语句如下:

```
CREATE TABLE tb_emp1
(
id      INT,
name    VARCHAR(25),
deptId  INT,
salary  FLOAT
);
```

以上语句执行后,便创建了一张名称为"tb_emp1"的数据表。

(4)查看数据表是否创建成功,SQL语句如下:

```
mysql> SHOW TABLES;
+---------------------+
| Tables_in_ test_db  |
+---------------------+
| tb_emp1             |
+---------------------+
```

可以看到,test_db数据库中已经有了数据表tb_emp1,说明数据表创建成功。

2.3.2 使用主键约束

主键又称主码,是表中一列或多列的组合。主键约束要求主键列的数据唯一,并且不为空。主键能够唯一地标识表中的一条记录,可以结合外键来定义不同数据表之间的关系,并且可以加快数据库查询的速度。主键和记录之间的关系如同身份证号和人之间的关系,它们是一一对应。主键分为两种类型:单字段主键和多字段联合主键。

1. 单字段主键

单字段主键由一个字段组成,SQL语句格式分为以下两种。

(1)在定义列的同时指定主键,语法格式如下:

```
字段名 数据类型 PRIMARY KEY [默认值]
```

【例2.5】定义数据表tb_emp 2,其主键为id,SQL语句如下:

```
CREATE TABLE tb_emp2
(
id          INT PRIMARY KEY,
name        VARCHAR(25),
deptId      INT,
```

```
salary    FLOAT
);
```

（2）在定义完所有列之后指定主键，语法规则如下：

```
[CONSTRAINT <约束名>] PRIMARY KEY [字段名]
```

【例2.6】定义数据表tb_emp 3，其主键为id，SQL语句如下：

```
CREATE TABLE tb_emp3
(
id INT,
name VARCHAR(25),
deptId INT,
salary FLOAT,
PRIMARY KEY(id)
);
```

上述两个例子执行后的结果是一样的，都会在id字段上设置主键约束。

2. 多字段联合主键

多字段联合主键由多个字段联合组成，语法格式如下：

```
PRIMARY KEY [字段1, 字段2,..., 字段n]
```

【例2.7】定义数据表tb_emp4，假设表中间没有主键id，为了唯一确定一个员工，可以把name、deptId联合起来作为主键，SQL语句如下：

```
CREATE TABLE tb_emp4
(
name VARCHAR(25),
deptId INT,
salary FLOAT,
PRIMARY KEY(name,deptId)
);
```

以上语句执行后，便创建了一张名称为"tb_emp4"的数据表，name字段和deptId字段组合在一起成为tb_emp4的多字段联合主键。

2.3.3 使用外键约束

外键用来在两张表的数据之间建立连接，可以是一列或者多列。一张表可以有一个或多个外键。外键对应的是参照完整性，一张表的外键可以为空值，若不为空值，则每一个外键值必须等于另一张表中主键的某个值。

外键是表中的一个字段，虽然可以不是本表的主键，但必须对应另外一张表的主键。外键的主要作用是保证数据引用的完整性，定义外键后，不允许删除另一张表中具有关联关系的行。例如，部门表tb_dept的主键是id，员工表tb_emp5中有一个键deptId与这个id关联。在这个例子中，部门表（tb_dept）是主表（父表），而员工表（tb_emp5）是从表（子表）。从表中的外键（deptId）与主表中的主键（id）建立了关联关系，确保了数据的完整性和一致性。

- 主表（父表）：对于两个具有关联关系的表而言，相关联字段中主键所在的那张表是主表。
- 从表（子表）：对于两个具有关联关系的表而言，相关联字段中外键所在的那张表是从表。

创建外键的语法格式如下：

```
[CONSTRAINT <外键名>] FOREIGN KEY 字段名1 [ ,字段名2,...]
REFERENCES <主表名> 主键列1 [ ,主键列2,...]
```

"外键名"为定义的外键约束的名称，一张表中不能有相同名称的外键；"字段名"表示子表需要添加外键约束的字段列；"主表名"表示被子表外键所依赖的表的名称；"主键列"表示主表中定义的主键列或者列组合。

【例2.8】定义数据表tb_emp5，并在tb_emp5表上创建外键约束。

（1）创建一张部门表tb_dept1，表结构如表2.2所示，SQL语句如下：

```
CREATE TABLE tb_dept1
(
id          INT PRIMARY KEY,
name        VARCHAR(22)  NOT NULL,
location    VARCHAR(50)
);
```

表 2.2　tb_dept1 表结构

字段名称	数据类型	备 注
id	INT	部门编号
name	VARCHAR（22）	部门名称
location	VARCHAR（50）	部门位置

（2）定义数据表tb_emp5，让它的键deptId作为外键关联到tb_dept1的主键id，SQL语句如下：

```
CREATE TABLE tb_emp5
(
id      INT PRIMARY KEY,
name    VARCHAR(25),
deptId  INT,
salary  FLOAT,
CONSTRAINT fk_emp_dept1 FOREIGN KEY(deptId) REFERENCES tb_dept1(id)
);
```

以上语句执行成功之后，在表tb_emp5上添加了名称为"fk_emp_dept1"的外键约束，外键名称为deptId，它依赖于表tb_dept1的主键id。

> **提示**　关联指的是在关系数据库中相关表之间的联系。它是通过相容或相同的属性或属性组来表示的。子表的外键必须关联父表的主键，且关联字段的数据类型必须匹配，如果类型不一样，则创建子表时，就会出现"ERROR 1005 (HY000): Can't create table 'database.tablename'(errno: 150)"错误。

2.3.4 使用非空约束

非空约束指字段的值不能为空。对于使用了非空约束的字段，如果用户在添加数据时没有指定值，则数据库系统会报错。

非空约束的语法格式如下：

```
字段名 数据类型 NOT NULL
```

【例2.9】定义数据表tb_emp6，指定员工的名称不能为空，SQL语句如下：

```
CREATE TABLE tb_emp6
(
id      INT PRIMARY KEY,
name    VARCHAR(25) NOT NULL,
deptId  INT,
salary  FLOAT
);
```

以上语句执行后，在tb_emp6中创建了一个name字段，其插入值不能为空。

2.3.5 使用唯一性约束

唯一性约束（Unique Constraint）要求该列唯一，允许为空，但只能出现一个空值。唯一性约束可以确保一列或者几列不出现重复值。

唯一性约束的语法格式有以下两种：

1. 在定义完列之后直接指定唯一约束

在定义完列之后直接指定唯一约束，语法格式如下：

```
字段名 数据类型 UNIQUE
```

【例2.10】定义数据表tb_dept2，指定部门的名称唯一，SQL语句如下：

```
CREATE TABLE tb_dept2
(
id          INT PRIMARY KEY,
name        VARCHAR(22) UNIQUE,
location    VARCHAR(50)
);
```

2. 在定义完所有列之后指定唯一约束

在定义完所有列之后指定唯一约束，语法格式如下：

```
[CONSTRAINT <约束名>] UNIQUE(<字段名>)
```

【例2.11】定义数据表tb_dept3，指定部门的名称唯一，SQL语句如下：

```
CREATE TABLE tb_dept3
(
id          INT PRIMARY KEY,
```

```
name         VARCHAR(22),
location     VARCHAR(50),
CONSTRAINT   STH UNIQUE(name)
);
```

UNIQUE和PRIMARY KEY的区别：一张表中可以有多个字段被声明为UNIQUE，但只能有一个PRIMARY KEY声明；声明为PRIMAY KEY的列不允许有空值，但声明为UNIQUE的字段允许空值的存在。

2.3.6 使用默认约束

默认约束（Default Constraint）指定某列的默认值。例如，当男性同学较多时，性别就默认为"男"。如果插入一条新的记录时没有为性别字段赋值，系统就会自动为这个字段赋值为"男"。

默认约束的语法格式如下：

```
字段名  数据类型 DEFAULT 默认值
```

【例2.12】定义数据表tb_emp7，指定员工的部门编号默认为1111，SQL语句如下：

```
CREATE TABLE tb_emp7
(
id       INT PRIMARY KEY,
name     VARCHAR(25) NOT NULL,
deptId   INT DEFAULT 1111,
salary   FLOAT
);
```

以上语句执行成功之后，表tb_emp7上的字段deptId拥有了一个默认值1111，新插入的记录如果没有指定部门编号，则默认为1111。请读者自行测试一下。

2.3.7 设置表的属性值自动增加

在数据库应用中，经常希望在每次插入新记录时，系统自动生成字段的主键值，可以通过为主键添加AUTO_INCREMENT关键字来实现。在MySQL中，AUTO_INCREMENT的初始值默认是1，每新增一条记录，字段值自动加1。一张表只能有一个字段使用AUTO_INCREMENT约束，且该字段必须为主键的一部分。AUTO_INCREMENT约束的字段可以是任何整数类型（TINYINT、SMALLIN、INT、BIGINT等）。

设置表的属性值自动增加的语法格式如下：

```
字段名 数据类型 AUTO_INCREMENT
```

【例2.13】定义数据表tb_emp8，指定员工的编号自动递增，SQL语句如下：

```
CREATE TABLE tb_emp8
(
id       INT PRIMARY KEY AUTO_INCREMENT,
name     VARCHAR(25) NOT NULL,
```

```
    deptId    INT,
    salary    FLOAT
);
```

上述语句执行后，会创建名称为"tb_emp8"的数据表，表tb_emp8中id字段的值在添加记录时会自动增加。即在插入记录的时候，默认的自增字段id的值从1开始，每添加一条新记录，该值自动加1。

例如，执行如下插入语句：

```
mysql> INSERT INTO tb_emp8 (name,salary)
    -> VALUES('Lucy',1000), ('Lura',1200),('Kevin',1500);
```

> **提示** 这里使用INSERT 声明向表中插入记录的方法，并不是SQL的标准语法，这种语法不一定被其他厂家的数据库支持，只能在MySQL中使用。

语句执行完后，表tb_emp8中增加了3条记录，这里并没有输入id的值，但系统已经自动添加该值。使用SELECT命令查看记录，结果如下：

```
mysql> SELECT * FROM tb_emp8;
+----+-------+--------+--------+
| id | name  | deptId | salary |
+----+-------+--------+--------+
|  1 | Lucy  | NULL   |   1000 |
|  2 | Lura  | NULL   |   1200 |
|  3 | Kevin | NULL   |   1500 |
+----+-------+--------+--------+
```

2.4 查看数据表结构

使用SQL语句创建好数据表之后，就可以查看表结构的定义，以确认表的定义是否正确。在MySQL中，查看表结构可以使用DESCRIBE和SHOW CREATE TABLE语句。本节将针对这两个语句分别进行详细讲解。

2.4.1 查看表基本结构的语句 DESCRIBE

DESCRIBE/DESC语句可以查看表的字段信息，其中包括字段名、字段数据类型、是否为主键、是否有默认值等。语法格式如下：

```
DESCRIBE 表名;
```

或者简写为：

```
DESC 表名;
```

【例2.14】分别使用DESCRIBE和DESC查看表tb_dept1和表tb_emp1的结构。

（1）查看tb_dept1表结构，SQL语句如下：

```
mysql> DESCRIBE tb_dept1;
+----------+-------------+------+-----+---------+-------+
| Field    | Type        | Null | Key | Default | Extra |
+----------+-------------+------+-----+---------+-------+
| id       | int         | NO   | PRI | NULL    |       |
| name     | varchar(22) | NO   |     | NULL    |       |
| location | varchar(50) | YES  |     | NULL    |       |
+----------+-------------+------+-----+---------+-------+
```

（2）查看tb_emp1表结构，SQL语句如下：

```
mysql> DESC tb_emp1;
+--------+-------------+------+-----+---------+-------+
| Field  | Type        | Null | Key | Default | Extra |
+--------+-------------+------+-----+---------+-------+
| id     | int         | YES  |     | NULL    |       |
| name   | varchar(25) | YES  |     | NULL    |       |
| deptId | int         | YES  |     | NULL    |       |
| salary | float       | YES  |     | NULL    |       |
+--------+-------------+------+-----+---------+-------+
```

其中，各个字段的含义分别解释如下：

- NULL：表示该列是否可以存储空值。
- Key：表示该列是否已编制索引。PRI表示该列是表主键的一部分；UNI表示该列是UNIQUE索引的一部分；MUL表示在列中允许某个给定值出现多次。
- Default：表示该列是否有默认值，有的话指定值是多少。
- Extra：表示可以获取的与给定列有关的附加信息，例如 AUTO_INCREMENT等。

2.4.2 查看表详细结构的语句 SHOW CREATE TABLE

SHOW CREATE TABLE语句可以用来显示创建表时所使用的CREATE TABLE语句，语法格式如下：

```
SHOW CREATE TABLE <表名\G>;
```

> **提示** 使用SHOW CREATE TABLE语句，不仅可以查看创建表时的详细语句，还可以查看存储引擎和字符编码。

如果不加"\G"参数，显示的结果可能非常混乱，加上"\G"参数，可使显示结果更加直观，易于查看。

【例2.15】使用SHOW CREATE TABLE查看表tb_emp1的详细信息，SQL语句如下：

```
mysql> SHOW CREATE TABLE tb_emp1;
```

```
+---------+-----------------------------------------------------------+
| Table   | Create Table                                              |
+---------+-----------------------------------------------------------+
| tb_emp1 | CREATE TABLE `tb_emp1` (
  `id` int DEFAULT NULL,
  `name` varchar(25) DEFAULT NULL,
  `deptId` int DEFAULT NULL,
  `salary` float DEFAULT NULL
) ENGINE=InnoDB DEFAULT CHARSET=utf8mb4 COLLATE=utf8mb4_0900_ai_ci |
+---------+-----------------------------------------------------------+
```

使用参数"\G"之后的结果如下：

```
mysql> SHOW CREATE TABLE tb_emp1\G
*************************** 1. row ***************************
       Table: tb_emp1
Create Table: CREATE TABLE `tb_emp1` (
  `id` int DEFAULT NULL,
  `name` varchar(25) DEFAULT NULL,
  `deptId` int DEFAULT NULL,
  `salary` float DEFAULT NULL
) ENGINE=InnoDB DEFAULT CHARSET=utf8mb4 COLLATE=utf8mb4_0900_ai_ci
```

2.5 修改数据表

修改数据表指的是修改数据库中已经存在的数据表的结构。MySQL使用ALTER TABLE语句修改数据表。常用的修改表的操作有修改表名、修改字段数据类型或字段名、增加或删除字段、修改字段的排列位置、更改表的存储引擎、删除表的外键约束等。本节将对与修改表有关的操作进行讲解。

2.5.1 修改表名

MySQL是通过ALTER TABLE语句来实现表名的修改的，具体的语法格式如下：

```
ALTER TABLE <旧表名> RENAME [TO] <新表名>;
```

其中，TO为可选参数，使用与否均不影响结果。

【例2.16】将数据表tb_dept3改名为tb_deptment3。

（1）执行修改表名操作之前，使用SHOW TABLES查看数据库中所有的表。

```
mysql> SHOW TABLES;
+--------------------+
| Tables_in_test_db  |
+--------------------+
| tb_dept1           |
| tb_dept2           |
```

```
| tb_dept3           |
...省略部分内容
```

（2）使用ALTER TABLE将表tb_dept3改名为tb_deptment3，SQL语句如下：

```
ALTER TABLE tb_dept3 RENAME tb_deptment3;
```

（3）语句执行之后，检验表tb_dept3是否改名成功。使用SHOW TABLES查看数据库中的表，结果如下：

```
mysql> SHOW TABLES;
+--------------------+
| Tables_in_test_db  |
+--------------------+
| tb_dept            |
| tb_dept2           |
| tb_deptment3       |
...省略部分内容
```

经过比较可以看到，数据表列表中已经有了名称为tb_deptment3的表，表名修改成功。

> **提示** 在修改表名称时，可以使用DESC命令查看表修改前后两张表的结构，修改表名并不修改表的结构，因此修改名称后的表和修改名称前的表的结构必然相同。

2.5.2 修改字段的数据类型

修改字段的数据类型，就是把字段的数据类型转换成另一种数据类型。在MySQL中，修改字段数据类型的语法格式如下：

```
ALTER TABLE <表名> MODIFY <字段名> <数据类型>
```

其中，"表名"指要修改数据类型的字段所在表的名称，"字段名"指需要修改的字段，"数据类型"指修改后的字段的新数据类型。

【例2.17】将数据表tb_dept1中name字段的数据类型由VARCHAR(22)修改成VARCHAR(30)。

（1）执行修改表名操作之前，使用DESC查看tb_dept1表结构，结果如下：

```
mysql> DESC tb_dept1;
+----------+-------------+------+-----+---------+-------+
| Field    | Type        | Null | Key | Default | Extra |
+----------+-------------+------+-----+---------+-------+
| id       | int         | NO   | PRI | NULL    |       |
| name     | varchar(22) | YES  |     | NULL    |       |
| location | varchar(50) | YES  |     | NULL    |       |
+----------+-------------+------+-----+---------+-------+
```

可以看到，name字段的数据类型为VARCHAR(22)。

(2) 修改数据类型，SQL语句如下：

```
ALTER TABLE tb_dept1 MODIFY name VARCHAR(30);
```

(3) 再次使用DESC查看tb_dept1表结构，结果如下：

```
mysql> DESC tb_dept1;
+----------+-------------+------+-----+---------+-------+
| Field    | Type        | Null | Key | Default | Extra |
+----------+-------------+------+-----+---------+-------+
| id       | int         | NO   | PRI | NULL    |       |
| name     | varchar(30) | YES  |     | NULL    |       |
| location | varchar(50) | YES  |     | NULL    |       |
+----------+-------------+------+-----+---------+-------+
```

可以发现，表tb_dept1中name字段的数据类型已经修改成了VARCHAR(30)，修改成功。

2.5.3 修改字段名

在MySQL中，修改表字段名的语法格式如下：

```
ALTER TABLE <表名> CHANGE <旧字段名> <新字段名> <新数据类型>;
```

其中，"旧字段名"指修改前的字段名；"新字段名"指修改后的字段名；"新数据类型"指修改后的数据类型，如果不需要修改字段的数据类型，将新数据类型设置成与原来一样即可，但数据类型不能为空。

【例2.18】将数据表tb_dept1中的location字段名称改为loc，数据类型保持不变，SQL语句如下：

```
ALTER TABLE tb_dept1 CHANGE location loc VARCHAR(50);
```

使用DESC查看表tb_dept1，结果如下：

```
mysql> DESC tb_dept1;
+-------+-------------+------+-----+---------+-------+
| Field | Type        | Null | Key | Default | Extra |
+-------+-------------+------+-----+---------+-------+
| id    | int         | NO   | PRI | NULL    |       |
| name  | varchar(30) | YES  |     | NULL    |       |
| loc   | varchar(50) | YES  |     | NULL    |       |
+-------+-------------+------+-----+---------+-------+
```

可以发现该字段的名称已经修改成功。

【例2.19】将数据表tb_dept1中的loc字段名称改为location，同时将数据类型变为VARCHAR(60)，SQL语句如下：

```
ALTER TABLE tb_dept1 CHANGE loc location VARCHAR(60);
```

使用DESC查看表tb_dept1，结果如下：

```
mysql> DESC tb_dept1;
```

```
+----------+-------------+------+-----+---------+-------+
| Field    | Type        | Null | Key | Default | Extra |
+----------+-------------+------+-----+---------+-------+
| id       | int         | NO   | PRI | NULL    |       |
| name     | varchar(30) | YES  |     | NULL    |       |
| location | varchar(60) | YES  |     | NULL    |       |
+----------+-------------+------+-----+---------+-------+
```

可以发现该字段的名称和数据类型均已经修改成功。

> **提示** CHANGE也可以只修改数据类型，实现和MODIFY同样的效果，方法是将SQL语句中的"新字段名"和"旧字段名"设置为相同的名称，只改变"数据类型"。由于不同类型的数据在机器中存储的方式及长度并不相同，修改数据类型可能会影响数据表中已有的数据记录，因此当数据表中已经有数据时，不要轻易修改数据类型。

2.5.4 添加字段

随着业务需求的变化，可能需要在已经存在的表中添加新的字段。一个完整字段包括字段名、数据类型、完整性约束。添加字段的语法格式如下：

```
ALTER TABLE <表名> ADD <新字段名> <数据类型>
[约束条件] [FIRST | AFTER 已存在字段名];
```

其中，"新字段名"为需要添加的字段的名称；"FIRST"为可选参数，其作用是将新添加的字段设置为表的第一个字段；"AFTER"为可选参数，其作用是将新添加的字段添加到指定的"已存在字段名"的后面。

> **提示** "FIRST"或"AFTER 已存在字段名"用于指定新增字段在表中的位置，如果SQL语句中没有这两个参数，则默认将新添加的字段放置在数据表的最后一列。

1. 添加无完整性约束条件的字段

【例2.20】在数据表tb_dept1中添加一个没有完整性约束的INT类型的字段managerId（部门经理编号），SQL语句如下：

```
ALTER TABLE tb_dept1 ADD managerId INT;
```

使用DESC查看表tb_dept1，结果如下：

```
mysql> DESC tb_dept1;
+-----------+-------------+------+-----+---------+-------+
| Field     | Type        | Null | Key | Default | Extra |
+-----------+-------------+------+-----+---------+-------+
| id        | int         | NO   | PRI | NULL    |       |
| name      | varchar(30) | YES  |     | NULL    |       |
| location  | varchar(60) | YES  |     | NULL    |       |
| managerId | int         | YES  |     | NULL    |       |
+-----------+-------------+------+-----+---------+-------+
```

可以发现,在表的最后一列添加了一个名为"managerId"的INT类型的字段。

2. 添加有完整性约束条件的字段

【例2.21】在数据表tb_dept1中添加一个不能为空的VARCHAR(12)类型的字段column1,SQL语句如下:

```
ALTER TABLE tb_dept1 ADD column1 VARCHAR(12) not null;
```

使用DESC查看表tb_dept1,结果如下:

```
mysql> DESC tb_dept1;
+-----------+-------------+------+-----+---------+-------+
| Field     | Type        | Null | Key | Default | Extra |
+-----------+-------------+------+-----+---------+-------+
| id        | int         | NO   | PRI | NULL    |       |
| name      | varchar(30) | YES  |     | NULL    |       |
| location  | varchar(60) | YES  |     | NULL    |       |
| managerId | int         | YES  |     | NULL    |       |
| column1   | varchar(12) | NO   |     | NULL    |       |
+-----------+-------------+------+-----+---------+-------+
```

可以发现,在表的最后一列添加了一个名为"column1"的VARCHAR(12)类型且不为空的字段。

3. 在表的第一列添加一个字段

【例2.22】在数据表tb_dept1中添加一个INT类型的字段column2,SQL语句如下:

```
ALTER TABLE tb_dept1 ADD column2 INT FIRST;
```

使用DESC查看表tb_dept1,结果如下:

```
mysql> DESC tb_dept1;
+-----------+-------------+------+-----+---------+-------+
| Field     | Type        | Null | Key | Default | Extra |
+-----------+-------------+------+-----+---------+-------+
| column2   | int         | YES  |     | NULL    |       |
| id        | int         | NO   | PRI | NULL    |       |
| name      | varchar(30) | YES  |     | NULL    |       |
| location  | varchar(60) | YES  |     | NULL    |       |
| managerId | int         | YES  |     | NULL    |       |
| column1   | varchar(12) | NO   |     | NULL    |       |
+-----------+-------------+------+-----+---------+-------+
```

可以发现,在表的第一列中添加了一个名为"column2"的INT类型字段。

4. 在表的指定列之后添加一个字段

【例2.23】在数据表tb_dept1的name列后添加一个INT类型的字段column3,SQL语句如下:

```
ALTER TABLE tb_dept1 ADD column3 INT AFTER name;
```

使用DESC查看表tb_dept1,结果如下:

```
mysql> DESC tb_dept1;
+-------------+-------------+------+-----+---------+-------+
| Field       | Type        | Null | Key | Default | Extra |
+-------------+-------------+------+-----+---------+-------+
| column2     | int         | YES  |     | NULL    |       |
| id          | int         | NO   | PRI | NULL    |       |
| name        | varchar(30) | YES  |     | NULL    |       |
| column3     | int         | YES  |     | NULL    |       |
| location    | varchar(60) | YES  |     | NULL    |       |
| managerId   | int         | YES  |     | NULL    |       |
| column1     | varchar(12) | NO   |     | NULL    |       |
+-------------+-------------+------+-----+---------+-------+
```

可以看到，tb_dept1表中增加了一个名称为column3的字段，其位置在指定的name字段后面，添加字段成功。

2.5.5 删除字段

删除字段是将数据表中的某个字段从表中移除，语法规则如下：

```
ALTER TABLE <表名> DROP <字段名>;
```

其中"字段名"指需要从表中删除的字段的名称。

【例2.24】删除数据表tb_dept1中的column2字段。

（1）删除字段之前，使用DESC查看tb_dept1表结构，结果如下：

```
mysql> DESC tb_dept1;
+-------------+-------------+------+-----+---------+------+
| Field       | Type        | Null | Key | Default | Extr |
+-------------+-------------+------+-----+---------+------+
| column2     | int         | YES  |     | NULL    |      |
| id          | int         | NO   | PRI | NULL    |      |
| name        | varchar(30) | YES  |     | NULL    |      |
| column3     | int         | YES  |     | NULL    |      |
| location    | varchar(60) | YES  |     | NULL    |      |
| managerId   | int         | YES  |     | NULL    |      |
| column1     | varchar(12) | NO   |     | NULL    |      |
+-------------+-------------+------+-----+---------+------+
```

（2）删除column2字段，SQL语句如下：

```
ALTER TABLE tb_dept1 DROP column2;
```

（3）再次使用DESC查看表tb_dept1，结果如下：

```
mysql> DESC tb_dept1;
+-------------+-------------+------+-----+---------+------+
| Field       | Type        | Null | Key | Default | Extr |
+-------------+-------------+------+-----+---------+------+
| id          | int         | NO   | PRI | NULL    |      |
| name        | varchar(30) | YES  |     | NULL    |      |
```

```
| column3     | int         | YES  |     | NULL    |       |
| location    | varchar(60) | YES  |     | NULL    |       |
| managerId   | int         | YES  |     | NULL    |       |
| column1     | varchar(12) | NO   |     | NULL    |       |
+-------------+-------------+------+-----+---------+-------+
```

可以看到，tb_dept1表中已经不存在名称为column2的字段，说明删除字段成功。

2.5.6 修改字段的排列位置

对于一张数据表来说，在创建的时候，表中字段的排列顺序就已经确定了，但表的结构并不是完全不可以改变的，可以通过ALTER TABLE来改变表中字段的相对位置。语法规则如下：

```
ALTER TABLE <表名> MODIFY <字段1> <数据类型> FIRST|AFTER <字段2>;
```

其中，"字段1"指要修改位置的字段；"数据类型"指"字段1"的数据类型；"FIRST"为可选参数，指将"字段1"修改为表的第一个字段；"AFTER 字段2"指将"字段1"插在"字段2"后面。

1. 修改字段为表的第一个字段

【例2.25】将数据表tb_dept1中的column1字段修改为表的第一个字段，SQL语句如下：

```
ALTER TABLE tb_dept1 MODIFY column1 VARCHAR(12) FIRST;
```

使用DESC查看表tb_dept1，结果如下：

```
mysql> DESC tb_dept1;
+-------------+-------------+------+-----+---------+-------+
| Field       | Type        | Null | Key | Default | Extra |
+-------------+-------------+------+-----+---------+-------+
| column1     | varchar(12) | YES  |     | NULL    |       |
| id          | int         | NO   | PRI | NULL    |       |
| name        | varchar(30) | YES  |     | NULL    |       |
| column3     | int         | YES  |     | NULL    |       |
| location    | varchar(60) | YES  |     | NULL    |       |
| managerId   | int         | YES  |     | NULL    |       |
+-------------+-------------  -+------+-----+---------+-------+
```

可以发现字段column1已经被移至表的第一列。

2. 修改字段到表的指定列之后

【例2.26】将数据表tb_dept1中的column1字段插到location字段后面，SQL语句如下：

```
ALTER TABLE tb_dept1 MODIFY column1 VARCHAR(12) AFTER location;
```

使用DESC查看表tb_dept1，结果如下：

```
mysql> DESC tb_dept1;
+-------------+-------------+------+-----+---------+-------+
| Field       | Type        | Null | Key | Default | Extra |
+-------------+-------------+------+-----+---------+-------+
```

```
| id        | int         | NO  | PRI | NULL |  |
| name      | varchar(30) | YES |     | NULL |  |
| column3   | int         | YES |     | NULL |  |
| location  | varchar(60) | YES |     | NULL |  |
| column1   | varchar(12) | YES |     | NULL |  |
| managerId | int         | YES |     | NULL |  |
+-----------+-------------+-----+-----+------+--+
```

可以看到，表tb_dept1中的字段column1已经被移至location字段之后。

2.5.7 删除表的外键约束

对于数据库中定义的外键，如果不再需要，可以将其删除。外键一旦删除，就会解除主表和从表间的关联关系。在MySQL中删除外键的语法格式如下：

```
ALTER TABLE <表名> DROP FOREIGN KEY <外键约束名>
```

其中，"外键约束名"指在定义表时CONSTRAINT关键字后面的参数，详细内容可参考2.3.3节。

【例2.27】删除数据表tb_emp9中的外键约束。

（1）创建表tb_emp9，其外键deptId关联tb_dept1表的主键id，SQL语句如下：

```
CREATE TABLE tb_emp9
(
id       INT PRIMARY KEY,
name     VARCHAR(25),
deptId   INT,
salary   FLOAT,
CONSTRAINT fk_emp_dept  FOREIGN KEY (deptId) REFERENCES tb_dept1(id)
);
```

（2）使用SHOW CREATE TABLE查看表tb_emp9的结构，结果如下：

```
mysql> SHOW CREATE TABLE tb_emp9 \G
*** 1. row ***
       Table: tb_emp9
Create Table: CREATE TABLE `tb_emp9` (
  `id` INT NOT NULL,
  `name` varchar(25) DEFAULT NULL,
  `deptId` INT DEFAULT NULL,
  `salary` float DEFAULT NULL,
  PRIMARY KEY (`id`),
  KEY `fk_emp_dept` (`deptId`),
  CONSTRAINT `fk_emp_dept` FOREIGN KEY (`deptId`) REFERENCES `tb_dept1` (`id`)
) ENGINE=InnoDB DEFAULT CHARSET=utf8mb4 COLLATE=utf8mb4_0900_ai_ci
1 row in set (0.00 sec)
```

可以看到，已经成功添加了表的外键。

（3）删除外键约束，SQL语句如下：

```
ALTER TABLE tb_emp9 DROP FOREIGN KEY fk_emp_dept;
```

上述语句执行完毕之后，将删除表tb_emp9的外键约束。

（4）使用SHOW CREATE TABLE再次查看表tb_emp9结构，结果如下：

```
mysql> SHOW CREATE TABLE tb_emp9 \G
*** 1. row ***
       Table: tb_emp9
Create Table: CREATE TABLE `tb_emp9` (
  `id` INT NOT NULL,
  `name` varchar(25) DEFAULT NULL,
  `deptId` INT DEFAULT NULL,
  `salary` float DEFAULT NULL,
  PRIMARY KEY (`id`),
  KEY `fk_emp_dept` (`deptId`)
) ENGINE=InnoDB DEFAULT CHARSET=utf8mb4 COLLATE=utf8mb4_0900_ai_ci
1 row in set (0.00 sec)
```

可以看到，表tb_emp9中已经不存在FOREIGN KEY，原有的名称为fk_emp_dept的外键约束已被删除。

2.6 删除数据表

删除数据表就是将数据库中已经存在的表从数据库中删除。注意，在删除表的同时，表的定义和表中所有的数据均会被删除。因此，在进行删除操作前，最好对表中的数据做一个备份，以免造成无法挽回的后果。本节将详细讲解数据表的删除方法。

2.6.1 删除没有被关联的表

在MySQL中，使用DROP TABLE可以一次性删除一张或多张没有被其他表关联的数据表。语法格式如下：

```
DROP TABLE [IF EXISTS]表1,表2,...,表n;
```

其中，"表n"指要删除的表的名称，可以同时删除多张表，只需将要删除的表名依次写在后面，相互之间用逗号隔开即可。如果要删除的数据表不存在，则MySQL会提示一条错误信息：ERROR 1051 (42S02): Unknown table '表名'。参数"IF EXISTS"用于在删除前判断要删除的表是否存在，加上该参数后，删除表时，如果表不存在，SQL语句也可以顺利执行，但是会发出警告。

在【例2.10】中，已经创建了名为"tb_dept2"的数据表。下面使用删除语句将该表删除。

【例2.28】删除数据表tb_dept2，SQL语句如下：

```
DROP TABLE IF EXISTS tb_dept2;
```

语句执行完毕之后，使用SHOW TABLES命令查看当前数据库中所有的表，结果如下：

```
mysql> SHOW TABLES;
+--------------------+
| Tables_in_test_db  |
+--------------------+
| tb_dept            |
| tb_deptment3       |
...省略部分内容
```

从结果中可以看到，数据表列表中已经不存在名称为"tb_dept2"的表，说明删除表操作成功。

2.6.2　删除被其他表关联的主表

在数据表之间存在外键关联的情况下，如果直接删除父表，结果会显示失败，原因是直接删除将破坏表的参照完整性。如果必须要删除父表，可以先删除与它关联的子表，再删除父表，只是这样就同时删除了两张表中的数据。有的情况下可能要保留子表，这时若要单独删除父表，只需将关联表的外键约束条件取消，然后就可以删除父表了，下面讲解这种方法。

在数据库中创建两张关联表，首先创建表tb_dept2，SQL语句如下：

```
CREATE TABLE tb_dept2
(
id       INT PRIMARY KEY,
name     VARCHAR(22),
location VARCHAR(50)
);
```

接下来创建表tb_emp，SQL语句如下：

```
CREATE TABLE tb_emp
(
id       INT PRIMARY KEY,
name     VARCHAR(25),
deptId   INT,
salary   FLOAT,
CONSTRAINT fk_emp_dept  FOREIGN KEY (deptId) REFERENCES tb_dept2(id)
);
```

使用SHOW CREATE TABLE命令查看表tb_emp的外键约束，结果如下：

```
mysql> SHOW CREATE TABLE tb_emp\G
*** 1. row ***
       Table: tb_emp
Create Table: CREATE TABLE `tb_emp` (
  `id` INT NOT NULL,
  `name` varchar(25) DEFAULT NULL,
  `deptId` INT DEFAULT NULL,
  `salary` float DEFAULT NULL,
  PRIMARY KEY (`id`),
  KEY `fk_emp_dept` (`deptId`),
  CONSTRAINT `fk_emp_dept` FOREIGN KEY (`deptId`) REFERENCES `tb_dept2` (`id`)
```

```
) ENGINE=InnoDB DEFAULT CHARSET=utf8mb4 COLLATE=utf8mb4_0900_ai_ci
```

可以看到，以上执行结果创建了两张关联表tb_dept2和tb_emp。其中，表tb_emp为子表，具有名称为fk_emp_dept的外键约束；表tb_dept2为父表，其主键id被子表tb_emp所关联。

【例2.29】删除被数据表tb_emp关联的数据表tb_dept2。

（1）首先试着直接删除父表tb_dept2，SQL语句如下：

```
mysql> DROP TABLE tb_dept2;
ERROR 3730 (HY000): Cannot drop table 'tb_dept2' referenced by a foreign key
constraINT 'fk_emp_dept' on table 'tb_emp'.
```

如前所述，在存在外键约束时，主表不能被直接删除。

（2）解除关联子表tb_emp的外键约束，SQL语句如下：

```
ALTER TABLE tb_emp DROP FOREIGN KEY fk_emp_dept;
```

语句成功执行后，将取消表tb_emp和表tb_dept2之间的关联关系。

（3）使用如下SQL语句将原来的父表tb_dept2删除：

```
DROP TABLE tb_dept2;
```

（4）通过"SHOW TABLES;"查看数据表列表：

```
mysql> show tables;
+---------------------+
| Tables_in_test_db   |
+---------------------+
| tb_dept             |
| tb_deptment3        |
...省略部分内容
```

可以看到，数据表列表中已经不存在名称为"tb_dept2"的表。

第 3 章
数据类型和运算符

数据库表由多列字段构成,每一个字段指定了不同的数据类型。指定了字段的数据类型,也就决定了向字段插入的数据内容。例如,当要插入数值的时候,可以将它们存储为整数类型,也可以将它们存储为字符串类型。不同的数据类型也决定了MySQL在存储它们时使用的方式,以及在使用它们时选择什么运算符号进行运算。本章将介绍MySQL中的数据类型和常见的运算符。

3.1 MySQL的数据类型

MySQL支持多种数据类型,主要有数值类型、日期/时间类型和字符串类型。

(1)数值类型:包括整数类型TINYINT、SMALLINT、MEDIUMINT、INT、BIGINT,浮点小数类型FLOAT、DOUBLE,以及定点小数类型DECIMAL。

(2)日期/时间类型:包括YEAR、TIME、DATE、DATETIME和TIMESTAMP。

(3)字符串类型:包括文本字符串类型CHAR、VARCHAR、TEXT、ENUM、SET,以及二进制字符串类型BIT、BINARY、VARBINARY、TINYBLOB、BLOB、MEDIUMBLOB、LONGBLOB。

3.1.1 整数类型

数值类型主要用来存储数字。MySQL提供了多种数值数据类型,不同的数据类型提供不同的取值范围,可以存储的值的范围越大,其所需要的存储空间也会越大。MySQL主要提供的整数类型有TINYINT、SMALLINT、MEDIUMINT、INT(INTEGER)、BIGINT,如表3.1所示。整数类型的属性字段可以添加AUTO_INCREMENT自增约束条件。

表 3.1　MySQL 中的整数类型

类型名称	说　　明	存储需求
TINYINT	很小的整数	1 字节
SMALLINT	小的整数	2 字节
MEDIUMINT	中等大小的整数	3 字节
INT（INTEGER）	普通大小的整数	4 字节
BIGINT	大整数	8 字节

从表3.1中可以看到，不同整数类型存储所需的字节数是不同的，占用字节数最小的是TINYINT类型，占用字节最大的是BIGINT类型，占用字节越多的类型所能表示的数值范围越大。根据占用字节数，可以求出每一种数据类型的取值范围。例如，TINYINT需要1字节（8 bits）来存储，那么TINYINT无符号数的最大值为2^8-1（255）、TINYINT有符号数的最大值为2^7-1（127）。其他整数类型的取值范围计算方法相同，如表3.2所示。

表 3.2　不同整数类型的取值范围

数据类型	有　符　号	无　符　号
TINYINT	−128~127	0~255
SMALLINT	−32768~32767	0~65535
MEDIUMINT	−8388608~8388607	0~16777215
INT（INTEGER）	−2147483648~2147483647	0~4294967295
BIGINT	−9223372036854775808~9223372036854775807	0~18446744073709551615

【例3.1】创建表tmp1，其中字段x、y、z、m、n的数据类型依次为TINYINT、SMALLINT、MEDIUMINT、INT、BIGINT，SQL语句如下：

```
CREATE TABLE tmp1 ( x TINYINT, y SMALLINT, z MEDIUMINT, m INT, n BIGINT );
```

语句执行成功之后，使用DESC查看表结构，结果如下：

```
mysql> DESC tmp1;
+-------+-----------+------+-----+---------+-------+
| Field | Type      | Null | Key | Default | Extra |
+-------+-----------+------+-----+---------+-------+
| x     | tinyint   | YES  |     | NULL    |       |
| y     | smallint  | YES  |     | NULL    |       |
| z     | mediumint | YES  |     | NULL    |       |
| m     | int       | YES  |     | NULL    |       |
| n     | bigint    | YES  |     | NULL    |       |
+-------+-----------+------+-----+---------+-------+
```

不同的整数类型有不同的取值范围，并且需要不同的存储空间，因此应该根据实际需要选择合适的数据类型，这样有利于提高查询的效率和节省存储空间。

3.1.2　小数类型

MySQL中使用浮点数和定点数来表示小数。浮点数类型有两种：单精度浮点类型（FLOAT）

和双精度浮点类型（DOUBLE）。定点数类型只有一种：DECIMAL。定点数类型都可以用(M,N)来表示。其中，M称为精度，表示总共的位数；N称为标度，表示小数的位数。表3.3列出了MySQL中的小数类型和存储需求。

表 3.3 MySQL 中的小数类型

类型名称	说 明	存储需求
FLOAT	单精度浮点数	4 字节
DOUBLE	双精度浮点数	8 字节
DECIMAL(M,D)，DEC	压缩的"严格"定点数	M+2 字节

DECIMAL类型不同于FLOAT和DOUBLE，它实际是以串存放的，可能的最大取值范围与DOUBLE一样，但其有效的取值范围由M和D的值决定。如果改变M而固定D，则其取值范围将随M的变大而变大。从表3.3中可以看到，DECIMAL的存储空间并不是固定的，而由其精度值M决定的，其占用M+2字节。

FLOAT类型的取值范围如下：

- 有符号的取值范围：$-3.402823466E+38 \sim -1.175494351E-38$。
- 无符号的取值范围：0和$1.175494351E-38 \sim 3.402823466E+38$。

DOUBLE类型的取值范围如下：

- 有符号的取值范围：$-1.7976931348623157E+308 \sim -2.2250738585072014E-308$。
- 无符号的取值范围：0和$2.2250738585072014E-308 \sim 1.7976931348623157E+308$。

【例3.2】创建表tmp2，其中字段x、y、z的数据类型依次为FLOAT、DOUBLE和DECIMAL(5,1)，SQL语句如下：

```
CREATE TABLE tmp2 ( x FLOAT, y DOUBLE, z DECIMAL(5,1) );
```

向表中插入数据5.12、5.15和5.123，SQL语句如下：

```
mysql>INSERT INTO tmp2 VALUES(5.12, 5.15, 5.123);
```

在插入数据时，MySQL给出了一个警告信息，使用SHOW WARNINGS;语句查看警告信息：

```
mysql> SHOW WARNINGS;
+-------+------+---------------------------------------+
| Level | Code | Message                               |
+-------+------+---------------------------------------+
| Note  | 1265 | Data truncated for column 'z' at row 1 |
+-------+------+---------------------------------------+
```

可以看到，给出z字段数值被截断的警告。使用SELECT * FROM tmp2语句查看数据输入结果，结果如下：

```
mysql> SELECT * FROM tmp2;
+------+------+------+
| x    | y    | z    |
+------+------+------+
```

```
| 5.1  | 5.2  | 5.1  |
+------+------+------+
```

3.1.3 日期与时间类型

MySQL中有多种表示日期的数据类型,主要有DATETIME、DATE、TIMESTAMP、TIME和YEAR。例如,当只记录年信息的时候,可以只使用YEAR类型,而没有必要使用DATE类型。每一个类型都有合法的取值范围,当指定确实不合法的值时,系统将"零"值插入数据库中。表3.4列出了MySQL中的日期与时间类型。

表 3.4 日期与时间数据类型

类型名称	日期格式	日期范围	存储需求
YEAR	YYYY	1901～2155	1字节
TIME	HH:MM:SS	-838:59:59 ～838:59:59	3字节
DATE	YYYY-MM-DD	1000-01-01～9999-12-3	3字节
DATETIME	YYYY-MM-DD HH:MM:SS	1000-01-01 00:00:00～9999-12-31 23:59:59	8字节
TIMESTAMP	YYYY-MM-DD HH:MM:SS	1970-01-01 00:00:01 UTC～2038-01-19 03:14:07 UTC	4字节

1. YEAR类型

YEAR是一个单字节类型,用于表示年,在存储时只需要1字节。可以使用以下3种格式指定YEAR值:

(1)以4位字符串或者4位数字格式表示的YEAR,范围为1901～2155。输入格式为"YYYY"或者YYYY。例如,输入'2010'或2010,插入数据库的值均为2010。

(2)以2位字符串格式表示的YEAR,范围为00～99。00～69和70～99范围的值分别被转换为2000～2069和1970～1999范围的YEAR值。"0"与"00"的作用相同。超过取值范围的值将被转换为2000。

(3)以2位数字表示的YEAR,范围为1～99。1～69和70～99范围的值分别被转换为2001～2069和1970～1999范围的YEAR值。注意:在这里0值将被转换为0000,而不是2000。

> **提示** 两位整数范围与两位字符串范围稍有不同。例如,要插入2000年,读者可能会使用数字格式的0表示YEAR,实际上,插入数据库的值为0000,而不是所希望的2000。只有使用字符串格式的'0'或'00',才可以被正确地解释为2000。非法YEAR值将被转换为0000。

【例3.3】创建数据表tmp3,定义数据类型为YEAR的字段y,向表中插入值2010、"2010"、"2166"。

(1)创建表tmp3,SQL语句如下:
```
CREATE TABLE tmp3(y YEAR );
```
(2)向表中插入数据:
```
mysql> INSERT INTO tmp3 values(2010),('2010');
```

（3）再次向表中插入数据：

```
mysql> INSERT INTO tmp3 values ('2166');
ERROR 1264 (22003): Out of range value for column 'y' at row 1
```

（4）上述语句执行之后，MySQL给出了一条错误提示，使用SHOW查看错误信息：

```
mysql> SHOW WARNINGS;
+-------+------+---------------------------------------------+
| Level | Code | Message                                     |
+-------+------+---------------------------------------------+
| Error | 1264 | Out of range value for column 'y' at row 1; |
+-------+------+---------------------------------------------+
```

可以看到，插入的第3个值2166超过了YEAR类型的取值范围，此时不能正常执行插入操作。

（5）查看插入结果：

```
mysql> SELECT * FROM tmp3;
+------+
| y    |
+------+
| 2010 |
| 2010 |
+------+
```

由结果可以看到，当插入值为数值类型的2010或者字符串类型的"2010"时，都正确地储存到了数据库中；而当插入值为"2166"时，由于它超出了YEAR类型的取值范围，因此不能插入。

【例3.4】向表tmp3中的y字段插入2位字符串表示的YEAR值，分别为"0""00""77"和"10"。

（1）删除表中的数据：

```
DELETE FROM tmp3;
```

（2）向表中插入数据：

```
INSERT INTO tmp3 values('0'),('00'),('77'),('10');
```

（3）查看插入结果：

```
mysql> SELECT * FROM tmp3;
+------+
| y    |
+------+
| 2000 |
| 2000 |
| 1977 |
| 2010 |
+------+
```

由结果可以看到，字符串"0"和"00"的作用相同，都被转换成了2000年；"77"被转换为1977；"10"被转换为2010。

【例3.5】向表tmp3中的y字段插入2位数字表示的YEAR值，分别为0、78和11。

（1）删除表中的数据：

```
DELETE FROM tmp3;
```

（2）向表中插入数据：

```
INSERT INTO tmp3 values(0),(78),(11);
```

（3）查看插入结果：

```
mysql> SELECT * FROM tmp3;
+------+
| y    |
+------+
| 0000 |
| 1978 |
| 2011 |
+------+
```

由结果可以看到，0被转换为0000，78被转换为1978，11被转换为2011。

2. TIME类型

TIME类型用在只需要时间信息时，在存储时需要3字节，格式为"HH:MM:SS"。其中，HH表示小时，MM表示分钟，SS表示秒。TIME类型的取值范围为-838:59:59～838:59:59，小时部分取值范围如此大的原因是，TIME类型不仅可以表示一天的时间（必须小于24小时），还可以表示某个事件过去的时间或两个事件之间的时间间隔(可以大于24小时,甚至可以为负)。可以使用以下两种格式指定TIME值：

（1）"D HH:MM:SS"格式的字符串。可以使用下面任何一种"非严格"的语法："HH:MM:SS""HH:MM""D HH:MM""D HH"或"SS"。这里的D表示日，可以取0~31的值。在插入数据库时，D被转换为小时保存，格式为"D*24 + HH"。

（2）"HHMMSS"格式的、没有间隔符的字符串或者HHMMSS格式的数值，假定是有意义的时间。例如，"101112"被理解为"10:11:12"，而"109712"是不合法的（它有一个没有意义的分钟部分），存储时将变为00:00:00。

> **提示** 为TIME列分配简写值时应注意：如果TIME值中没有冒号，那么MySQL在解释值时，假定最右边的两位表示秒（MySQL解释TIME值为过去的时间而不是当天的时间）。例如，读者可能认为"1112"和1112都表示11:12:00（11点12分），但MySQL将它们解释为00:11:12（11分12秒）。同样地，"12"和12被解释为00:00:12。相反，如果TIME值中使用冒号，则肯定被看作当天的时间，也就是说，"11:12"表示11:12:00，而不是00:11:12。

【例3.6】创建数据表tmp4，定义数据类型为TIME的字段t，向表中插入值"10:05:05" "23:23" "2 10:10" "3 02"和"10"。

（1）创建表tmp4：

```
CREATE TABLE tmp4( t TIME );
```

（2）向表中插入数据：

```
mysql> INSERT INTO tmp4 values('10:05:05 '), ('23:23'), ('2 10:10'), ('3 02'), ('10');
```

（3）查看插入结果：

```
mysql> SELECT * FROM tmp4;
+----------+
| t        |
+----------+
| 10:05:05 |
| 23:23:00 |
| 58:10:00 |
| 74:00:00 |
| 00:00:10 |
+----------+
```

由结果可以看到，"10:05:05"被转换为10:05:05；"23:23"被转换为23:23:00；"2 10:10"被转换为58:10:00，"3 02"被转换为74:00:00；"10"被转换成00:00:10。

> **提示** 在使用"D HH"格式时，小时一定要使用双位数值，如果是小于10的小时数，应在前面加0。

【例3.7】向表tmp4中插入值"101112"、111213、"0"和107010。

（1）删除表中的数据：

```
DELETE FROM tmp4;
```

（2）向表中插入数据：

```
mysql>INSERT INTO tmp4 values('101112'),(111213),( '0');
```

（3）再向表中插入数据：

```
mysql>INSERT INTO tmp4 values ( 107010);
ERROR 1292 (22007): Incorrect time value: '107010' for column 't' at row 1
```

可以看到，在插入数据时，MySQL给出了一个错误提示信息，使用"SHOW WARNINGS;"语句查看错误信息，结果如下：

```
mysql> show warnings;
+---------+------+----------------------------------------------------+
| Level   | Code | Message                                            |
+---------+------+----------------------------------------------------+
```

```
| Error    | 1292 |Incorrect time value: '107010' for column 't' at row 1|
+---------+------+-------------------------------------------------------+
```

可以看到，第二次插入数据的时候，数据超出了范围，原因是107010的分钟部分超过了60（分钟部分是不会超过60的）。

（4）查看插入结果：

```
mysql> SELECT * FROM tmp4;
+----------+
| t        |
+----------+
| 10:11:12 |
| 11:12:13 |
| 00:00:00 |
+----------+
```

由结果可以看到，"101112"被转换为10:11:12；111213被转换为11:12:13；"0"被转换为00:00:00；107010因为是不合法的值，所以不能插入。

也可以使用系统日期函数向TIME字段列插入值。

【例3.8】向表tmp4中插入系统当前时间。

（1）删除表中的数据：

```
DELETE FROM tmp4;
```

（2）向表中插入数据：

```
mysql> INSERT INTO tmp4 values (CURRENT_TIME) ,(NOW());
```

（3）查看插入结果：

```
mysql> SELECT * FROM tmp4;
+----------+
| t        |
+----------+
| 08:43:51 |
| 08:43:51 |
+----------+
```

由结果可以看到，系统当前的日期已插入TIME类型的t列。因为读者输入语句的时间不确定，所以获取的值可能与这里的不同，但都是系统当前的时间值。

3. DATE类型

DATE类型用在仅需要日期值时，没有时间部分，在存储时需要3字节。日期格式为"YYYY-MM-DD"。其中，YYYY表示年，MM表示月，DD表示日。在给DATE类型的字段赋值时，可以使用字符串类型或者数字类型，只要符合DATE的日期格式即可。

（1）以"YYYY-MM-DD"或者"YYYYMMDD"字符串格式表示的日期，取值范围为1000-01-01～9999-12-3。例如，输入"2012-12-31"或者"20121231"，插入数据库的日期都为2012-12-31。

（2）以"YY-MM-DD"或者"YYMMDD"字符串格式表示的日期，这里YY表示两位的年值。MySQL使用以下规则解释两位年值：00～69范围的年值被转换为2000～2069；70～99范围的年值被转换为1970～1999。例如，输入"12-12-31"，插入数据库的日期为2012-12-31；输入"981231"，插入数据的日期为1998-12-31。

（3）以YY-MM-DD或者YYMMDD数字格式表示的日期。与前面类似，00~69范围的年值被转换为2000～2069，70～99范围的年值被转换为1970～1999。例如，输入12-12-31，插入数据库的日期为2012-12-31；输入981231，插入数据库的日期为1998-12-31。

（4）使用CURRENT_DATE或者NOW()，插入系统当前日期。

【例3.9】创建数据表tmp5，定义数据类型为DATE的字段d，向表中插入"YYYY-MM-DD"和"YYYYMMDD"字符串格式的日期。

（1）创建表tmp5：

```
MySQL> CREATE TABLE tmp5(d DATE);
```

（2）向表中插入"YYYY-MM-DD"和"YYYYMMDD"格式的日期：

```
MySQL> INSERT INTO tmp5 values('1998-08-08'),('19980808'),('20101010');
```

（3）查看插入结果：

```
MySQL> SELECT * FROM tmp5;
+------------+
| d          |
+------------+
| 1998-08-08 |
| 1998-08-08 |
| 2010-10-10 |
+------------+
```

可以看到，各个不同类型的日期值都被正确地插入数据表中。

【例3.10】向表tmp5中插入"YY-MM-DD"和"YYMMDD"字符串格式的日期。

（1）删除表中的数据：

```
DELETE FROM tmp5;
```

（2）向表中插入"YY-MM-DD"和"YYMMDD"格式的日期：

```
mysql> INSERT INTO tmp5 values ('99-09-09'),( '990909'), ('000101') ,('111111');
```

（3）查看插入结果：

```
mysql> SELECT * FROM tmp5;
+------------+
| d          |
+------------+
| 1999-09-09 |
| 1999-09-09 |
| 2000-01-01 |
```

```
| 2011-11-11 |
+------------+
```

【例3.11】向表tmp5中插入YYYYMMDD和YYMMDD数字格式日期。

（1）删除表中的数据：

`DELETE FROM tmp5;`

（2）向表中插入YYYYMMDD和YYMMDD格式的日期：

`mysql> INSERT INTO tmp5 values (19990909),(990909), ( 000101) ,( 111111);`

（3）查看插入结果：

```
mysql> SELECT * FROM tmp5;
+------------+
| d          |
+------------+
| 1999-09-09 |
| 1999-09-09 |
| 2000-01-01 |
| 2011-11-11 |
+------------+
```

【例3.12】向表tmp5中插入系统当前日期。

（1）删除表中的数据：

`DELETE FROM tmp5;`

（2）向表中插入系统当前日期：

`mysql> INSERT INTO tmp5 values( CURRENT_DATE() ),( NOW() );`

（3）查看插入结果：

```
mysql> SELECT * FROM tmp5;
+------------+
| d          |
+------------+
| 2024-08-09 |
| 2024-08-09 |
+------------+
```

CURRENT_DATE只返回当前日期值，不包括时间部分；NOW()函数返回日期和时间值，在保存到数据库时，只保留日期部分。

4. DATETIME类型

DATETIME类型用在需要同时包含日期和时间信息时，在存储时需要8字节。日期格式为"YYYY-MM-DD HH:MM:SS"。其中，YYYY表示年，MM表示月，DD表示日，HH表示小时，MM表示分钟，SS表示秒。在给DATETIME类型的字段赋值时，可以使用字符串类型或者数字类型，只要符合DATETIME的日期格式即可。

（1）以"YYYY-MM-DD HH:MM:SS"或者"YYYYMMDDHHMMSS"字符串格式表示的值，取值范围为1000-01-01 00:00:00～9999-12-3 23:59:59。例如，输入"2012-12-31 05:05:05"或者"20121231050505"，插入数据库的DATETIME值都为2012-12-31 05: 05: 05。

（2）以"YY-MM-DD HH:MM:SS"或者"YYMMDDHHMMSS"字符串格式表示的日期，这里YY表示两位的年值。与前面相同，00～69范围的年值被转换为2000～2069，70～99范围的年值被转换为1970～1999。例如，输入"12-12-31 05:05:05"，插入数据库的DATETIME为2012-12-31 05:05:05；输入"980505050505"，插入数据库的DATETIME为1998-05-05 05:05:05。

（3）以YYYYMMDDHHMMSS或者YYMMDDHHMMSS数字格式表示的日期和时间。例如，输入20121231050505，插入数据库的DATETIME为2012-12-31 05:05:05；输入981231050505，插入数据的DATETIME为1998-12-31 05:05:05。

【例3.13】创建数据表tmp6，定义数据类型为DATETIME的字段dt，向表中插入"YYYY-MM-DD HH:MM:SS"和"YYYYMMDDHHMMSS"字符串格式日期和时间值。

（1）创建表tmp6：

```
CREATE TABLE tmp6( dt DATETIME );
```

（2）向表中插入"YYYY-MM-DD HH:MM:SS"和"YYYYMMDDHHMMSS"格式的日期：

```
mysql> INSERT INTO tmp6 values('1998-08-08 08:08:08'),('19980808080808'),
('20101010101010');
```

（3）查看插入结果：

```
mysql> SELECT * FROM tmp6;
+---------------------+
| dt                  |
+---------------------+
| 1998-08-08 08:08:08 |
| 1998-08-08 08:08:08 |
| 2010-10-10 10:10:10 |
+---------------------+
```

可以看到，各个不同类型的日期值都被正确地插入数据表中。

【例3.14】向表tmp6中插入"YY-MM-DD HH:MM:SS"和"YYMMDDHHMMSS"字符串格式的日期和时间值。

（1）删除表中的数据：

```
DELETE FROM tmp6;
```

（2）向表中插入"YY-MM-DD HH:MM:SS"和"YYMMDDHHMMSS"格式的日期：

```
mysql> INSERT INTO tmp6 values('99-09-09 09:09:09'),('990909090909'),
('101010101010');
```

（3）查看插入结果：

```
mysql> SELECT * FROM tmp6;
```

```
+---------------------+
| dt                  |
+---------------------+
| 1999-09-09 09:09:09 |
| 1999-09-09 09:09:09 |
| 2010-10-10 10:10:10 |
+---------------------+
```

【例3.15】向表tmp6中插入YYYYMMDDHHMMSS 和YYMMDDHHMMSS数字格式的日期和时间值。

(1) 删除表中的数据:

```
DELETE FROM tmp6;
```

(2) 向表中插入YYYYMMDDHHMMSS和YYMMDDHHMMSS数字格式的日期和时间：

```
mysql> INSERT INTO tmp6 values(19990909090909), (101010101010);
```

(3) 查看插入结果:

```
mysql> SELECT * FROM tmp6;
+---------------------+
| dt                  |
+---------------------+
| 1999-09-09 09:09:09 |
| 2010-10-10 10:10:10 |
+---------------------+
```

【例3.16】向表tmp6中插入系统当前日期和时间。

(1) 删除表中的数据:

```
DELETE FROM tmp6;
```

(2) 向表中插入系统当前日期和时间：

```
mysql> INSERT INTO tmp6 values( NOW() );
```

(3) 查看插入结果:

```
mysql> SELECT * FROM tmp6;
+---------------------+
| dt                  |
+---------------------+
| 2024-08-09 17:07:30 |
+---------------------+
```

NOW()函数返回当前系统的日期和时间，格式为"YYYY-MM-DD HH:MM:SS"。

提示 MySQL允许"不严格"语法：任何标点符号都可以用作日期部分或时间部分之间的间隔符。例如，"98-12-31 11:30:45" "98.12.31 11+30+45" "98/12/31 11*30*45" 和 "98@12@31 11^30^45" 是等价的，这些值都可以被正确地插入数据库中。

5. TIMESTAMP类型

TIMESTAMP的显示格式与DATETIME相同，显示宽度固定在19个字符，日期格式为YYYY-MM-DD HH:MM:SS，在存储时需要4字节。TIMESTAMP的取值范围小于DATETIME的取值范围，为"1970-01-01 00:00:01"UTC～"2038-01-19 03:14:07"UTC。其中，UTC（Coordinated Universal Time）为世界标准时间，因此插入的数据要保证在合法的取值范围内。

【例3.17】创建数据表tmp7，定义数据类型为TIMESTAMP的字段ts，向表中插入值"19950101010101""950505050505""1996-02-02 02:02:02""97@03@03 03@03@03"和121212121212、NOW()。

（1）创建数据表tmp7：

```
CREATE TABLE tmp7( ts TIMESTAMP);
```

（2）向表中插入数据：

```
INSERT INTO tmp7 values ('19950101010101'),
('950505050505'),
('1996-02-02 02:02:02'),
('97@03@03 03@03@03'),
(121212121212),
( NOW() );
```

（3）查看插入结果：

```
mysql>SELECT * FROM tmp7;
+---------------------+
| ts                  |
+---------------------+
| 1995-01-01 01:01:01 |
| 1995-05-05 05:05:05 |
| 1996-02-02 02:02:02 |
| 1997-03-03 03:03:03 |
| 2012-12-12 12:12:12 |
| 2024-08-09 17:08:25 |
+---------------------+
```

由结果可以看到，"19950101010101"被转换为1995-01-01 01:01:01；"950505050505"被转换为1995-05-05 05:05:05；"1996-02-02 02:02:02"被转换为1996-02-02 02:02:02；"97@03@03 03@03@03"被转换为1997-03-03 03:03:03；121212121212被转换为2012-12-12 12:12:12；NOW()被转换为系统当前日期和时间2024-08-09 17:08:25。

> 提示 TIMESTAMP与DATETIME除了存储字节和支持的范围不同之外，还有一个最大的区别就是：DATETIME在存储日期数据时，按实际输入的格式存储，即输入什么就存储什么，与时区无关；而TIMESTAMP值的存储是以UTC（世界标准时间）格式保存的，存储时对当前时区进行转换，检索时再转换回当前时区。查询时，不同时区显示的时间值是不同的。

【例3.18】向表tmp7中插入当前日期，查看插入值，更改时区为东十区，再次查看插入值。

（1）删除表中的数据：

```
DELETE FROM tmp7;
```

（2）向表中插入系统当前日期：

```
mysql> INSERT INTO tmp7 values( NOW() );
```

（3）查看当前时区下的日期值：

```
mysql> SELECT * FROM tmp7;
+---------------------+
| ts                  |
+---------------------+
| 2024-08-09 17:12:20 |
+---------------------+
```

查询结果为插入时的日期和时间。

（4）读者所在时区一般为东八区，下面修改当前时区为东十区：

```
mysql> set time_zone='+10:00';
```

（5）再次查看插入时的日期值：

```
mysql> SELECT * FROM tmp7;
+---------------------+
| ts                  |
+---------------------+
| 2024-08-09 19:12:20 |
+---------------------+
```

由结果可以看到，因为东十区时间比东八区快2个小时，因此查询的结果经过时区转换之后，显示的值增加了2个小时。对应地，每减少1个时区，则查询显示的日期中的小时数减1。

> **提示** 如果为一个DATETIME或TIMESTAMP对象分配一个DATE值，那么结果值的时间部分将被设置为"00:00:00"，因为DATE值未包含时间信息。如果为一个DATE对象分配一个DATETIME或TIMESTAMP值，那么结果值的时间部分将被删除，因为DATE值未包含时间信息。

3.1.4 文本字符串类型

字符串类型用来存储字符串数据，除此之外，还可以存储其他数据，比如图片和声音的二进制数据。MySQL支持两类字符型数据：文本字符串和二进制字符串。本小节主要讲解文本字符串类型。

文本字符串不仅可以进行区分或者不区分大小写的字符串比较，还可以进行模式匹配查找。表3.5列出了MySQL中的文本字符串类型。

表 3.5 MySQL 中文本字符串类型

类型名称	说明	存储需求
CHAR(M)	固定长度非二进制字符串	M 字节，1≤M≤255
VARCHAR(M)	变长非二进制字符串	L+1 字节，在此 L≤M 和 1≤M≤255
TINYTEXT	非常小的非二进制字符串	L+1 字节，在此 $L<2^8$
TEXT	小的非二进制字符串	L+2 字节，在此 $L<2^{16}$
MEDIUMTEXT	中等大小的非二进制字符串	L+3 字节，在此 $L<2^{24}$
LONGTEXT	大的非二进制字符串	L+4 字节，在此 $L<2^{32}$
ENUM	枚举类型，只能有一个枚举字符串值	1 或 2 字节，取决于枚举值的数目（最大值为65535）
SET	一个设置，字符串对象可以有 0 个或多个 SET 成员	1、2、3、4 或 8 字节，取决于集合成员的数量（最多为 64 个成员）

VARCHAR和TEXT类型与下一小节要讲的BLOB都是变长类型，其存储需求取决于列值的实际长度（在表3.5中用L表示），而不是取决于类型的最大可能尺寸。例如，一个VARCHAR(10)列能保存最大长度为10个字符的字符串，实际的存储需要是字符串的长度L加上1字节（记录字符串的长度）。对于字符串"abcd"，L是4字节而存储要求是5字节。下面具体介绍这些数据类型的作用以及如何在查询中使用这些类型。

1. CHAR和VARCHAR类型

CHAR(M) 为固定长度字符串，在定义时指定字符串列长，在保存时在右侧填充空格，以达到指定的长度。M表示列长度，其范围是0~255个字符。例如，CHAR(4)定义了一个固定长度的字符串列，其包含的字符个数最大为4。当检索到CHAR值时，尾部的空格将被删除。

VARCHAR(M) 是长度可变的字符串，M表示最大列长度，其范围是0~65535。VARCHAR的最大实际长度由最长的行的大小和使用的字符集确定，而其实际占用的空间为字符串的实际长度加1。例如，VARCHAR(50)定义了一个最大长度为50的字符串，如果插入的字符串只有10个字符，则实际存储的字符串为10个字符和一个字符串结束字符。VARCHAR在值保存和检索时尾部的空格仍保留。

【例3.19】将不同字符串保存到CHAR(4)列和VARCHAR(4)列，并说明CHAR和VARCHAR之间的差别，如表3.6所示。

表 3.6 CHAR(4)与 VARCHAR(4)存储区别

插入值	CHAR(4)	存储需求	VARCHAR(4)	存储需求
" "	" "	4 字节	" "	1 字节
"ab"	"ab "	4 字节	"ab"	3 字节
"abc"	"abc "	4 字节	"abc"	4 字节
"abcd"	"abcd"	4 字节	"abcd"	5 字节
"abcdef"	"abcd"	4 字节	"abcd"	5 字节

对比结果可以看到，CHAR(4)定义了固定长度为4的列，不管存入的数据长度为多少，所占用的空间均为4字节；VARCHAR(4)定义的列所占的字节数为实际长度加1。

查询时，CHAR(4)和VARCHAR(4)的值并不一定相同，如【例3.20】所示。

【例3.20】创建表tmp8，定义字段ch和vch的数据类型分别为CHAR(4)、VARCHAR(4)，向表中插入数据"ab "。

（1）创建表tmp8：
```
CREATE TABLE tmp8(ch CHAR(4), vch VARCHAR(4));
```
（2）插入数据：
```
INSERT INTO tmp8 VALUES('ab ', 'ab ');
```
（3）查询插入结果：
```
mysql> SELECT concat('(', ch, ')'), concat('(',vch,')') FROM tmp8;
+----------------------+---------------------+
| concat('(', ch, ')') | concat('(',vch,')') |
+----------------------+---------------------+
| (ab)                 | (ab )               |
+----------------------+---------------------+
1 row in set (0.00 sec)
```

从查询结果可以看到，ch字段在保存"ab "时将末尾的两个空格删除了，而vch字段保留了末尾的两个空格。

> 提示：在表3.6中，最后一行的字符串只有在使用"不严格"模式时，才会被截断插入；如果MySQL运行在"严格"模式下，则超过列长度的值不会被保存，并且会出现错误信息"ERROR 1406(22001): Data too long for column"，即字符串长度超过指定长度，无法插入。

2. TEXT类型

TEXT用于保存非二进制字符串，如文章内容、评论等。当保存或查询TEXT列的值时，不删除尾部空格。TEXT类型分为4种：TINYTEXT、TEXT、MEDIUMTEXT和LONGTEXT。不同的TEXT类型的数据长度不同。

（1）TINYTEXT的最大长度为255（2^8-1）个字符。

（2）TEXT的最大长度为65535（$2^{16}-1$）个字符。

（3）MEDIUMTEXT的最大长度为16777215（$2^{24}-1$）个字符。

（4）LONGTEXT的最大长度为4294967295（$2^{32}-1$）个或4GB字符。

3. ENUM类型

ENUM是一个字符串对象，其值为表创建时在列规定中枚举的一列值。语法格式如下：

```
字段名 ENUM('值1','值2',...,'值n')
```

其中，"字段名"指将要定义的字段，"值n"指枚举列表中的第n个值。ENUM类型的字段在取值时，只能在指定的枚举列表中取，而且一次只能取一个。如果创建的成员中有空格，则其尾部的空格将自动被删除。ENUM值在内部用整数表示，并且每个枚举值均有一个索引值：

列表值所允许的成员值从1开始编号，MySQL存储的就是这个索引编号。枚举最多可以有65535个元素。

例如，定义ENUM类型的列('first', 'second', 'third')，该列可以取的值和每个值的索引如表3.7所示。

表 3.7　ENUM 类型的取值范围

值	索　引
NULL	NULL
''	0
first	1
second	2
third	3

ENUM值依照列索引顺序排列，并且空字符串排在非空字符串前，NULL排在其他所有的枚举值前。这一点也可以从表3.7中看到。

在这里，有一个方法可以查看列成员的索引值，如【例3.21】所示。

【例3.21】创建表tmp9，定义ENUM类型的列enm('first', 'second', 'third')，查看列成员的索引值。

（1）创建表tmp9：

```
CREATE TABLE tmp9( enm ENUM('first','second','third') );
```

（2）插入各个列值：

```
INSERT INTO tmp9 values('first'),('second') ,('third') , (NULL);
```

（3）查看索引值：

```
mysql> SELECT enm, enm+0 FROM tmp9;
+--------+-------+
| enm    | enm+0 |
+--------+-------+
| first  |     1 |
| second |     2 |
| third  |     3 |
| NULL   |  NULL |
+--------+-------+
```

可以看到，这里的索引值和表3.7中的相同。

> **提示**　ENUM列总有一个默认值：如果将ENUM列声明为NULL，则NULL为该列的一个有效值，并且默认值为NULL；如果ENUM列被声明为NOT NULL，其默认值为允许的值列表中的第以个元素。

【例3.22】创建表tmp10，定义INT类型的字段soc，ENUM类型的字段level，并且取值列表为('excellent','good', 'bad')，向表tmp10中插入数据(70,'good')、(90,1)、(75,2)、(50,3)、(100,'best')。

（1）创建数据表：

```
CREATE TABLE tmp10 (soc INT, level enum('excellent', 'good','bad') );
```

（2）插入数据：

```
INSERT INTO tmp10 values(70,'good'), (90,1),(75,2),(50,3);
```

（3）再次插入数据：

```
mysql>INSERT INTO tmp10 values (100,'best');
ERROR 1265 (01000): Data truncated for column 'level' at row 1
```

这里系统提示错误信息，可以看到，由于字符串值'best'不在ENUM列表中，因此阻止将它插入表中。

（4）查询插入结果：

```
mysql> SELECT * FROM tmp10;
+------+-----------+
| soc  | level     |
+------+-----------+
|   70 | good      |
|   90 | excellent |
|   75 | good      |
|   50 | bad       |
+------+-----------+
```

由结果可以看到，因为ENUM列表中的值在MySQL中都是以编号序列存储的，所以插入列表中的值"good"与插入其对应序号"2"的结果是相同的，而"best"不是列表中的值，所以不能进行插入操作。

4. SET类型

SET是一个字符串对象，可以有0个或多个值。SET列最多可以有64个成员，其值为表创建时规定的一个列值。指定包括多个SET成员的SET列值时，各成员之间用逗号（,）隔开。语法格式如下：

```
SET('值1','值2',...,'值n')
```

与ENUM类型相同，SET值在内部用整数表示，列表中每一个值都有一个索引编号。当创建表时，SET成员值的尾部空格将自动被删除。与ENUM类型不同的是，ENUM类型的字段只能从定义的多个列值中选择一个值插入，而SET类型的列可从定义的多个列值中选择多个字符的联合。

如果插入SET字段中的列值有重复，则MySQL自动删除重复的值；插入SET字段的值的顺序并不重要，MySQL会在将这些值存入数据库时按照定义的顺序显示；如果插入了不正确的值，默认情况下，MySQL将忽视这些值，并给出警告。

【例3.23】创建表tmp11，定义SET类型的字段s，取值列表为('a', 'b', 'c', 'd')，插入数据('a')、('a,b,a')、('c,a,d')、('a,x,b,y')。

（1）创建表tmp11：

```
CREATE TABLE tmp11 ( s SET('a', 'b', 'c', 'd'));
```

（2）插入数据：

```
INSERT INTO tmp11 values('a'),( 'a,b,a'),('c,a,d');
```

（3）再次插入数据：

```
mysql>INSERT INTO tmp11 values ('a,x,b,y');
ERROR 1265 (01000): Data truncated for column 's' at row 1
```

由于插入了SET列不支持的值，因此MySQL给出错误提示。

（4）查看插入结果：

```
mysql> SELECT * FROM tmp11;
+-------+
| s     |
+-------+
| a     |
| a,b   |
| a,c,d |
+-------+
```

从结果中可以看到，对于SET来说，如果插入的值为重复的，则只取一个，例如插入"a,b,a"，则结果为"a,b"；如果插入了不按顺序排列的值，则自动按顺序插入，例如插入"c,a,d"，结果为"a,c,d"；如果插入了不正确的值，那么该值将被阻止插入，例如插入"a,x,b,y"，则提示错误并阻止插入。

3.1.5 二进制字符串类型

前面讲解了存储文本的字符串类型，这一小节将讲解MySQL中存储二进制数据的字符串类型。MySQL中的二进制字符串类型有BIT、BINARY、VARBINARY、TINYBLOB、BLOB、MEDIUMBLOB和LONGBLOB，如表3.8所示。

表 3.8　MySQL 中的二进制字符串类型

类型名称	说　　明	存储需求
BIT(M)	位字段类型	大约(M+7)/8 字节
BINARY(M)	固定长度二进制字符串	M 字节
VARBINARY(M)	可变长度二进制字符串	M+1 字节
TINYBLOB(M)	非常小的 BLOB	L+1 字节，在此 L<2^8
BLOB(M)	小的 BLOB	L+2 字节，在此 L<2^{16}
MEDIUMBLOB(M)	中等大小的 BLOB	L+3 字节，在此 L<2^{24}
LONGBLOB(M)	非常大的 BLOB	L+4 字节，在此 L<2^{32}

1. BIT类型

BIT(M)是位字段类型。M表示每个值的位数，范围为1~64。如果M被省略，则默认为1。

如果为BIT(M)列分配的值的长度小于M位，就在值的左边用0填充。例如，为BIT(6)列分配一个值b'101'，其效果与分配b'000101'相同。BIT数据类型用来保存位字段值。例如，以二进制的形式保存数据13（13的二进制形式为1101），则需要位数至少为4位的BIT类型，即可以定义列类型为BIT(4)，大于二进制1111的数据是不能插入BIT(4)类型的字段中的。

【例3.24】 创建表tmp12，定义BIT(4)类型的字段b，向表中插入数据2、9、15。

（1）创建表tmp12：

```
CREATE TABLE tmp12( b BIT(4) );
```

（2）插入数据：

```
mysql> INSERT INTO tmp12 VALUES(2), (9), (15);
```

（3）查询插入结果：

```
mysql> SELECT BIN(b+0) FROM tmp12;
+----------+
| BIN(b+0) |
+----------+
| 10       |
| 1001     |
| 1111     |
+----------+
```

b+0表示将二进制的结果转换为对应数字的值，BIN()函数将数字转换为二进制。从结果中可以看到，成功地将3个数插入表中。

> **提示** 默认情况下，MySQL不可以插入超出该列允许范围的值，因而要确保插入的数据值在指定的范围内。

2. BINARY和VARBINARY类型

BINARY和VARBINARY类型类似于CHAR和VARCHAR，不同的是它们包含二进制字节字符串。其语法格式如下：

```
列名称 BINARY(M) 或者VARBINARY(M)
```

BINARY类型的长度是固定的，指定长度之后，如果插入值长度没有达到指定长度，则将在它们右边填充"\0"以达到指定长度。例如，指定列数据类型为BINARY(3)，当插入"a"时，存储的内容实际为"a\0\0"；当插入"ab"时，实际存储的内容为"ab\0"。不管存储的内容是否达到指定的长度，其存储空间均为指定的值M。

VARBINARY类型的长度是可变的，指定好长度之后，插入值的长度可以在0到最大值之间。例如，指定列数据类型为VARBINARY(20)，如果插入值的长度只有10，则实际存储空间为10加1，即实际占用的空间为字符串的实际长度加1。

【例3.25】 创建表tmp13，定义BINARY(3)类型的字段b和VARBINARY(3)类型的字段vb，并向表中插入数据'5'，比较两个字段的存储空间。

(1)创建表tmp13：

```
CREATE TABLE tmp13(b binary(3), vb varbinary(3));
```

(2)插入数据：

```
INSERT INTO tmp13 VALUES(5,5);
```

(3)查看两个字段存储数据的长度：

```
mysql> SELECT length(b), length(vb) FROM tmp13;
+-----------+------------+
| length(b) | length(vb) |
+-----------+------------+
|         3 |          1 |
+-----------+------------+
```

可以看到，b字段的数据长度为3，而vb字段的数据长度仅为插入的一个字符的长度1。

3. BLOB类型

BLOB是一个二进制大对象，用来存储可变数量的数据。BLOB类型分为4种：TINYBLOB、BLOB、MEDIUMBLOB和LONGBLOB，如表3.9所示。

表3.9 BLOB 类型的存储范围

数据类型	存储范围
TINYBLOB	最大长度为 255（2^8-1）B
BLOB	最大长度为 65535（$2^{16}-1$）B
MEDIUMBLOB	最大长度为 16777215（$2^{24}-1$）B
LONGBLOB	最大长度为 4294967295（$2^{32}-1$）B 或 4GB

BLOB列存储的是二进制字符串（字节字符串），TEXT列存储的是非二进制字符串（字符字符串）。BLOB列没有字符集，并且排序和比较基于列值字节的数值；TEXT列有一个字符集，并且根据字符集对值进行排序和比较。

3.2 如何选择数据类型

MySQL提供了大量的数据类型，为了优化存储、提高数据库性能，在任何情况下均应使用最精确的类型，即在所有可以表示该列值的类型中，选择存储量最少的那种类型。

1. 整数和浮点数

如果不需要小数部分，就使用整数来保存数据；如果需要表示小数部分，就使用浮点数类型。对于浮点数据列，存入的数值会对该列定义的小数位进行四舍五入。例如，某列的值的范围为1~99999，若只需存储整数，则MEDIUMINT UNSIGNED是最好的类型；若需要存储小数，则使用FLOAT类型。

浮点类型包括FLOAT和DOUBLE类型。DOUBLE类型的精度比FLOAT类型的精度更高，因此当存储精度要求较高时，应选择DOUBLE类型。

2. 浮点数和定点数

浮点数FLOAT、DOUBLE相对于定点数DECIMAL的优势是：在长度一定的情况下，浮点数能表示更大的数据范围。由于浮点数容易产生误差，因此当对精确度要求比较高时，建议使用DECIMAL来存储。DECIMAL在MySQL中是以字符串存储的，用于定义货币等对精确度要求较高的数据。在数据迁移中，float(M,D)是非标准SQL定义，数据库迁移可能会出现问题，最好不要这样使用。另外，两个浮点数在进行减法和比较运算时也容易出问题，因此在进行计算的时候一定要小心。进行数值比较时，最好使用DECIMAL类型。

3. 日期与时间类型

MySQL对于不同种类的日期和时间有很多数据类型，比如YEAR和TIME。如果只需要记录年份，则使用YEAR类型即可；如果只记录时间，则使用TIME类型。

如果同时需要记录日期和时间，则可以使用TIMESTAMP或者DATETIME类型。由于DATETIME的取值范围大于TIMESTAMP，因此存储范围较大的日期时最好使用DATETIME。

TIMESTAMP有一个DATETIME不具备的属性——默认情况下，当插入一条记录但没有指定TIMESTAMP这个列值时，MySQL会把TIMESTAMP列设为当前的时间。因此，当需要在插入记录的同时插入当前时间，使用TIMESTAMP会更方便。另外，TIMESTAMP在空间上比DATETIME更有效。

4. CHAR与VARCHAR之间的特点与选择

CHAR和VARCHAR的区别如下：

- CHAR是固定长度字符，VARCHAR是可变长度字符。
- CHAR会自动删除插入数据的尾部空格，VARCHAR不会删除插入数据的尾部空格。
- 因为CHAR是固定长度，所以它的处理速度比VARCHAR的要快，但也因此浪费了存储空间。对于存储量不大但在速度上有要求的字段，可以使用CHAR类型，反之可以使用VARCHAR类型。

存储引擎对于选择CHAR和VARCHAR的影响：

- 对于MyISAM存储引擎：最好使用固定长度的数据列代替可变长度的数据列。这样可以使整张表静态化，从而使数据检索更快，用空间换时间。
- 对于InnoDB存储引擎：最好使用可变长度的数据列。因为InnoDB数据表的存储格式不分固定长度和可变长度，所以使用CHAR不一定比使用VARCHAR更好，但由于VARCHAR是按照实际的长度存储的，比较节省空间，因此比较节省磁盘I/O和数据存储总量。

5. ENUM和SET

ENUM只能取单值，它的数据列表是一个枚举集合。它的合法取值列表最多允许有65535个成员。因此，在需要从多个值中选取一个时，可以使用ENUM。例如，性别字段适合定义为ENUM类型，每次只能从"男"或"女"中取一个值。

SET可取多值。它的合法取值列表最多允许有64个成员。空字符串也是一个合法的SET值。在需要取多个值的时候，适合使用SET类型，比如要存储一个人的兴趣爱好，最好使用SET类型。

ENUM和SET的值是以字符串形式出现的，但在MySQL内部，是以数值的形式存储它们的。

6. BLOB和TEXT

BLOB是二进制字符串，TEXT是非二进制字符串，两者均可存放大容量的信息。BLOB主要存储图片、音频信息等，而TEXT只能存储纯文本文件。

3.3 运 算 符

运算符连接表达式中的各个操作数，其作用是指明对操作数所进行的运算。运用运算符可以更加灵活地使用表中的数据。本节将介绍MySQL中的运算符的特点和使用方法。

3.3.1 运算符概述

运算符是告诉MySQL执行特定算术或逻辑操作的符号。MySQL内部的运算符很丰富，主要有四大类，分别是算术运算符、比较运算符、逻辑运算符、位运算符。

1. 算术运算符

算术运算符用于各类数值运算，包括加（+）、减（–）、乘（*）、除（/）和求余（或称模运算，%）。

2. 比较运算符

比较运算符用于比较运算，包括大于（>）、小于（<）、等于（=）、大于或等于（>=）、小于或等于（<=）、不等于（!=），以及IN、BETWEEN AND、IS NULL、GREATEST、LEAST、LIKE、REGEXP等。

3. 逻辑运算符

逻辑运算符的求值结果均为1（TRUE）或0（FALSE），这类运算符有逻辑非（NOT 或者!）、逻辑与（AND 或者&&）、逻辑或（OR 或者||）、逻辑异或（XOR）。

4. 位运算符

位运算符包括位与（&）、位或（|）、位非（~）、位异或（^）、左移（<<）、右移（>>）6种，参与运算的操作数按二进制位进行运算。

接下来将对MySQL中的4种运算符的使用进行详细介绍。

3.3.2 算术运算符

算术运算符是MySQL中最基本的运算符，如表3.10所示。

表3.10 MySQL 中的算术运算符

运算符	作用
+	加法运算
–	减法运算
*	乘法运算
/	除法运算，返回商
%	求余运算，返回余数

下面分别讨论不同算术运算符的使用方法。

【例3.26】创建表tmp14，定义数据类型为INT的字段num，插入值64，对num值进行加法和减法运算。

（1）创建表tmp14：

```
CREATE TABLE tmp14 ( num INT);
```

（2）向字段num插入数据64：

```
INSERT INTO tmp14 value(64);
```

（3）对num值进行加法和减法运算：

```
mysql> SELECT num, num+10, num-3+5, num+5-3, num+36.5 FROM tmp14;
+-------+--------+---------+---------+----------+
| num   | num+10 | num-3+5 | num+5-3 | num+36.5 |
+-------+--------+---------+---------+----------+
|   64  |   74   |   66    |   66    |  100.5   |
+-------+--------+---------+---------+----------+
```

由计算结果可以看到，可以对num字段的值进行加法和减法的运算，而且"+"和"–"的优先级相同，先加后减或者先减后加的结果是相同的。

【例3.27】对表tmp14中的num值进行乘法和除法运算，SQL语句如下：

```
mysql> SELECT num, num *2, num /2, num/3, num%3 FROM tmp14;
+-------+--------+---------+---------+-------+
| num   | num *2 | num /2  | num/3   | num%3 |
+-------+--------+---------+---------+-------+
|   64  |  128   | 32.0000 | 21.3333 |   1   |
+-------+--------+---------+---------+-------+
```

由计算结果可以看到，对num值进行除法运算的时候，由于64无法被3整除，MySQL对num/3求商的结果保存到了小数点后面4位，结果为21.3333；64除以3的余数为1，因此取余运算num%3的结果为1。

在进行数学运算时，除数为0的除法是没有意义的，因此除法运算中的除数不能为0，如果被0除，则返回结果为NULL。

【例3.28】 用num除以0，SQL语句如下：

```
mysql> SELECT num, num / 0, num %0 FROM tmp14;
+------+---------+---------+
| num  | num / 0 | num %0  |
+------+---------+---------+
|  64  |  NULL   |  NULL   |
+------+---------+---------+
```

由计算结果可以看到，除数为0时，对num值进行求商或者求余运算的结果均为NULL。

3.3.3 比较运算符

一个比较运算符的结果总是1、0或者是NULL。比较运算符经常在SELECT的查询条件子句中使用，用来查询满足指定条件的记录。MySQL中的比较运算符如表3.11所示。

表 3.11 MySQL 中的比较运算符

运 算 符	作 用
=	等于
<=>	安全等于
<> (!=)	不等于
<=	小于或等于
>=	大于或等于
>	大于
IS NULL	判断一个值是否为 NULL
IS NOT NULL	判断一个值是否不为 NULL
LEAST	当有两个或多个参数时，返回最小值
GREATEST	当有两个或多个参数时，返回最大值
BETWEEN AND	判断一个值是否落在两个值之间
ISNULL	与 IS NULL 作用相同
IN	判断一个值是否为 IN 列表中的任意一个值
NOT IN	判断一个值是不是 IN 列表中的任意一个值
LIKE	通配符匹配
REGEXP	正则表达式匹配

下面分别讨论不同比较运算符的使用方法。

1. 等于运算符（=）

等号（=）用来判断数字、字符串和表达式是否相等：如果相等，则返回值为1；否则返回值为0。

【例3.29】 使用"="进行相等判断，SQL语句如下：

```
mysql> SELECT 1=0, '2'=2, 2=2,'0.02'=0, 'b'='b', (1+3) = (2+2),NULL=NULL;
+-----+-----+-----+--------+-------+-------------+-----------+
| 1=0 |'2'=2| 2=2 |'0.02'=0|'b'='b'| (1+3) = (2+2)| NULL=NULL |
+-----+-----+-----+--------+-------+-------------+-----------+
|   0 |   1 |   1 |      0 |     1 |           1 |      NULL |
+-----+-----+-----+--------+-------+-------------+-----------+
```

由结果可以看到，在进行判断时，2=2和'2'=2的返回值相同，都为1。因为在进行判断时，MySQL自动进行了转换，把字符'2'转换成了数字2；'b'='b'为相同的字符比较，因此返回值为1；表达式1+3和表达式2+2的结果都为4，因此结果相等，返回值为1；由于"="不能用于NULL的判断，因此返回值为NULL。

数值比较时有如下规则：

（1）若有一个或两个参数为NULL，则比较运算的结果为NULL。
（2）若同一个比较运算中的两个参数都是字符串，则按照字符串进行比较。
（3）若两个参数均为整数，则按照整数进行比较。
（4）若用字符串和数字进行相等判断，则MySQL可以自动将字符串转换为数字。

2. 安全等于运算符（<=>）

这个操作符和"="操作符执行相同的比较操作，不过"<=>"可以用来判断NULL。在两个操作数均为NULL时，其返回值为1，而不为NULL；当一个操作数为NULL时，其返回值为0，而不为NULL。

【例3.30】 使用"<=>"进行相等判断，SQL语句如下：

```
mysql> SELECT 1<=>0, '2'<=>2, 2<=>2,'0.02'<=>0, 'b'<=>'b', (1+3) <=> (2+1),
NULL<=>NULL;
+------+-------+-----+----------+---------+--------------+-----------+
|1<=>0 |'2'<=>2|2<=>2|'0.02'<=>0|'b'<=>'b'|(1+3) <=> (2+1)| NULL<=>NULL|
+------+-------+-----+----------+---------+--------------+-----------+
|   0  |    1  |   1 |       0  |      1  |           0  |         1 |
+------+-------+-----+----------+---------+--------------+-----------+
```

由结果可以看到，"<=>"在执行比较操作时，其作用和"="相似，唯一的区别是"<=>"可以用来对NULL进行判断，两者都为NULL时返回值为1。

3. 不等于运算符（<>或者!=）

"<>"或者"!="用于进行数字、字符串、表达式不相等的判断：如果不相等，则返回值为1；否则返回值为0。这两个运算符不能用于判断NULL。

【例3.31】 使用"<>"和"!="进行不相等判断，SQL语句如下：

```
mysql> SELECT 'good'<>'god', 1<>2, 4!=4, 5.5!=5, (1+3)!=(2+1),NULL<>NULL;
+---------------+------+------+--------+-------------+------------+
| 'good'<>'god' | 1<>2 | 4!=4 | 5.5!=5 | (1+3)!=(2+1)| NULL<>NULL |
```

```
+---------------+------+------+--------+----------------+------------+
|             1 |    1 |    0 |      1 |              1 |       NULL |
+---------------+------+------+--------+----------------+------------+
```

由结果可以看到,两个不等于运算符的作用相同,都可以进行数字、字符串、表达式的比较判断。

4. 小于或等于运算符(<=)

"<="用来判断左边的操作数是否小于或等于右边的操作数:如果小于或等于,则返回值为1;否则返回值为0。"<="不能用于判断NULL。

【例3.32】使用"<="进行比较判断,SQL语句如下:

```
mysql>SELECT 'good'<='god', 1<=2, 4<=4, 5.5<=5, (1+3) <= (2+1),NULL<=NULL;
+---------------+------+------+--------+-----------------+------------+
| 'good'<='god' | 1<=2 | 4<=4 | 5.5<=5 | (1+3) <= (2+1)  | NULL<=NULL |
+---------------+------+------+--------+-----------------+------------+
|             0 |    1 |    1 |      0 |               0 |       NULL |
+---------------+------+------+--------+-----------------+------------+
```

由结果可以看到,左边操作数小于或等于右边时,返回值为1,例如对于4<=4,其判断结果为真,所以返回值为1;当左边操作数大于右边时,返回值为0,例如对于'good'<='god','good'第3个位置的o字符在字母表中的顺序大于'god'中第3个位置的d字符,因此判断结果为假,返回值为0;比较NULL时,返回NULL。

5. 小于运算符(<)

"<"运算符用来判断左边的操作数是否小于右边的操作数:如果小于,则返回值为1;否则返回值为0。"<"不能用于判断NULL。

【例3.33】使用"<"进行比较判断,SQL语句如下:

```
mysql> SELECT 'good'<'god', 1<2, 4<4, 5.5<5, (1+3) < (2+1),NULL<NULL;
+--------------+-----+-----+-------+----------------+-----------+
| 'good'<'god' | 1<2 | 4<4 | 5.5<5 | (1+3) < (2+1)  | NULL<NULL |
+--------------+-----+-----+-------+----------------+-----------+
|            0 |   1 |   0 |     0 |              0 |      NULL |
+--------------+-----+-----+-------+----------------+-----------+
```

从结果中可以看到,当左边操作数小于右边操作数时,返回值为1,例如1<2的返回值1;当左边操作数大于右边操作数时,返回值为0,例如'good'<'god'的返回值为0;比较NULL时返回NULL。

6. 大于或等于运算符(>=)

">="运算符用来判断左边的操作数是否大于或等于右边的操作数:如果大于或等于,则返回值为1;否则返回值为0。">="不能用于判断NULL。

【例3.34】使用 ">=" 进行比较判断，SQL语句如下：

```
MySQL> SELECT 'good'>='god', 1>=2, 4>=4, 5.5>=5, (1+3) >= (2+1),NULL>=NULL;
+---------------+------+------+--------+----------------+------------+
| 'good'>='god' | 1>=2 | 4>=4 | 5.5>=5 | (1+3) >= (2+1) | NULL>=NULL |
+---------------+------+------+--------+----------------+------------+
|             1 |    0 |    1 |      1 |              1 |       NULL |
+---------------+------+------+--------+----------------+------------+
```

由结果可以看到，左边操作数大于或等于右边操作数时，返回值为1，例如4>=4的返回值为1；当左边操作数小于右边操作数时，返回值为0，例如1>=2的返回值为0；比较NULL时返回NULL。

7. 大于运算符（>）

">" 运算符用来判断左边的操作数是否大于右边的操作数：如果大于，则返回值为1；否则返回值为0。 ">" 不能用于判断NULL。

【例3.35】使用 ">" 进行比较判断，SQL语句如下：

```
mysql> SELECT 'good'>'god', 1>2, 4>4, 5.5>5, (1+3) > (2+1),NULL>NULL;
+--------------+-----+-----+-------+---------------+-----------+
| 'good'>'god' | 1>2 | 4>4 | 5.5>5 | (1+3) > (2+1) | NULL>NULL |
+--------------+-----+-----+-------+---------------+-----------+
|            1 |   0 |   0 |     1 |             1 |      NULL |
+--------------+-----+-----+-------+---------------+-----------+
```

由结果可以看到，左边操作数大于右边操作数时，返回值为1，例如5.5>5的返回值为1；左边操作数小于右边操作数时，返回值为0，例如1>2的返回值为0；比较NULL时返回NULL。

8. IS NULL(ISNULL)和IS NOT NULL运算符

IS NULL和ISNULL检验一个值是否为NULL：如果为NULL，则返回值为1；否则返回值为0。IS NOT NULL检验一个值是否为非NULL：如果是非NULL，则返回值为1；否则返回值为0。

【例3.36】使用IS NULL、ISNULL和IS NOT NULL判断NULL值和非NULL值，SQL语句如下：

```
mysql> SELECT NULL IS NULL, ISNULL(NULL),ISNULL(10), 10 IS NOT NULL;
+--------------+--------------+------------+----------------+
| NULL IS NULL | ISNULL(NULL) | ISNULL(10) | 10 IS NOT NULL |
+--------------+--------------+------------+----------------+
|            1 |            1 |          0 |              1 |
+--------------+--------------+------------+----------------+
```

由结果可以看到，IS NULL和ISNULL的作用相同，只是格式不同；ISNULL和IS NOT NULL的返回值正好相反。

9. BETWEEN AND运算符

BETWEEN AND运算符语法格式为：

```
expr BETWEEN min AND max
```

假如expr大于或等于min且小于或等于max，则BETWEEN的返回值为1，否则返回值为0。

【例3.37】 使用BETWEEN AND运算符进行值区间判断，SQL语句如下：

```
mysql> SELECT 4 BETWEEN 2 AND 5, 4 BETWEEN 4 AND 6,12 BETWEEN 9 AND 10;
+-------------------+-------------------+--------------------+
| 4 BETWEEN 2 AND 5 | 4 BETWEEN 4 AND 6 | 12 BETWEEN 9 AND 10 |
+-------------------+-------------------+--------------------+
|                 1 |                 1 |                  0 |
+-------------------+-------------------+--------------------+

mysql> SELECT 'x' BETWEEN 'f' AND 'g', 'b' BETWEEN 'a' AND 'c';
+-------------------------+-------------------------+
| 'x' BETWEEN 'f' AND 'g' | 'b' BETWEEN 'a' AND 'c' |
+-------------------------+-------------------------+
|                       0 |                       1 |
+-------------------------+-------------------------+
```

由结果可以看到，当4在端点值区间内或者等于其中一个端点值时，BETWEEN AND表达式返回值为1；12并不在指定区间内，因此返回值为0；对于字符串类型的比较，按字母表中字母顺序进行比较，"x"不在指定的字母区间内，因此返回值为0，而"b"位于指定字母区间内，因此返回值为1。

10. LEAST运算符

LEAST运算符语法格式为：

```
LEAST(值1,值2,...,值n)
```

其中，"值n"表示参数列表中有n个值。在有两个或多个参数的情况下，返回最小值。假如任意一个自变量为NULL，则LEAST()的返回值为NULL。

【例3.38】 使用LEAST运算符进行大小判断，SQL语句如下：

```
mysql> SELECT least(2,0), least(20.0,3.0,100.5),
least('a','c','b'),least(10,NULL);
+-----------+----------------------+-------------------+----------------+
|least(2,0) |least(20.0,3.0,100.5) |least('a','c','b') |least(10,NULL)  |
+-----------+----------------------+-------------------+----------------+
|         0 |                  3.0 | a                 |           NULL |
+-----------+----------------------+-------------------+----------------+
```

由结果可以看到，当参数是整数或者浮点数时，LEAST将返回其中最小的值；当参数为字符串时，返回字母表中顺序最靠前的字符；当比较值列表中有NULL时，不能判断大小，返回值为NULL。

11. GREATEST运算符

GREATEST运算符语法格式为：

```
GREATEST(值1, 值2,...,值n)
```

其中，"值n"表示参数列表中有n个值。当有两个或多个参数时，返回值为最大值。假如任意一个自变量为NULL，则GREATEST()的返回值为NULL。

【例3.39】使用GREATEST运算符进行大小判断，SQL语句如下：

```
mysql> SELECT greatest(2,0), greatest(20.0,3.0,100.5),
greatest('a','c','b'),greatest(10,NULL);
+---------------+--------------------------+-----------------------+-------------------+
| greatest(2,0) | greatest(20.0,3.0,100.5) | greatest('a','c','b') | greatest(10,NULL) |
+---------------+--------------------------+-----------------------+-------------------+
|             2 |                    100.5 | c                     |              NULL |
+---------------+--------------------------+-----------------------+-------------------+
```

由结果可以看到，当参数是整数或者浮点数时，GREATEST将返回其中最大的值；当参数为字符串时，返回字母表中顺序最靠后的字符；当比较值列表中有NULL时，不能判断大小，返回值为NULL。

12. IN、NOT IN运算符

IN运算符用来判断操作数是否为IN列表中的一个值：如果是，则返回值为1；否则返回值为0。

NOT IN运算符用来判断表达式是否为IN列表中一个值：如果不是，则返回值为1；否则返回值为0。

【例3.40】使用IN、NOT IN运算符进行判断，SQL语句如下：

```
mysql> SELECT 2 IN (1,3,5,'thks'), 'thks' IN (1,3,5,'thks');
+---------------------+--------------------------+
| 2 IN (1,3,5,'thks') | 'thks' IN (1,3,5,'thks') |
+---------------------+--------------------------+
|                   0 |                        1 |
+---------------------+--------------------------+
mysql> SELECT 2 NOT IN (1,3,5,'thks'), 'thks' NOT IN (1,3,5,'thks');
+-------------------------+------------------------------+
| 2 NOT IN (1,3,5,'thks') | 'thks' NOT IN (1,3,5,'thks') |
+-------------------------+------------------------------+
|                       1 |                            0 |
+-------------------------+------------------------------+
```

由结果可以看到，IN和NOT IN的返回值正好相反。

在左侧表达式为NULL，或者在IN列表中找不到匹配项并且表中一个表达式为NULL的情况下，IN的返回值均为NULL。

【例3.41】存在NULL时的IN查询，SQL语句如下：

```
mysql> SELECT NULL IN (1,3,5,'thks'),10 IN (1,3,NULL,'thks');
+------------------------+-------------------------+
| NULL IN (1,3,5,'thks') | 10 IN (1,3,NULL,'thks') |
+------------------------+-------------------------+
|                   NULL |                    NULL |
+------------------------+-------------------------+
```

IN()语法也可用于在SELECT语句中进行嵌套子查询，在后面的章节中会详细介绍。

13. LIKE运算符

LIKE运算符用来匹配字符串，语法格式为：

```
expr LIKE 匹配条件
```

如果expr满足匹配条件，则返回值为1；如果不匹配，则返回值为0。expr或匹配条件中任何一个为NULL，则返回值为NULL。

LIKE运算符在进行匹配时，可以使用下面两种通配符：

（1）"%"：匹配任何数目的字符，甚至包括零字符。
（2）"_"：只能匹配一个字符。

【例3.42】使用LIKE运算符进行字符串匹配运算，SQL语句如下：

```
mysql> SELECT 'stud' LIKE 'stud', 'stud' LIKE 'stu_','stud' LIKE '%d','stud' LIKE 't_ _ _', 's' LIKE NULL;
+------------------+------------------+----------------+--------------------+--------------+
|'stud' LIKE 'stud'|'stud' LIKE 'stu_'|'stud' LIKE '%d'|'stud' LIKE 't_ _ _'| 's' LIKE NULL|
+------------------+------------------+----------------+--------------------+--------------+
|        1         |        1         |       1        |         0          |    NULL      |
+------------------+------------------+----------------+--------------------+--------------+
```

由结果可以看到，指定匹配的字符串为"stud"。"stud"表示直接匹配"stud"字符串，满足匹配条件，返回1；"stu_"表示匹配以stu开头的字符长度为4个字符的字符串，"stud"正好是4个字符，满足匹配条件，因此返回1；"%d"表示匹配以字母"d"结尾的字符串，"stud"满足匹配条件，因此返回1；"t_ _ _"表示匹配以't'开头的字符长度为4的字符串，"stud"不满足匹配条件，因此返回0；当字符's'与NULL匹配时，结果为NULL。

14. REGEXP运算符

REGEXP运算符用来匹配字符串，语法格式为：

```
expr REGEXP 匹配条件
```

如果expr满足匹配条件，则返回1；如果不满足，则返回0。若expr或匹配条件任意一个为NULL，则返回值为NULL。

REGEXP运算符在进行匹配时，常用的通配符有下面5种：

（1）"^"：匹配以该字符后面的字符开头的字符串。
（2）"$"：匹配以该字符后面的字符结尾的字符串。
（3）"."：匹配任何一个单字符。
（4）"[...]"：匹配方括号内的任何字符。例如，"[abc]"匹配"a""b"或"c"。为了命名字符的范围，使用一个"-"。"[a-z]"匹配任何字母，而"[0-9]"匹配任何数字。
（5）"*"：匹配0个或多个在它前面的字符。例如，"x*"匹配任何数量的"x"字符，"[0-9]*"匹配任何数量的数字，而"*"匹配任何数量的任何字符。

【例3.43】 使用REGEXP运算符进行字符串匹配运算，SQL语句如下：

```
mysql> SELECT 'ssky' REGEXP '^s', 'ssky' REGEXP 'y$', 'ssky' REGEXP '.sky', 'ssky' REGEXP '[ab]';
+--------------------+--------------------+----------------------+---------------------+
|'ssky' REGEXP '^s' |'ssky' REGEXP 'y$' |'ssky' REGEXP '.sky' |'ssky' REGEXP '[ab]'|
+--------------------+--------------------+----------------------+---------------------+
|         1          |         1          |          1           |          0          |
+--------------------+--------------------+----------------------+---------------------+
```

由结果可以看到，指定匹配的字符串为"ssky"。"^s"表示匹配任何以字母"s"开头的字符串，因此满足匹配条件，返回1；"y$"表示匹配任何以字母"y"结尾的字符串，因此满足匹配条件，返回1；".sky"表示匹配任何以"sky"结尾、字符长度为4的字符串，因此满足匹配条件，返回1；"[ab]"匹配任何包含字母a或者b的字符串，指定字符串中既没有字母a也没有字母b，因此不满足匹配条件，返回0。

> **提示** 正则表达式是一个可以进行复杂查询的强大工具。相对于LIKE字符串匹配，它可以使用更多的通配符类型，查询结果更加灵活。读者可以参考相关的书籍或资料，详细学习正则表达式的写法，这里就不再详细介绍了。在后面的章节中，将会介绍如何使用正则表达式查询表中的记录。

3.3.4 逻辑运算符

在SQL中，所有逻辑运算符的求值所得结果均为TRUE、FALSE或NULL。在MySQL中，它们体现为1（TRUE）、0（FALSE）和NULL。MySQL中的逻辑运算符如表3.12所示。

表 3.12 MySQL 中的逻辑运算符

运 算 符	作 用
NOT 或者 ！	逻辑非
AND 或者 &&	逻辑与
OR 或者 ||	逻辑或
XOR	逻辑异或

接下来，分别讨论不同的逻辑运算符的使用方法。

1. NOT

逻辑非运算符"NOT"表示当操作数为0时，所得值为1；当操作数为非零值时，所得值为0；当操作数为NULL时，所得的返回值为NULL。

【例3.44】 使用逻辑非运算符"NOT"进行逻辑判断，SQL语句如下：

```
mysql> SELECT NOT 10, NOT (1-1), NOT -5, NOT NULL, NOT 1 + 1;
+--------+-----------+--------+----------+-----------+
| NOT 10 | NOT (1-1) | NOT -5 | NOT NULL | NOT 1 + 1 |
+--------+-----------+--------+----------+-----------+
```

```
|      0 |      1 |     0 |      NULL |        0 |
+--------+--------+-------+-----------+----------+
```

2. AND

逻辑与运算符"AND"表示当所有操作数均为非零值并且不为NULL时，计算所得结果为1；当一个或多个操作数为0时，所得结果为0；其余情况返回值为NULL。

【例3.45】 使用逻辑与运算符"AND"进行逻辑判断，SQL语句如下：

```
mysql> SELECT 1 AND -1,1 AND 0,1 AND NULL, 0 AND NULL;
+----------+---------+------------+------------+
| 1 AND -1 | 1 AND 0 | 1 AND NULL | 0 AND NULL |
+----------+---------+------------+------------+
|        1 |       0 |       NULL |          0 |
+----------+---------+------------+------------+
```

由结果可以看到，"1 AND -1"中没有0或者NULL，因此结果为1；"1 AND 0"中有操作数0，因此结果为0；"1 AND NULL"中虽然有NULL，但是没有操作数0，因此结果为NULL。

> **提示** "AND"运算符可以有多个操作数，需要注意的是：当有多个操作数进行运算时，AND两边一定要使用空格隔开，否则会影响结果的正确性。

3. OR

逻辑或运算符"OR"表示当两个操作数均为非NULL值且任意一个操作数为非零值时，结果为1，否则结果为0；当有一个操作数为NULL，且另一个操作数为非零值时，结果为1，否则结果为NULL；当两个操作数均为NULL时，所得结果为NULL。

【例3.46】 使用逻辑或运算符"OR"进行逻辑判断，SQL语句如下：

```
mysql> SELECT 1 OR -1 OR 0, 1 OR 2,1 OR NULL, 0 OR NULL, NULL OR NULL;
+--------------+--------+-----------+-----------+--------------+
| 1 OR -1 OR 0 | 1 OR 2 | 1 OR NULL | 0 OR NULL | NULL OR NULL |
+--------------+--------+-----------+-----------+--------------+
|            1 |      1 |         1 |      NULL |         NULL |
+--------------+--------+-----------+-----------+--------------+
```

由结果可以看到，"1 OR -1 OR 0"中没有NULL，并且包含有非零的值1和-1，因此结果为1；"1 OR 2"中没有操作数0，因此结果为1；"1 OR NULL"中虽然有NULL，但是有操作数1，因此结果为1；"0 OR NULL"中没有非零值，并且有NULL，因此结果为NULL；"NULL OR NULL"中只有NULL，因此结果为NULL。

4. XOR

逻辑异或运算符"XOR"表示当任意一个操作数为NULL时，返回值为NULL；对于非NULL的操作数，如果两个操作数都是非零值或者都是零值，则返回结果为0，如果一个为零值、另一个为非零值，则返回结果为1。

【例3.47】 使用逻辑异或运算符"XOR"进行逻辑判断，SQL语句如下：

```
mysql> SELECT 1 XOR 1, 0 XOR 0, 1 XOR 0, 1 XOR NULL, 1 XOR 1 XOR 1;
+---------+---------+---------+------------+---------------+
| 1 XOR 1 | 0 XOR 0 | 1 XOR 0 | 1 XOR NULL | 1 XOR 1 XOR 1 |
+---------+---------+---------+------------+---------------+
|       0 |       0 |       1 |       NULL |             1 |
+---------+---------+---------+------------+---------------+
```

由结果可以看到，在"1 XOR 1"和"0 XOR 0"中，运算符两边的操作数都为非零值或者都是零值，因此返回0；在"1 XOR 0"中，两边的操作数一个为零值、一个为非零值，因此结果为1；在"1 XOR NULL"中，有一个操作数为NULL，因此结果为NULL；在"1 XOR 1 XOR 1"中，有多个操作数且运算符相同，因此从左到右依次计算，"1 XOR 1"的结果为0，再与1进行异或运算，最终结果为1。

> **提示** a XOR b的计算等同于(a AND (NOT b))或者((NOT a)AND b)。

3.3.5 位运算符

位运算符是在二进制数上进行计算的运算符。位运算符会先将操作数变成二进制数，然后进行位运算，最后将计算结果从二进制数变回十进制数。MySQL中提供的位运算符有位或（|）、位与（&）、位异或（^）、位左移（<<）、位右移（>>）和位取反（~）运算符，如表3.13所示。

表 3.13　MySQL 中的位运算符

运 算 符	作 用
|	位或
&	位与
^	位异或
<<	位左移
>>	位右移
~	位取反，反转所有比特

接下来，分别讨论不同的位运算符的使用方法。

1. 位或运算符（|）

位或运算的实质是将参与运算的几个操作数按照对应的二进制数逐位进行逻辑或运算。对应的二进制位有一个或两个为1，则该位的运算结果为1，否则为0。

【例3.48】 使用位或运算符进行运算，SQL语句如下：

```
mysql> SELECT 10 | 15, 9 | 4 | 2;
+---------+-----------+
| 10 | 15 | 9 | 4 | 2 |
+---------+-----------+
```

```
|       15 |       15 |
+----------+----------+
```

10的二进制值为1010,15的二进制值为1111,按位或运算之后,结果为1111,即十进制整数15;9的二进制值为1001,4的二进制值为0100,2的二进制值为0010,按位或运算之后,结果为1111,即十进制整数15。其结果为一个64位无符号整数。

2. 位与运算符（&）

位与运算的实质是将参与运算的几个操作数按照对应的二进制数逐位进行逻辑与运算。对应的二进制位都为1,则该位的运算结果为1,否则为0。

【例3.49】使用位与运算符进行运算,SQL语句如下:

```
mysql> SELECT 10 & 15, 9 &4& 2;
+---------+---------+
| 10 & 15 | 9 &4& 2 |
+---------+---------+
|      10 |       0 |
+---------+---------+
```

10的二进制值为1010,15的二进制值为1111,按位与运算之后,结果为1010,即十进制整数10;9的二进制值为1001,4的二进制值为0100,2的二进制值为0010,按位与运算之后,结果为0000,即十进制整数0。其结果为一个64位无符号整数。

3. 位异或运算符（^）

位异或运算的实质是将参与运算的两个操作数按照对应的二进制数逐位进行逻辑异或运算。对应位的二进制数不同时,对应位的结果才为1。如果两个对应位的数都为0或者都为1,则对应位的结果为0。

【例3.50】使用位异或运算符进行运算,SQL语句如下:

```
mysql> SELECT 10 ^ 15, 1 ^0, 1 ^ 1;
+---------+------+-------+
| 10 ^ 15 | 1 ^0 | 1 ^ 1 |
+---------+------+-------+
|       5 |    1 |     0 |
+---------+------+-------+
```

10的二进制值为1010,15的二进制值为1111,按位异或运算之后,结果为0101,即十进制整数5;1的二进制值为0001,0的二进制值为0000,按位异或运算之后,结果为0001,即十进制整数1;1和1本身二进制位完全相同,因此结果为0。

4. 位左移运算符（<<）

位左移运算符<<使指定的二进制值的所有位都左移指定的位数。左移指定位数之后,左边高位的数值将被移出并丢弃,右边低位空出的位置用0补齐。语法格式为:

```
expr<<n
```

其中，n指定expr要移位的位数。

【例3.51】使用位左移运算符进行运算，SQL语句如下：

```
mysql> SELECT 1<<2, 4<<2;
+------+------+
| 1<<2 | 4<<2 |
+------+------+
|    4 |   16 |
+------+------+
```

1的二进制值为0000 0001，左移两位之后变成0000 0100，即十进制整数4；十进制4左移两位之后变成0001 0000，即整十进制数16。

5. 位右移运算符（>>）

位右移运算符使指定的二进制值的所有位都右移指定的位数。右移指定位数之后，右边低位的数值将被移出并丢弃，左边高位空出的位置用0补齐。语法格式为：

```
expr>>n
```

其中，n指定expr要移位的位数。

【例3.52】使用位右移运算符进行运算，SQL语句如下：

```
mysql> SELECT 1>>1, 16>>2;
+------+-------+
| 1>>1 | 16>>2 |
+------+-------+
|    0 |     4 |
+------+-------+
```

1的二进制值为0000 0001，右移1位之后变成0000 0000，即十进制整数0；16的二进制值为0001 0000，右移两位之后变成0000 0100，即十进制整数4。

6. 位取反运算符（~）

位取反运算的实质是将参与运算的操作数按照对应的二进制数逐位反转，即1取反后变为0，0取反后变为1。

【例3.53】使用位取反运算符进行运算，SQL语句如下：

```
mysql> SELECT 5 & ~1;
+--------+
| 5 & ~1 |
+--------+
|      4 |
+--------+
```

在逻辑运算"5&~1"中，由于位取反运算符（~）的级别高于位与运算符（&），因此先对1进行取反操作，取反之后，除了最低位为0外，其他位都为1，即1110；然后与十进制数值5进行与运算，结果为0100，即十进制整数4。

提示 MySQL经过位运算之后的数值是一个64位的无符号整数，1的二进制值表示为最右边位为1、其他位均为0，取反操作之后，除了最低位为0外，其他位均变为1。

可以使用BIN()函数查看1取反之后的结果，SQL语句如下：

```
mysql> SELECT BIN(~1);
+------------------------------------------------------------------+
| BIN(~1)                                                          |
+------------------------------------------------------------------+
| 1111111111111111111111111111111111111111111111111111111111111110 |
+------------------------------------------------------------------+
```

这样，读者就可以明白【例3.53】是如何计算的了。

3.3.6 运算符的优先级

运算符的优先级决定了不同的运算符在表达式中计算的先后顺序。表3.14列出了MySQL中的各类运算符及其优先级。

表3.14 运算符按优先级由低到高排列

优先级	运算符
最低	=（赋值运算），:=
	||，OR
	XOR
	&&，AND
	NOT
	BETWEEN, CASE, WHEN, THEN, ELSE
	=（比较运算），<=>, >=, >, <=, <, <>, !=, IS, LIKE, REGEXP, IN
	|
	&
	<<, >>
	-, +
	*, /（DIV），%（MOD）
	^
最高	-（负号），~（位反转）
	!

可以看到，不同运算符的优先级是不同的。一般情况下，级别高的运算符先进行计算，如果级别相同，MySQL按表达式的顺序从左到右依次计算。当然，在无法确定优先级的情况下，可以使用圆括号来改变优先级，这样会使计算过程更加清晰。

第 4 章 MySQL函数

MySQL提供了众多功能强大、方便易用的函数。使用这些函数，可以极大地提高用户对数据库的管理效率。MySQL中的函数包括数学函数、字符串函数、日期和时间函数、条件判断函数、系统信息函数、加密函数和其他函数。本章将介绍这些函数的功能和用法。

4.1 MySQL函数简介

函数用于对输入参数返回一个具有特定关系的值。MySQL中提供了大量函数，以方便用户对数据库进行管理以及对数据进行查询和操作。通过对数据进行处理，数据库的功能变得更加强大，可以更加灵活地满足不同用户的需求。MySQL中的函数从功能方面主要分为数学函数、字符串函数、日期和时间函数、条件判断函数、系统信息函数、加密函数和其他函数。

4.2 数学函数

数学函数主要用于处理数值数据。主要的数学函数有绝对值函数、三角函数（包括正弦函数、余弦函数、正切函数、余切函数等）、对数函数、随机数函数等。当有错误产生时，数学函数将会返回NULL。本节将介绍各种数学函数的功能和用法。

4.2.1 绝对值函数 ABS(x)和返回圆周率的函数 PI()

ABS(X)返回X的绝对值。

【例4.1】求2、-3.3和-33的绝对值,SQL语句如下:

```
mysql>SELECT ABS(2), ABS(-3.3), ABS(-33);
+--------+-----------+---------+
| ABS(2) | ABS(-3.3) | ABS(-33)|
+--------+-----------+---------+
|      2 |       3.3 |      33 |
+--------+-----------+---------+
```

正数的绝对值是其本身,因此2的绝对值为2;负数的绝对值是其相反数,因此-3.3的绝对值为3.3,-33的绝对值为33。

PI()返回圆周率的值,默认显示的小数位数是6位。

【例4.2】返回圆周率值,SQL语句如下:

```
mysql> SELECT pi();
+----------+
| pi()     |
+----------+
| 3.141593 |
+----------+
```

返回结果保留了7位有效数字。

4.2.2 平方根函数 SQRT(x)和求余函数 MOD(x,y)

SQRT(x)返回非负数x的平方根。

【例4.3】求9、40和-49的平方根,SQL语句如下:

```
mysql> SELECT SQRT(9), SQRT(40), SQRT(-49);
+---------+--------------------+-----------+
| SQRT(9) | SQRT(40)           | SQRT(-49) |
+---------+--------------------+-----------+
|       3 |  6.324555320336759 |      NULL |
+---------+--------------------+-----------+
```

9的平方根为3,40的平方根为6.324555320336759;而负数没有平方根,因此-49返回的结果为NULL。

MOD(x,y)返回x被y除后的余数。MOD()对带有小数部分的数值也起作用,它返回除法运算后的精确余数。

【例4.4】对(31,8)、(234, 10)、(45.5,6)进行求余运算,SQL语句如下:

```
mysql> SELECT MOD(31,8),MOD(234, 10),MOD(45.5,6);
+-----------+--------------+-------------+
| MOD(31,8) | MOD(234, 10) | MOD(45.5,6) |
+-----------+--------------+-------------+
|         7 |            4 |         3.5 |
+-----------+--------------+-------------+
```

4.2.3 获取整数的函数 CEIL(x)、CEILING(x)和 FLOOR(x)

CEIL(x)和CEILING(x)的意义相同,返回不小于x的最小整数值,返回值转换为一个BIGINT。

【例4.5】使用CEILING(x)的函数返回最小整数,SQL语句如下:

```
mysql> SELECT CEIL(-3.35),CEILING(3.35);
+-------------+---------------+
| CEIL(-3.35) | CEILING(3.35) |
+-------------+---------------+
|          -3 |             4 |
+-------------+---------------+
```

−3.35为负数,不小于−3.35的最小整数为−3,因此返回值为−3;不小于3.35的最小整数为4,因此返回值为4。

FLOOR(x)返回不大于x的最大整数值,返回值转换为一个BIGINT。

【例4.6】使用FLOOR(x)函数返回最大整数,SQL语句如下:

```
mysql> SELECT FLOOR(-3.35), FLOOR(3.35);
+--------------+-------------+
| FLOOR(-3.35) | FLOOR(3.35) |
+--------------+-------------+
|           -4 |           3 |
+--------------+-------------+
```

−3.35为负数,不大于−3.35的最大整数为−4,因此返回值为−4;不大于3.35的最大整数为3,因此返回值为3。

4.2.4 获取随机数的函数 RAND()和 RAND(x)

RAND(x)返回一个随机浮点值v,范围在0到1之间(0 ≤ v ≤ 1.0)。若已指定一个整数参数x,则它被用作种子值,用来产生重复序列。

【例4.7】使用不带参数的RAND()函数产生随机数,SQL语句如下:

```
mysql> SELECT RAND(),RAND(),RAND();
+--------------------+--------------------+--------------------+
| RAND()             | RAND()             | RAND()             |
+--------------------+--------------------+--------------------+
| 0.1380485446546679 | 0.45428662510056667 | 0.8572875222724746 |
+--------------------+--------------------+--------------------+
```

可以看到,不带参数的RAND()每次产生的随机数值是不同的。

【例4.8】使用带参数的RAND(x)函数产生随机数,SQL语句如下:

```
mysql> SELECT RAND(10),RAND(10),RAND(11);
+--------------------+--------------------+--------------------+
| RAND(10)           | RAND(10)           | RAND(11)           |
```

```
+--------------------+--------------------+--------------------+
| 0.6570515219653505 | 0.6570515219653505 | 0.907234631392392  |
+--------------------+--------------------+--------------------+
```

可以看到，当RAND(x)的参数x相同时，将产生相同的随机数；当参数x不同时，产生不同的随机数。

4.2.5　函数 ROUND(x)、ROUND(x,y)和 TRUNCATE(x,y)

ROUND(x)返回最接近于参数x的整数，即对x值进行四舍五入。

【例4.9】 使用ROUND(x)函数对操作数进行四舍五入操作，SQL语句如下：

```
mysql> SELECT ROUND(-1.14),ROUND(-1.67), ROUND(1.14),ROUND(1.66);
+--------------+--------------+-------------+-------------+
| ROUND(-1.14) | ROUND(-1.67) | ROUND(1.14) | ROUND(1.66) |
+--------------+--------------+-------------+-------------+
|           -1 |           -2 |           1 |           2 |
+--------------+--------------+-------------+-------------+
```

可以看到，对参数x进行四舍五入处理之后，只保留了各个值的整数部分。

ROUND(x,y)返回最接近于参数x的数，若y为正值，则将保留到小数点后面y位；若y为负值，则将保留到小数点左边y位。

【例4.10】 使用ROUND(x,y)函数对操作数进行四舍五入操作，结果保留小数点后面指定y位，SQL语句如下：

```
mysql> SELECT ROUND(1.38, 1), ROUND(1.38, 0), ROUND(232.38, -1), ROUND(232.38,-2);
+----------------+----------------+-------------------+------------------+
| ROUND(1.38, 1) | ROUND(1.38, 0) | ROUND(232.38, -1) | ROUND(232.38,-2) |
+----------------+----------------+-------------------+------------------+
|            1.4 |              1 |               230 |              200 |
+----------------+----------------+-------------------+------------------+
```

ROUND(1.38, 1)保留小数点后面1位，四舍五入的结果为1.4；ROUND(1.38, 0) 保留小数点后面0位，即返回四舍五入后的整数值； ROUND(23.38, -1)和ROUND (232.38,-2)分别保留小数点左边1位和2位。

> **提示** y值为负数时,保留的小数点左边的相应位数直接保存为0,不进行四舍五入。

TRUNCATE(x,y)返回被舍去至小数点后y位的数值x。若y为0，则结果不带有小数部分；若y为负数，则截去（归零数值）x自小数点左起第y位开始的所有低位的值。

【例4.11】 使用TRUNCATE(x,y)函数对操作数进行截取操作，结果保留小数点后面指定y位，SQL语句如下：

```
mysql> SELECT TRUNCATE(1.31,1), TRUNCATE(1.99,1), TRUNCATE(1.99,0),
TRUNCATE(19.99,-1);
```

```
+---------------+---------------+---------------+------------------+
|TRUNCATE(1.31,1)|TRUNCATE(1.99,1)|TRUNCATE(1.99,0)| TRUNCATE(19.99,-1)|
+---------------+---------------+---------------+------------------+
|          1.3  |          1.9  |          1    |           10     |
+---------------+---------------+---------------+------------------+
```

TRUNCATE(1.31,1)和TRUNCATE(1.99,1)都保留小数点后1位数字，返回值分别为1.3和1.9；TRUNCATE(1.99,0)返回整数部分1；TRUNCATE(19.99,-1)截去自小数点左边第1位开始的所有值，并将对应的整数部分置0，结果为10。

> **提示** ROUND(x,y)函数在截取值的时候会进行四舍五入，而TRUNCATE(x,y)直接截取值，并不进行四舍五入。

4.2.6 符号函数 SIGN(x)

SIGN(x)返回参数的符号，x的值分别为负、0或正时，返回结果依次为-1、0或1。

【例4.12】使用SIGN(x)函数返回参数的符号，SQL语句如下：

```
mysql> SELECT SIGN(-21),SIGN(0), SIGN(21);
+-----------+---------+---------+
| SIGN(-21) | SIGN(0) | SIGN(21) |
+-----------+---------+---------+
|        -1 |       0 |       1 |
+-----------+---------+---------+
```

SIGN(-21)返回-1；SIGN(0)返回0；SIGN(21)返回1。

4.2.7 幂运算函数 POW(x,y)、POWER(x,y)和 EXP(x)

POW(x,y)或者POWER(x,y)函数返回x的y次乘方的结果值。

【例4.13】使用POW(x,y)和POWER(x,y)函数进行乘方运算，SQL语句如下：

```
mysql> SELECT POW(2,2), POWER(2,2),POW(2,-2), POWER(2,-2);
+----------+------------+-----------+-------------+
| POW(2,2) | POWER(2,2) | POW(2,-2) | POWER(2,-2) |
+----------+------------+-----------+-------------+
|        4 |          4 |      0.25 |        0.25 |
+----------+------------+-----------+-------------+
```

可以看到，POW和POWER的结果是相同的，POW(2,2)和POWER(2,2)返回2的2次方，结果都是4；POW(2,-2)和POWER(2,-2)都返回2的-2次方，结果为4的倒数，即0.25。

EXP(x)返回e的x次乘方的结果值。

【例4.14】使用EXP(x)函数计算e的乘方，SQL语句如下：

```
mysql> SELECT EXP(3),EXP(-3),EXP(0);
+---------------------------+---------------------------+--------+
```

```
| EXP(3)               | EXP(-3)              | EXP(0) |
+----------------------+----------------------+--------+
| 20.085536923187668   | 0.049787068367863944 |   1    |
+----------------------+----------------------+--------+
```

EXP(3)计算以e为底的3次方，结果为20.085536923187668；EXP(-3)计算以e为底的-3次方，结果为0.049787068367863944；EXP(0)计算以e为底的0次方，结果为1。

4.2.8 对数运算函数 LOG(x)和 LOG10(x)

LOG(x)返回x的自然对数，即x相对于基数e的对数。

【例4.15】使用LOG(x)函数计算自然对数，SQL语句如下：

```
mysql> SELECT LOG(3), LOG(-3);
+--------------------+---------+
| LOG(3)             | LOG(-3) |
+--------------------+---------+
| 1.0986122886681098 |  NULL   |
+--------------------+---------+
```

对数定义域不能为负数，因此LOG(-3)的返回结果为NULL。
LOG10(x)返回x的基数为10的对数。

【例4.16】使用LOG10(x)计算以10为基数的对数，SQL语句如下：

```
mysql> SELECT LOG10(2), LOG10(100), LOG10(-100);
+--------------------+------------+-------------+
| LOG10(2)           | LOG10(100) | LOG10(-100) |
+--------------------+------------+-------------+
| 0.3010299956639812 |     2      |    NULL     |
+--------------------+------------+-------------+
```

10的2次方等于100，因此LOG10(100)的返回结果为2；LOG10(-100)的定义域为负，因此返回NULL。

4.2.9 角度与弧度相互转换的函数 RADIANS(x)和 DEGREES(x)

RADIANS(x)将参数x由角度转换为弧度。

【例4.17】使用RADIANS(x)将角度转换为弧度，SQL语句如下：

```
mysql> SELECT RADIANS(90),RADIANS(180);
+--------------------+-------------------+
| RADIANS(90)        | RADIANS(180)      |
+--------------------+-------------------+
| 1.5707963267948966 | 3.141592653589793 |
+--------------------+-------------------+
```

DEGREES(x)将参数x由弧度转换为角度。

【例4.18】 使用DEGREES(x)将弧度转换为角度，SQL语句如下：

```
mysql> SELECT DEGREES(PI()), DEGREES(PI() / 2);
+---------------+-------------------+
| DEGREES(PI()) | DEGREES(PI() / 2) |
+---------------+-------------------+
|           180 |                90 |
+---------------+-------------------+
```

4.2.10　正弦函数 SIN(x)和反正弦函数 ASIN(x)

SIN(x)返回x的正弦值，其中x为弧度值。

【例4.19】 使用SIN(x)函数计算正弦值，SQL语句如下：

```
mysql> SELECT SIN(1), ROUND(SIN(PI()));
+--------------------+------------------+
| SIN(1)             | ROUND(SIN(PI())) |
+--------------------+------------------+
| 0.8414709848078965 |                0 |
+--------------------+------------------+
```

ASIN(x)返回x的反正弦值，即正弦为x的值。若x不在−1～1的范围之内，则返回NULL。

【例4.20】 使用ASIN(x)函数计算反正弦值，SQL语句如下：

```
mysql> SELECT ASIN(0.8414709848078965), ASIN(3);
+--------------------------+---------+
| ASIN(0.8414709848078965) | ASIN(3) |
+--------------------------+---------+
|                        1 |    NULL |
+--------------------------+---------+
```

对比【例4.19】和【例4.20】可以看到，函数ASIN(x)和SIN(x)互为反函数；ASIN(3)中的参数3超出了正弦值的范围，因此返回NULL。

4.2.11　余弦函数 COS(x)和反余弦函数 ACOS(x)

COS(x)返回x的余弦值，其中x为弧度值。

【例4.21】 使用COS(x)函数计算余弦值，SQL语句如下：

```
mysql> SELECT COS(0),COS(PI()),COS(1);
+--------+-----------+--------------------+
| COS(0) | COS(PI()) | COS(1)             |
+--------+-----------+--------------------+
|      1 |        -1 | 0.5403023058681398 |
+--------+-----------+--------------------+
```

ACOS(x)返回x的反余弦值，即余弦是x的值。若x不在−1~1的范围之内，则返回NULL。

【例4.22】使用ACOS(x)函数计算反余弦值，SQL语句如下：

```
mysql> SELECT ACOS(1),ACOS(0), ROUND(ACOS(0.5403023058681398));
+---------+--------------------+---------------------------------+
| ACOS(1) | ACOS(0)            | ROUND(ACOS(0.5403023058681398)) |
+---------+--------------------+---------------------------------+
|       0 | 1.5707963267948966 |                               1 |
+---------+--------------------+---------------------------------+
```

对比【例4.21】和【例4.22】可以看到，函数ACOS(x)和COS(x)互为反函数。

4.2.12　正切函数、反正切函数和余切函数

TAN(x)返回x的正切值，其中x为给定的弧度值。

【例4.23】使用TAN(x)函数计算正切值，SQL语句如下：

```
mysql> SELECT TAN(0.3), ROUND(TAN(PI()/4));
+---------------------+--------------------+
| TAN(0.3)            | ROUND(TAN(PI()/4)) |
+---------------------+--------------------+
| 0.30933624960962325 |                  1 |
+---------------------+--------------------+
```

ATAN(x)返回x的反正切值，即正切为x的值。

【例4.24】使用ATAN(x)函数计算反正切值，SQL语句如下：

```
mysql> SELECT ATAN(0.30933624960962325), ATAN(1);
+---------------------------+--------------------+
| ATAN(0.30933624960962325) | ATAN(1)            |
+---------------------------+--------------------+
|                       0.3 | 0.7853981633974483 |
+---------------------------+--------------------+
```

对比【例4.23】和【例4.24】可以看到，函数ATAN(x)和TAN(x)互为反函数。

COT(x)返回x的余切值，其中x为给定的弧度值。

【例4.25】使用COT(x)函数计算余切值，SQL语句如下：

```
mysql> SELECT COT(0.3), 1/TAN(0.3),COT(PI() / 4);
+--------------------+--------------------+--------------------+
| COT(0.3)           | 1/TAN(0.3)         | COT(PI() / 4)      |
+--------------------+--------------------+--------------------+
| 3.2327281437658275 | 3.2327281437658275 | 1.0000000000000002 |
+--------------------+--------------------+--------------------+
```

由结果可以看到，函数COT(x)和TAN(x)互为倒函数。

4.3 字符串函数

字符串函数主要用来处理数据库中的字符串数据。MySQL中的字符串函数有计算字符串长度函数、字符串合并函数、字符串替换函数、字符串比较函数、查找指定字符串位置函数等。本节将介绍各种字符串函数的功能和用法。

4.3.1 计算字符串的字符数的函数和计算字符串长度的函数

CHAR_LENGTH(str)的返回值为字符串str所包含的字符个数。一个多字节字符算作一个单字符。

【例4.26】使用CHAR_LENGTH()函数计算字符串中的字符个数，SQL语句如下：

```
mysql> SELECT CHAR_LENGTH('date'), CHAR_LENGTH('egg');
+---------------------+--------------------+
| CHAR_LENGTH('date') | CHAR_LENGTH('egg') |
+---------------------+--------------------+
|                   4 |                  3 |
+---------------------+--------------------+
```

LENGTH(str)的返回值为字符串的字节长度，使用utf8（UNICODE的一种变长字符编码，又称万国码）编码字符集时，一个汉字是3字节，一个数字或字母是1字节。

【例4.27】使用LENGTH()函数计算字符串长度，SQL语句如下：

```
mysql> SELECT LENGTH('date'), LENGTH('egg');
+----------------+---------------+
| LENGTH('date') | LENGTH('egg') |
+----------------+---------------+
|              4 |             3 |
+----------------+---------------+
```

可以看到，LENGTH()函数的计算结果与CHAR_LENGTH()相同，因为英文字符的个数和所占的字节数相同，一个字符占1字节。

4.3.2 合并字符串函数 CONCAT(s1,s2,...)、CONCAT_WS(x,s1,s2,...)

CONCAT(s1,s2,...)的返回结果为所有参数相连而产生的新字符串，或许有一个或多个参数。如果任何一个参数为NULL，则返回值为NULL；如果所有参数均为非二进制字符串，则结果为非二进制字符串；如果参数中含有二进制字符串，则结果为一个二进制字符串。

【例4.28】使用CONCAT()函数连接字符串，SQL语句如下：

```
mysql> SELECT CONCAT('My SQL', '9.0'),CONCAT('My',NULL, 'SQL');
+-------------------------+--------------------------+
| CONCAT('My SQL', '9.0') | CONCAT('My',NULL, 'SQL') |
```

```
+------------------------+------------------------+
| My SQL9.0              | NULL                   |
+------------------------+------------------------+
```

CONCAT('My SQL', '9.0')返回两个字符串相连之后的新字符串；CONCAT('My',NULL,'SQL')中有一个参数为NULL，因此返回结果为NULL。

在CONCAT_WS(x,s1,s2,...)中，CONCAT_WS代表CONCAT With Separator，是CONCAT()的特殊形式。第一个参数x是其他参数的分隔符，分隔符放在要连接的两个字符串之间。分隔符可以是一个字符串，也可以是其他参数。如果分隔符为NULL，则结果为NULL。函数会忽略任何分隔符参数后的NULL。

【例4.29】使用CONCAT_WS()函数连接带分隔符的字符串，SQL语句如下：

```
mysql> SELECT CONCAT_WS('-', '1st','2nd', '3rd'), CONCAT_WS('*', '1st', NULL, '3rd');
+------------------------------------+------------------------------------+
| CONCAT_WS('-', '1st','2nd', '3rd') | CONCAT_WS('*', '1st', NULL, '3rd') |
+------------------------------------+------------------------------------+
| 1st-2nd-3rd                        | 1st*3rd                            |
+------------------------------------+------------------------------------+
```

CONCAT_WS('-', '1st','2nd', '3rd')使用分隔符"-"将3个字符串连接成一个字符串，结果为"1st-2nd-3rd"；CONCAT_WS('*', '1st', NULL, '3rd')使用分隔符"*"将两个字符串连接成一个字符串，同时忽略NULL。

4.3.3 替换字符串的函数 INSERT(s1,x,len,s2)

NSERT(s1,x,len,s2)返回一个新的字符串，其内容是将字符串s1中从位置x开始的len个字符替换为字符串s2的内容。如果x超过字符串s1的长度，则返回值为原始字符串；如果len的长度超出了字符串s1的长度，则从位置x开始替换到s1的末尾；若任何一个参数为NULL，则返回值为NULL。

【例4.30】使用INSERT()函数进行字符串替换操作，SQL语句如下：

```
MySQL> SELECT INSERT('Quest', 2, 4, 'What') AS col1, INSERT('Quest', -1, 4, 'What') AS col2, INSERT('Quest', 3, 100, 'What') AS col3;
+-------+-------+--------+
| col1  | col2  | col3   |
+-------+-------+--------+
| QWhat | Quest | QuWhat |
+-------+-------+--------+
```

第一个函数INSERT('Quest', 2, 4, 'What')将"Quest"从第2个字符开始长度为4的字符串替换为What，结果为"QWhat"；第二个函数INSERT('Quest', -1, 4, 'What')中的起始位置−1超出了字符串长度，直接返回原字符；第三个函数INSERT('Quest', 3, 100, 'What')的替换长度超出了原字符串长度，则从第3个字符开始，截取后面所有的字符，并替换为指定字符What，结果为"QuWhat"。

4.3.4 字母大小写转换函数

LOWER (str)或者LCASE (str)可以将字符串str中的大写字母全部转换为小写字母。

【例4.31】使用LOWER()函数或者LCASE()函数将字符串中所有大写字母转换为小写字母，SQL语句如下：

```
mysql> SELECT LOWER('BEAUTIFUL'), LCASE('Well');
+--------------------+----------------+
| LOWER('BEAUTIFUL') | LCASE('Well')  |
+--------------------+----------------+
| beautiful          | well           |
+--------------------+----------------+
```

由结果可以看到，原来所有字母为大写的，都被转换为小写，如"BEAUTIFUL"转换之后为"beautiful"；大小写字母混合的字符串，小写不变，大写字母转换为小写字母，如"WelL"转换之后为"well"。

UPPER(str)或者UCASE(str)可以将字符串str中的小写字母全部转换为大写字母。

【例4.32】使用UPPER()函数或者UCASE()函数将字符串中所有小写字母转换为大写字母，SQL语句如下：

```
mysql> SELECT UPPER('black'), UCASE('BLacK');
+----------------+----------------+
| UPPER('black') | UCASE('BLacK') |
+----------------+----------------+
| BLACK          | BLACK          |
+----------------+----------------+
```

由结果可以看到，原来所有字母为小写的，全部转换为大写，如"black"转换之后为"BLACK"；大小写字母混合的字符串，大写不变，小写字母转换为大写字母，如"BLacK"转换之后为"BLACK"。

4.3.5 获取指定长度的字符串的函数 LEFT(s,n)和 RIGHT(s,n)

LEFT(s,n)返回字符串s中的最左边n个字符。

【例4.33】使用LEFT()函数返回字符串中左边的字符，SQL语句如下：

```
mysql> SELECT LEFT('football', 5);
+---------------------+
| LEFT('football', 5) |
+---------------------+
| footb               |
+---------------------+
```

LEFT()函数返回字符串"football"从左边开始的长度为5的子字符串，结果为"footb"。

RIGHT(s,n)返回字符串str中的最右边的n个字符。

【例4.34】使用RIGHT()函数返回字符串中右边的字符,SQL语句如下:

```
MySQL> SELECT RIGHT('football', 4);
+----------------------+
| RIGHT('football', 4) |
+----------------------+
| ball                 |
+----------------------+
```

RIGHT()函数返回字符串"football"从右边开始的长度为4的子字符串,结果为"ball"。

4.3.6 填充字符串的函数 LPAD(s1,len,s2)和 RPAD(s1,len,s2)

LPAD(s1,len,s2)返回一个新字符串,新字符串的内容是将字符串s1的左侧用字符串s2填充至len个字符长度。如果s1的长度本身已经大于或等于len,则返回值被缩短至len个字符。

【例4.35】使用LPAD()函数对字符串进行填充操作,SQL语句如下:

```
MySQL> SELECT LPAD('hello',4,'?'), LPAD('hello',10,'?');
+---------------------+----------------------+
| LPAD('hello',4,'?') | LPAD('hello',10,'?') |
+---------------------+----------------------+
| hell                | ?????hello           |
+---------------------+----------------------+
```

字符串"hello"的长度大于4,不需要填充,因此LPAD('hello',4,'??')只返回被缩短的长度为4的子字符串"hell";字符串"hello"的长度小于10,因此LPAD('hello',10,'??')的返回结果为"?????hello",左侧填充"?",长度为10。

RPAD(s1,len,s2)返回一个新字符串,新字符串的内容是将字符串s1的右侧用字符串s2填充至len个字符长度。如果s1的长度本身已经大于或等于len,则返回值被缩短至len个字符。

【例4.36】使用RPAD()函数对字符串进行填充操作,SQL语句如下:

```
mysql> SELECT RPAD('hello',4,'?'), RPAD('hello',10,'?');
+---------------------+----------------------+
| RPAD('hello',4,'?') | RPAD('hello',10,'?') |
+---------------------+----------------------+
| hell                | hello?????           |
+---------------------+----------------------+
```

字符串"hello"的长度大于4,不需要填充,因此RPAD('hello',4,'?')只返回被缩短的长度为4的子串"hell";字符串"hello"的长度小于10,因此RPAD('hello',10,'?')的返回结果为"hello?????",右侧填充"?",长度为10。

4.3.7 删除空格的函数 LTRIM(s)、RTRIM(s)和 TRIM(s)

LTRIM(s)返回被删除了左侧空格的字符串s。

【例4.37】使用LTRIM()函数删除字符串左边的空格，SQL语句如下：

```
mysql> SELECT '( book )',CONCAT('(',LTRIM(' book '),')');
+------------+----------------------------------+
| ( book )   | CONCAT('(',LTRIM(' book '),')')  |
+------------+----------------------------------+
| ( book )   | (book )                          |
+------------+----------------------------------+
```

LTRIM()只删除字符串左边的空格，而右边的空格不会被删除，因此" book "在删除左边空格之后的结果为"book "。

RTRIM(s)返回被删除了右侧空格的字符串s。

【例4.38】使用RTRIM()函数删除字符串右边的空格，SQL语句如下：

```
mysql> SELECT '( book )',CONCAT('(', RTRIM (' book '),')');
+------------+-----------------------------------+
| ( book )   | CONCAT('(', RTRIM (' book '),')') |
+------------+-----------------------------------+
| ( book )   | ( book)                           |
+------------+-----------------------------------+
```

RTRIM()只删除字符串右边的空格，左边的空格不会被删除，因此" book "在删除右边空格之后的结果为" book"。

TRIM(s)删除字符串s两侧的空格。

【例4.39】使用TRIM()函数删除字符串两侧的空格，SQL语句如下：

```
mysql> SELECT '( book )',CONCAT('(', TRIM(' book '),')');
+------------+---------------------------------+
| ( book )   | CONCAT('(', TRIM(' book '),')') |
+------------+---------------------------------+
| ( book )   | (book)                          |
+------------+---------------------------------+
```

可以看到，函数执行之后，字符串" book "两边的空格都被删除，结果为"book"。

4.3.8 删除指定字符串的函数 TRIM(s1 FROM s)

TRIM(s1 FROM s)删除字符串s两端所有的子字符串s1。s1为可选项，在未指定情况下，默认删除s两端的空格。

【例4.40】使用TRIM(s1 FROM s)函数删除字符串两端指定的字符，SQL语句如下：

```
mysql> SELECT TRIM('xy' FROM 'xyxboxyokxxyxy') ;
+----------------------------------+
| TRIM('xy' FROM 'xyxboxyokxxyxy') |
+----------------------------------+
| xboxyokx                         |
+----------------------------------+
```

TRIM(s1 FROM s)函数会删除字符串"xyxboxyokxxyxy"两端的重复字符串"xy",而中间的"xy"并不删除,因此结果为"xboxyokx"。

4.3.9 重复生成字符串的函数 REPEAT(s,n)

REPEAT(s,n)返回一个由重复的字符串s组成的新字符串,字符串s的数目等于n。若n≤0,则返回一个空字符串;若s或n为NULL,则返回NULL。

【例4.41】 使用REPEAT()函数重复生成相同的字符串,SQL语句如下:

```
mysql> SELECT REPEAT('mysql', 3);
+--------------------+
| REPEAT('mysql', 3) |
+--------------------+
| mysqlmysqlmysql    |
+--------------------+
```

REPEAT('mysql', 3)函数返回由3个重复的"mysql"字符串组成的新字符串。

4.3.10 空格函数 SPACE(n)和替换函数 REPLACE(s,s1,s2)

SPACE(n)返回一个由n个空格组成的字符串。

【例4.42】 使用SPACE()函数生成由空格组成的字符串,SQL语句如下:

```
mysql> SELECT CONCAT('(', SPACE(6), ')' );
+-----------------------------+
| CONCAT('(', SPACE(6), ')' ) |
+-----------------------------+
| (      )                    |
+-----------------------------+
```

SPACE(6)返回的字符串由6个空格组成。

REPLACE(s,s1,s2)使用字符串s2替代字符串s中所有的字符串s1。

【例4.43】 使用REPLACE()函数进行字符串替代操作,SQL语句如下:

```
mysql> SELECT REPLACE('xxx.mysql.com', 'x', 'w');
+------------------------------------+
| REPLACE('xxx.mysql.com', 'x', 'w') |
+------------------------------------+
| www.mysql.com                      |
+------------------------------------+
```

REPLACE('xxx.mysql.com', 'x', 'w')将"xxx.mysql.com"字符串中的"x"字符替换为"w"字符,结果为"www.mysql.com"。

4.3.11 比较字符串大小的函数 STRCMP(s1,s2)

STRCMP(s1,s2)函数用于比较字符串s1和s2的大小，若两个字符串s1和s2完全相同，则返回0；若根据当前的字符排序规则，字符串s1小于字符串s2，则返回-1；其他情况返回1。

【例4.44】使用STRCMP()函数比较字符串大小，SQL语句如下：

```
mysql> SELECT STRCMP('txt', 'txt2'),STRCMP('txt2', 'txt'), STRCMP('txt', 'txt');
+-----------------------+-----------------------+----------------------+
| STRCMP('txt', 'txt2') | STRCMP('txt2', 'txt') | STRCMP('txt', 'txt') |
+-----------------------+-----------------------+----------------------+
|                    -1 |                     1 |                    0 |
+-----------------------+-----------------------+----------------------+
```

"txt"小于"txt2"，因此STRCMP('txt', 'txt2')的返回结果为-1，STRCMP('txt2', 'txt')的返回结果为1；"txt"与"txt"相等，因此STRCMP('txt', 'txt')的返回结果为0。

4.3.12 获取子串的函数 SUBSTRING(s,n,len)和 MID(s,n,len)

SUBSTRING(s,n,len)函数返回一个从字符串s的第n位开始的、长度为len的子字符串。如果n为负值，则子字符串的起始位置是从字符串s的末尾往前数的第n个字符；如果n的值超出了字符串s的长度范围，则返回一个空字符串；如果len的值大于剩余子字符串的长度，则返回从n开始到字符串末尾的子字符串。

【例4.45】使用SUBSTRING()函数获取指定位置处的子字符串，SQL语句如下：

```
MySQL> SELECT SUBSTRING('breakfast',5) AS col1, SUBSTRING('breakfast',5,3) AS col2,SUBSTRING('lunch', -3) AS col3,SUBSTRING('lunch', -5, 3) AS col4;
+-------+-------+------+------+
| col1  | col2  | col3 | col4 |
+-------+-------+------+------+
| kfast | kfa   | nch  | lun  |
+-------+-------+------+------+
```

SUBSTRING('breakfast',5)返回从字符串"breakfast"第5个位置开始到结尾的子字符串，结果为"kfast"；SUBSTRING('breakfast',5,3)返回从字符串"breakfast"第5个位置开始、长度为3的子字符串，结果为"kfa"；SUBSTRING('lunch', -3)返回从字符串"breakfast"结尾开始往前数第3个位置到字符串结尾的子字符串，结果为"nch"；SUBSTRING('lunch', -5, 3)返回从字符串"breakfast"结尾开始往前数第5个位置开始、长度为3的子字符串，结果为"lun"。

MID(s,n,len)与SUBSTRING(s,n,len)的作用相同。

【例4.46】使用MID()函数获取指定位置处的子字符串，SQL语句如下：

```
MySQL> SELECT MID('breakfast',5) as col1, MID('breakfast',5,3) as col2,MID('lunch', -3) as col3, MID('lunch', -5, 3) as col4;
+-------+-------+------+------+
| col1  | col2  | col3 | col4 |
```

```
+-------+-------+------+------+
| kfast | kfa   | nch  | lun  |
+-------+-------+------+------+
```

可以看到MID()和SUBSTRING()的结果是一样的。

> **提示** 如果对len使用的是一个小于1的值，则结果始终为空字符串。

4.3.13 匹配子字符串开始位置的函数

LOCATE(str1,str)、POSITION(str1 IN str)和INSTR(str, str1)这3个函数的作用相同，都返回子字符串str1在字符串str中的开始位置。

【例4.47】 使用LOCATE()、POSITION()、INSTR()函数查找字符串中指定子字符串的开始位置，SQL语句如下：

```
mysql> SELECT LOCATE('ball','football'),POSITION('ball'IN 'football'),INSTR
('football', 'ball');
+--------------------------+-------------------------------+----------------------+
| LOCATE('ball','football') | POSITION('ball'IN 'football') | INSTR ('football','ball') |
+--------------------------+-------------------------------+----------------------+
|                        5 |                             5 |                    5 |
+--------------------------+-------------------------------+----------------------+
```

子字符串"ball"在字符串"football"中是从第5个位置开始的，因此3个函数的返回结果都为5。

4.3.14 字符串逆序的函数 REVERSE(s)

REVERSE(s)将字符串s反转，返回的字符串的顺序和字符串s的顺序相反。

【例4.48】 使用REVERSE()函数反转字符串，SQL语句如下：

```
mysql> SELECT REVERSE('abc');
+----------------+
| REVERSE('abc') |
+----------------+
| cba            |
+----------------+
```

可以看到，字符串"abc"经过REVERSE()函数处理之后，所有字母顺序被反转，结果为"cba"。

4.3.15 返回指定位置的字符串的函数

ELT(N,字符串1,字符串2,字符串3,...,字符串N)函数返回指定位置的字符串，如果N的值等于1，则返回第1个字符串（字符串1）；如果N的值等于2，则返回第2个字符串（字符串2）；以此类推。如果N的值小于1或大于参数的总数，则返回NULL。

【例4.49】使用ELT()函数返回指定位置的字符串，SQL语句如下：

```
mysql> SELECT ELT(3,'1st','2nd','3rd'), ELT(3,'net','os');
+--------------------------+-------------------+
| ELT(3,'1st','2nd','3rd') | ELT(3,'net','os') |
+--------------------------+-------------------+
| 3rd                      | NULL              |
+--------------------------+-------------------+
```

由结果可以看到，ELT(3,'1st','2nd','3rd')返回第3个字符串"3rd"；ELT(3,'net','os')指定返回的字符串位置超出了参数个数，因此返回NULL。

4.3.16 返回指定字符串位置的函数 FIELD(s,s1,s2,…,sn)

FIELD(s,s1,s2,…,sn)返回字符串s在字符串列表s1,s2,…,sn中第一次出现的位置，在找不到s的情况下，返回0。如果s为NULL，则返回值为0，原因是NULL不能同任何值进行同等比较。

【例4.50】使用FIELD()函数返回指定字符串第一次出现的位置，SQL语句如下：

```
mysql> SELECT FIELD('Hi', 'hihi', 'Hey', 'Hi', 'bas') as col1, FIELD('Hi', 'Hey',
'Lo', 'Hilo', 'foo') as col2;
+------+------+
| col1 | col2 |
+------+------+
|    3 |    0 |
+------+------+
```

在FIELD('Hi', 'hihi', 'Hey', 'Hi', 'bas')函数中，字符串"Hi"出现在字符串列表的第3个位置，因此返回结果为3；在FIELD('Hi', 'Hey', 'Lo', 'Hilo', 'foo')函数中，列表中没有字符串"Hi"，因此返回结果为0。

4.3.17 返回子字符串位置的函数 FIND_IN_SET(s1,s2)

FIND_IN_SET(s1,s2)返回字符串s1在字符串列表s2中出现的位置，字符串列表是一个由多个逗号（,）分开的字符串组成的列表。如果s1不在s2中或s2为空字符串，则返回值为0；如果任意一个参数为NULL，则返回值为NULL。如果这个函数的第一个参数中包含一个逗号，则该函数将无法正常运行。

【例4.51】使用FIND_IN_SET()函数返回子字符串在字符串列表中的位置，SQL语句如下：

```
mysql> SELECT FIND_IN_SET('Hi','hihi,Hey,Hi,bas');
+-------------------------------------+
| FIND_IN_SET('Hi','hihi,Hey,Hi,bas') |
+-------------------------------------+
|                                   3 |
+-------------------------------------+
```

虽然FIND_IN_SET()和FIELD()两个函数的格式不同，但作用类似，都可以返回指定字符串在字符串列表中的位置。

4.3.18 选取字符串的函数 MAKE_SET(x,s1,s2,...,sn)

MAKE_SET(x,s1,s2,...,sn)函数按x的二进制数从s1,s2,...,sn中选取字符串。例如5的二进制值是0101，这个二进制从右往左的第1位和第3位是1，所以选取s1和s3。s1,s2,...,sn中的NULL不会被添加到结果中。

【例4.52】使用MAKE_SET根据二进制位选取指定字符串，SQL语句如下：

```
mysql> SELECT MAKE_SET(1,'a','b','c') as col1, MAKE_SET(1 | 4,'hello','nice',
'world') as col2, MAKE_SET(1 | 4,'hello','nice',NULL,'world') as col3, MAKE_SET(0,'a',
'b','c') as col4;
+------+-------------+-------+------+
| col1 | col2        | col3  | col4 |
+------+-------------+-------+------+
| a    | hello,world | hello |      |
+------+-------------+-------+------+
```

1的二进制值为0001，4的二进制值为0100，1与4进行或操作之后的二进制值为0101，从右到左的第1位和第3位为1。MAKE_SET(1,'a','b','c')返回第1个字符串；MAKE_SET(1 | 4,'hello','nice','world')返回从左侧开始的由第1个和第3个字符串组成的新字符串；NULL不会添加到结果中，因此MAKE_SET(1 | 4,'hello','nice',NULL,'world')只返回第1个字符串"hello"；MAKE_SET(0,'a','b','c')返回空字符串。

4.4 日期和时间函数

日期和时间函数主要用来处理日期和时间值，一般的日期函数除了使用DATE类型的参数外，也可以使用DATETIME或者TIMESTAMP类型的参数，但会忽略这些值的时间部分。相同的是，以TIME类型值为参数的函数，可以接收TIMESTAMP类型的参数，但会忽略日期部分。许多日期函数可以同时接收数字和字符串类型的参数。本节将介绍各种日期和时间函数的功能和用法。

4.4.1 获取当前日期的函数和获取当前时间的函数

CURDATE()和CURRENT_DATE()函数的作用相同，将当前日期按照"YYYY-MM-DD"或YYYYMMDD格式返回，具体格式根据函数是在字符串语境还是数字语境中而定。

【例4.53】使用日期函数获取系统当前日期，SQL语句如下：

```
mysql> SELECT CURDATE(),CURRENT_DATE(), CURDATE() + 0;
+------------+----------------+---------------+
| CURDATE()  | CURRENT_DATE() | CURDATE() + 0 |
+------------+----------------+---------------+
| 2024-07-16 | 2024-07-16     |      20240716 |
+------------+----------------+---------------+
```

可以看到，两个函数的作用相同，返回了相同的系统当前日期，"CURDATE()＋0"将当前日期值转换为数值型。

CURTIME()和CURRENT_TIME()函数的作用相同，将当前时间以"HH:MM:SS"或HHMMSS格式返回，具体格式根据函数是在字符串语境还是数字语境中而定。

【例4.54】使用时间函数获取系统当前时间，SQL语句如下：

```
mysql> SELECT CURTIME(),CURRENT_TIME(),CURTIME() + 0;
+-----------+----------------+----------------+
| CURTIME() | CURRENT_TIME() | CURTIME() + 0  |
+-----------+----------------+----------------+
| 18:03:22  | 18:03:22       |         180322 |
+-----------+----------------+----------------+
```

可以看到，两个函数的作用相同，都返回了相同的系统当前时间，"CURTIME () + 0"将当前时间值转换为数值型。

4.4.2 获取当前日期和时间的函数

CURRENT_TIMESTAMP()、LOCALTIME()、NOW()和SYSDATE()这4个函数的作用相同，均返回当前日期和时间值，格式为"YYYY-MM-DD HH:MM:SS"或YYYYMMDDHHMMSS，具体格式根据函数是在字符串语境还是数字语境中而定。

【例4.55】使用日期时间函数获取当前系统日期和时间，SQL语句如下：

```
mysql>SELECT CURRENT_TIMESTAMP(),LOCALTIME(),NOW(),SYSDATE();
+-----------+-------------+-------------+-------------------+
| CURRENT_TIMESTAMP() | LOCALTIME() | NOW() | SYSDATE() |
+-----------+---  -------+-------------+-------------------+
| 2024-07-16 12:06:09 | 2024-07-16 12:06:09 | 2024-07-16 12:06:09 | 2024-07-16 12:06:09 |
+-----------+-------------+-------------+-------------------+
```

可以看到，4个函数返回的结果是相同的。

4.4.3 UNIX 时间戳函数

如果不带任何参数调用UNIX_TIMESTAMP()函数，它会返回一个无符号整数，代表从"1970-01-01 00:00:00"GMT开始至当前时间的秒数。其中，GMT（Greenwich Mean Time）为格林尼治标准时间。如果用date参数调用UNIX_TIMESTAMP()函数，它会返回date所表示的时间点距离"1970-01-01 00:00:00"GMT的秒数。date参数可以是DATE字符串、DATETIME字符串、TIMESTAMP类型或者是当地时间的YYMMDD或YYYYMMDD格式的数字。

【例4.56】使用UNIX_TIMESTAMP函数返回UNIX格式的时间戳，SQL语句如下：

```
mysql> SELECT UNIX_TIMESTAMP(), UNIX_TIMESTAMP(NOW()), NOW();
+------------------+-----------------------+---------------------+
| UNIX_TIMESTAMP() | UNIX_TIMESTAMP(NOW()) | NOW()               |
```

```
+------------------+----------------------+---------------------+
| 1721882134       |           1721882134 | 2024-07-25 12:35:34 |
+------------------+----------------------+---------------------+
```

FROM_UNIXTIME(date)函数把UNIX时间戳转换为普通格式的时间，与UNIX_TIMESTAMP (date)函数互为反函数。

【例4.57】使用FROM_UNIXTIME函数将UNIX时间戳转换为普通格式时间，SQL语句如下：

```
mysql> SELECT FROM_UNIXTIME('1721882134');
+-----------------------------+
| FROM_UNIXTIME('1721882134') |
+-----------------------------+
| 2024-07-25 12:35:34.000000  |
+-----------------------------+
```

可以看到，FROM_UNIXTIME('1721882134')与【例4.56】中UNIX_TIMESTAMP(NOW())的结果正好相反，即两个函数互为反函数。

4.4.4　返回UTC日期的函数和返回UTC时间的函数

UTC_DATE()函数返回当前UTC（世界标准时间）日期值，其格式为"YYYY-MM-DD"或YYYYMMDD，具体格式取决于函数是用在字符串语境还是数字语境中。

【例4.58】使用UTC_DATE()函数返回当前UTC日期值，SQL语句如下：

```
mysql> SELECT UTC_DATE(), UTC_DATE() + 0;
+------------+----------------+
| UTC_DATE() | UTC_DATE() + 0 |
+------------+----------------+
| 2024-07-25 |       20240725 |
+------------+----------------+
```

UTC_DATE()函数的返回值为当前时区的日期值。

UTC_TIME()返回当前UTC时间值，其格式为"HH:MM:SS"或HHMMSS，具体格式取决于函数是用在字符串语境还是数字语境中。

【例4.59】使用UTC_TIME()函数返回当前UTC时间值，SQL语句如下：

```
mysql> SELECT UTC_TIME(), UTC_TIME() + 0;
+------------+----------------+
| UTC_TIME() | UTC_TIME() + 0 |
+------------+----------------+
| 10:08:22   |         100822 |
+------------+----------------+
```

UTC_TIME()函数返回当前时区的时间值。

4.4.5 获取月份的函数 MONTH(date)和 MONTHNAME(date)

MONTH(date)函数返回date对应的月份，值的范围为1~12。

【例4.60】使用MONTH()函数返回指定日期中的月份，SQL语句如下：

```
mysql> SELECT MONTH('2020-02-13');
+---------------------+
| MONTH('2020-02-13') |
+---------------------+
|                   2 |
+---------------------+
```

MONTHNAME(date)函数返回date对应月份的英文全名。

【例4.61】使用MONTHNAME()函数返回指定日期中月份的名称，SQL语句如下：

```
mysql> SELECT MONTHNAME('2018-02-13');
+-------------------------+
| MONTHNAME('2018-02-13') |
+-------------------------+
| February                |
+-------------------------+
```

4.4.6 获取星期的函数 DAYNAME(d)、DAYOFWEEK(d)和 WEEKDAY(d)

DAYNAME(d)函数返回日期d所对应的星期几的英文名称，例如Sunday、Monday等。

【例4.62】使用DAYNAME()函数返回指定日期所对应的星期几的英文名称，SQL语句如下：

```
mysql> SELECT DAYNAME('2018-10-10');
+-----------------------+
| DAYNAME('2018-10-10') |
+-----------------------+
| Wednesday             |
+-----------------------+
```

可以看到，2018年10月10日是星期三，因此返回结果为Wednesday。

DAYOFWEEK(d)函数返回日期d所对应的周索引，1表示星期日，2表示星期一……7表示星期六。

【例4.63】使用DAYOFWEEK()函数返回日期对应的周索引，SQL语句如下：

```
mysql> SELECT DAYOFWEEK('2018-10-10');
+-------------------------+
| DAYOFWEEK('2018-10-10') |
+-------------------------+
|                       4 |
+-------------------------+
```

由【例4.62】可知,2018年10月10日为周三,其对应的周索引值为4,因此本例结果为4。WEEKDAY(d)函数返回日期d所对应的周索引,0表示星期一,1表示星期二……6表示星期日。

【例4.64】使用WEEKDAY()函数返回日期对应的周索引,SQL语句如下:

```
mysql>SELECT WEEKDAY('2018-10-10 22:23:00'), WEEKDAY('2018-11-11');
+--------------------------------+------------------------+
| WEEKDAY('2018-10-10 22:23:00') | WEEKDAY('2018-11-11')  |
+--------------------------------+------------------------+
|                              2 |                      6 |
+--------------------------------+------------------------+
```

可以看到,WEEKDAY()和DAYOFWEEK()函数都返回指定日期在某一周内的位置,只是索引编号不同。

4.4.7 获取星期数的函数 WEEK(d)和 WEEKOFYEAR(d)

WEEK(d)计算日期d是一年中的第几周。WEEK(d, Mode)的双参数形式允许指定该星期是否起始于周日或周一,以及返回值的范围是否为0~53或1~53。若Mode参数被省略,则使用系统自变量default_week_format的值,具体可参考表4.1。

表 4.1 WEEK()函数中 Mode 参数的取值

Mode	一周的第一天	范围	Week 1 为第一周…
0	周日	0~53	本年度中有一个周日
1	周一	0~53	本年度中有 3 天以上
2	周日	1~53	本年度中有一个周日
3	周一	1~53	本年度中有 3 天以上
4	周日	0~53	本年度中有 3 天以上
5	周一	0~53	本年度中有一个周一
6	周日	1~53	本年度中有 3 天以上
7	周一	1~53	本年度中有一个周一

【例4.65】使用WEEK()函数查询指定日期是一年中的第几周,SQL语句如下:

```
mysql>SELECT WEEK('2018-02-20'),WEEK('2018-02-20',0), WEEK('2018-02-20',1);
+--------------------+----------------------+----------------------+
| WEEK('2018-02-20') | WEEK('2018-02-20',0) | WEEK('2018-02-20',1) |
+--------------------+----------------------+----------------------+
|                  7 |                    7 |                    8 |
+--------------------+----------------------+----------------------+
```

可以看到,WEEK('2018-02-20')使用一个参数,其第二个参数为default_week_format的默认值,MySQL中该值为0,指定一周的第一天为周日,因此和WEEK('2018-02-20',0)的返回结果相同;WEEK('2018-02-20',1)中第二个参数为1,指定一周的第一天为周一,返回值为8。可以看到,第二个参数不同,返回的结果也不同。使用不同的参数的原因是不同地区和国家的习惯不同,每周的第一天并不相同。

WEEKOFYEAR(d)计算某天位于一年中的第几周,范围是1~53,相当于WEEK(d,3)。

【例4.66】使用WEEKOFYEAR()查询指定日期是一年中的第几周,SQL语句如下:

```
mysql> SELECT WEEK('2018-01-20',3), WEEKOFYEAR('2018-01-20');
+----------------------+--------------------------+
| WEEK('2018-01-20',3) | WEEKOFYEAR('2018-01-20') |
+----------------------+--------------------------+
|                    3 |                        3 |
+----------------------+--------------------------+
```

可以看到,两个函数返回结果相同。

4.4.8 获取天数的函数 DAYOFYEAR(d)和 DAYOFMONTH(d)

DAYOFYEAR(d)函数返回日期d是一年中的第几天,范围是1~366。

【例4.67】使用DAYOFYEAR()函数返回指定日期在一年中的位置,SQL语句如下:

```
mysql> SELECT DAYOFYEAR('2018-02-20');
+-------------------------+
| DAYOFYEAR('2018-02-20') |
+-------------------------+
|                      51 |
+-------------------------+
```

1月有31天,再加上2月的20天,因此返回结果为51。

DAYOFMONTH(d)函数返回日期d是一个月中的第几天,范围是1~31。

【例4.68】使用DAYOFMONTH()函数返回指定日期在一个月中的位置,SQL语句如下:

```
mysql> SELECT DAYOFMONTH('2018-08-20');
+--------------------------+
| DAYOFMONTH('2018-08-20') |
+--------------------------+
|                       20 |
+--------------------------+
```

4.4.9 获取年份、季度、小时、分钟和秒钟的函数

YEAR(date)函数返回date对应的年份,范围是1970~2069。

【例4.69】使用YEAR()函数返回指定日期对应的年份,SQL语句如下:

```
mysql>SELECT YEAR('18-02-03'),YEAR('96-02-03');
+------------------+------------------+
| YEAR('18-02-03') | YEAR('96-02-03') |
+------------------+------------------+
|             2018 |             1996 |
+------------------+------------------+
```

> 提示 00~69转换为2000~2069，70~99转换为1970~1999。

QUARTER(date)函数返回date对应的一年中的季度值，范围是1~4。

【例4.70】使用QUARTER()函数返回指定日期对应的季度，SQL语句如下：

```
mysql> SELECT QUARTER('18-04-01');
+---------------------+
| QUARTER('18-04-01') |
+---------------------+
|                   2 |
+---------------------+
```

MINUTE(time)函数返回time对应的分钟数，范围是0~59。

【例4.71】使用MINUTE()函数返回指定时间的分钟值，SQL语句如下：

```
mysql> SELECT MINUTE('18-02-03 10:10:03');
+-----------------------------+
| MINUTE('18-02-03 10:10:03') |
+-----------------------------+
|                          10 |
+-----------------------------+
```

SECOND(time)函数返回time对应的秒数，范围是0~59。

【例4.72】使用SECOND()函数返回指定时间的秒值，SQL语句如下：

```
mysql> SELECT SECOND('10:05:03');
+--------------------+
| SECOND('10:05:03') |
+--------------------+
|                  3 |
+--------------------+
```

4.4.10 获取日期的指定值的函数 EXTRACT(type FROM date)

EXTRACT(type FROM date)函数所使用的时间间隔类型说明符与DATE_ADD()和DATE_SUB()的相同，但它作用是从日期中提取一部分，而不是执行日期运算。

【例4.73】使用EXTRACT()函数提取日期或者时间值，SQL语句如下：

```
mysql> SELECT EXTRACT(YEAR FROM '2018-07-02') AS col1, EXTRACT(YEAR_MONTH FROM
'2018-07-12 01:02:03') AS col2, EXTRACT(DAY_MINUTE FROM '2018-07-12 01:02:03') AS col3;
+------+--------+--------+
| col1 | col2   | col3   |
+------+--------+--------+
| 2018 | 201807 | 120102 |
+------+--------+--------+
```

type值为YEAR时，只返回年值，结果为2018；type值为YEAR_MONTH时，返回年与月份，结果为201807；type值为DAY_MINUTE时，返回日、小时和分钟值，结果为120102。

4.4.11 时间和秒数转换的函数

TIME_TO_SEC(time)函数用于将时间值转换为秒数值。转换公式为：小时×3600+分钟×60+秒。

【例4.74】使用TIME_TO_SEC()函数将时间值转换为秒数值，SQL语句如下：

```
mysql> SELECT TIME_TO_SEC('23:23:00');
+-------------------------+
| TIME_TO_SEC('23:23:00') |
+-------------------------+
|                   84180 |
+-------------------------+
```

SEC_TO_TIME(seconds)函数用于将秒数值转换为时间值，时间值的格式为"HH:MM:SS"或HHMMSS，具体格式根据该函数是用在字符串语境还是数字语境中而定。

【例4.75】使用SEC_TO_TIME()函数将秒数值转换为时间格式，SQL语句如下：

```
mysql> SELECT SEC_TO_TIME(2345),SEC_TO_TIME(2345)+0, TIME_TO_SEC('23:23:00'),
SEC_TO_TIME(84180);
+------------------+--------------------+-------------------------+-------------------+
|SEC_TO_TIME(2345) |SEC_TO_TIME(2345)+0 |TIME_TO_SEC('23:23:00')  |SEC_TO_TIME(84180) |
+------------------+--------------------+-------------------------+-------------------+
|00:39:05          |               3905 |                   84180 | 23:23:00          |
+------------------+--------------------+-------------------------+-------------------+
```

可以看到，SEC_TO_TIME()函数的返回值加上0值之后变成了数值格式；TIME_TO_SEC和SEC_TO_TIME互为反函数。

4.4.12 计算日期和时间的函数

计算日期和时间的函数有DATE_ADD()、ADDDATE()、DATE_SUB()、SUBDATE()、ADDTIME()、SUBTIME()和DATE_DIFF()。

在DATE_ADD(date,INTERVAL expr type)和DATE_SUB(date,INTERVAL expr type)中，date是一个DATETIME或DATE值，用来指定起始时间；expr是一个表达式，用来指定从起始日期添加或减去的时间间隔值，对于负值的时间间隔，expr可以以一个负号（-）开头；type为关键词，表明了表达式被解释的方式。表4.2说明了type和expr参数的关系。

表 4.2　MySQL 中计算日期和时间的格式

type 值	预期的 expr 格式
MICROSECOND	MICROSECONDS
SECOND	SECONDS
MINUTE	MINUTES
HOUR	HOURS
DAY	DAYS

（续表）

type 值	预期的 expr 格式
WEEK	WEEKS
MONTH	MONTHS
QUARTER	QUARTERS
YEAR	YEARS
SECOND_MICROSECOND	'SECONDS.MICROSECONDS'
MINUTE_MICROSECOND	'MINUTES.MICROSECONDS'
MINUTE_SECOND	'MINUTES:SECONDS'
HOUR_MICROSECOND	'HOURS.MICROSECONDS'
HOUR_SECOND	'HOURS:MINUTES:SECONDS'
HOUR_MINUTE	'HOURS:MINUTES'
DAY_MICROSECOND	'DAYS.MICROSECONDS'
DAY_SECOND	'DAYS HOURS:MINUTES:SECONDS'
DAY_MINUTE	'DAYS HOURS:MINUTES'
DAY_HOUR	'DAYS HOURS'
YEAR_MONTH	'YEARS-MONTHS'

若date参数是一个DATE值，则计算只会包括YEAR、MONTH和DAY部分（没有时间部分），其结果是一个DATE值；否则，结果将是一个DATETIME值。

DATE_ADD(date,INTERVAL expr type)和ADDDATE(date,INTERVAL expr type)函数的作用相同，执行日期的加运算。

【例4.76】使用DATE_ADD()和ADDDATE()函数执行日期加操作，SQL语句如下：

```
mysql> SELECT DATE_ADD('2010-12-31 23:59:59', INTERVAL 1 SECOND) AS col1,
ADDDATE('2010-12-31 23:59:59', INTERVAL 1 SECOND) AS col2, DATE_ADD('2010-12-31
23:59:59', INTERVAL '1:1' MINUTE_SECOND) AS col3;
+---------------------+---------------------+---------------------+
| col1                | col2                | col3                |
+---------------------+---------------------+---------------------+
| 2011-01-01 00:00:00 | 2011-01-01 00:00:00 | 2011-01-01 00:01:00 |
+---------------------+---------------------+---------------------+
```

由结果可以看到，DATE_ADD('2010-12-31 23:59:59', INTERVAL 1 SECOND)和ADDDATE('2010-12-31 23:59:59', INTERVAL 1 SECOND)两个函数的执行结果是相同的，将时间增加1秒后返回，结果都为"2011-01-01 00:00:00"；DATE_ADD('2010-12-31 23:59:59', INTERVAL '1:1' MINUTE_SECOND)日期运算类型是MINUTE_SECOND，将指定时间增加1分1秒后返回，结果为"2011-01-01 00:01:00"。

DATE_SUB(date,INTERVAL expr type)和SUBDATE(date,INTERVAL expr type)函数的作用相同，均用于执行日期的减运算。

【例4.77】使用DATE_SUB()和SUBDATE()函数执行日期减操作，SQL语句如下：

```
mysql> SELECT DATE_SUB('2011-01-02', INTERVAL 31 DAY) AS col1,
SUBDATE('2011-01-02', INTERVAL 31 DAY) AS col2, DATE_SUB('2011-01-01 00:01:00',
INTERVAL '0 0:1:1' DAY_SECOND) AS col3;
+------------+------------+---------------------+
| col1       | col2       | col3                |
+------------+------------+---------------------+
| 2010-12-02 | 2010-12-02 | 2010-12-31 23:59:59 |
+------------+------------+---------------------+
```

由结果可以看到，DATE_SUB('2011-01-02', INTERVAL 31 DAY)和SUBDATE('2011-01-02', INTERVAL 31 DAY)两个函数的执行结果是相同的，将日期值减少31天后返回，结果都为"2010-12-02"；DATE_SUB('2011-01-01 00:01:00',INTERVAL '0 0:1:1' DAY_SECOND)函数将指定日期减少1天，时间减少1分1秒后返回，结果为"2010-12-31 23:59:59"。

> **提示** DATE_ADD()和DATE_SUB()在指定修改的时间段时，也可以指定负值，负值代表相减，即返回以前的日期和时间。

ADDTIME(date,expr)函数将expr值添加到date，并返回修改后的值。其中，date是一个日期或者日期时间表达式，而expr是一个时间表达式。

【例4.78】使用ADDTIME()进行时间加操作，SQL语句如下：

```
mysql>SELECT ADDTIME('2000-12-31 23:59:59','1:1:1'), ADDTIME('02:02:02',
'02:00:00');
+----------------------------------------+-------------------------------+
| ADDTIME('2000-12-31 23:59:59','1:1:1') | ADDTIME('02:02:02','02:00:00')|
+----------------------------------------+-------------------------------+
| 2001-01-01 01:01:00                    | 04:02:02                      |
+----------------------------------------+-------------------------------+
```

可以看到，将"2000-12-31 23:59:59"的时间部分值增加1小时1分钟1秒后变为"2001-01-01 01:01:00"；"02:02:02"增加2小时后变为"04:02:02"。

SUBTIME(date,expr)函数将date减去expr值，并返回修改后的值。其中，date是一个日期或者日期时间表达式，而expr是一个时间表达式。

【例4.79】使用SUBTIME()函数执行时间减操作，SQL语句如下：

```
mysql> SELECT SUBTIME('2000-12-31 23:59:59','1:1:1'),
SUBTIME('02:02:02','02:00:00');
+----------------------------------------+-------------------------------+
| SUBTIME('2000-12-31 23:59:59','1:1:1') | SUBTIME('02:02:02','02:00:00')|
+----------------------------------------+-------------------------------+
| 2000-12-31 22:58:58                    | 00:02:02                      |
+----------------------------------------+-------------------------------+
```

可以看到，将"2000-12-31 23:59:59"的时间部分值减少1小时1分钟1秒后变为"2000-12-31 22:58:58"；"02:02:02"减少2小时后变为"00:02:02"。

DATEDIFF(date1,date2)返回起始时间date1和结束时间date2之间的天数。date1和date2为日期或日期时间表达式,计算中只用到这些值的日期部分。

【例4.80】使用DATEDIFF()函数计算两个日期的间隔天数,SQL语句如下:

```
mysql> SELECT DATEDIFF('2010-12-31 23:59:59','2010-12-30') AS col1,
DATEDIFF('2010-11-30 23:59:59','2010-12-31') AS col2;
+------+------+
| col1 | col2 |
+------+------+
|    1 |  -31 |
+------+------+
```

DATEDIFF()函数返回date1-date2后的值,因此DATEDIFF('2010-12-31 23:59:59','2010-12-30')的返回值为1,DATEDIFF('2010-11-30 23:59:59','2010-12-31')的返回值为-31。

4.4.13 将日期和时间格式化的函数

DATE_FORMAT(date,format)函数根据format指定的格式显示date值。format格式如表4.3所示。

表4.3 DATE_FORMAT 时间日期格式

说 明 符	说 明
%a	工作日的缩写名称(Sun,...,Sat)
%b	月份的缩写名称(Jan,...,Dec)
%c	月份,数字形式(0,...,12)
%D	以英文后缀表示月中的几号(1st, 2nd...)
%d	该月日期,数字形式(00,...,31)
%e	该月日期,数字形式(0,...,31)
%f	微秒(000000,...,999999)
%H	以2位数表示24小时(00,...,23)
%h,%I	以2位数表示12小时(01,...,12)
%i	分钟,数字形式(00,...,59)
%j	一年中的天数(001,...,366)
%k	以24(0,...,23)小时表示时间
%l	以12(1,...,12)小时表示时间
%M	月份名称(January,...,December)
%m	月份,数字形式(00,...,12)
%p	上午(AM)或下午(PM)
%r	时间,12小时制(小时:分钟:秒数,后加AM或PM)
%S,%s	以2位数形式表示秒(00,...,59)
%T	时间,24小时制(小时:分钟:秒数)
%U	周(00,...,53),其中周日为每周的第一天
%u	周(00,...,53),其中周一为每周的第一天
%V	周(01,...,53),其中周日为每周的第一天;和%X同时使用

(续表)

说明符	说明
%v	周（01,...,53），其中周一为每周的第一天；和 %x 同时使用
%W	周中每日的名称（周日,...,周六）
%w	一周中的每日（0=周日,...,6=周六）
%X	该周的年份，其中周日为每周的第一天；数字形式，4位数；和%V 同时使用
%x	该周的年份，其中周一为每周的第一天；数字形式，4位数；和%v 同时使用
%Y	4位数形式表示年份
%y	2位数形式表示年份
%%	标识符%

【例4.81】使用DATE_FORMAT()函数格式化输出日期和时间值，SQL语句如下：

```
mysql> SELECT DATE_FORMAT('1997-10-04 22:23:00', '%W %M %Y') AS col1,
DATE_FORMAT('1997-10-04 22:23:00','%D %y %a %d %m %b %j') AS col2;
+----------------------+-------------------------+
| col1                 | col2                    |
+----------------------+-------------------------+
| Saturday October 1997 | 4th 97 Sat 04 10 Oct 277 |
+----------------------+-------------------------+

mysql> SELECT DATE_FORMAT('1997-10-04 22:23:00', '%H:%i:%s') AS col3,
DATE_FORMAT('1999-01-01', '%X %V') AS col4;
+----------+---------+
| col3     | col4    |
+----------+---------+
| 22:23:00 | 1998 52 |
+----------+---------+
```

可以看到"1997-10-04 22:23:00"分别按照不同参数转换成了不同格式的日期值和时间值。

TIME_FORMAT(time,format)函数根据表达式format的要求显示时间time。表达式format指定了显示的格式。因为TIME_FORMAT(time,format)只处理时间，所以format只使用时间格式。

【例4.82】使用TIME_FORMAT()函数格式化输入时间值，SQL语句如下：

```
mysql>SELECT TIME_FORMAT('16:00:00', '%H %k %h %I %l');
+-------------------------------------------+
| TIME_FORMAT('16:00:00', '%H %k %h %I %l') |
+-------------------------------------------+
| 16 16 04 04 4                             |
+-------------------------------------------+
```

可以看到，"16:00:00"按照不同的参数转换为不同格式的时间值。

GET_FORMAT(val_type, format_type)返回日期和时间字符串的显示格式，val_type表示日期数据类型，包括DATE、DATETIME和TIME；format_type表示格式化显示类型，包括EUR、INTERVAL、ISO、JIS、USA。GET_FORMAT根据两个值类型组合返回的字符串显示格式如表4.4所示。

表 4.4 GET_FORMAT 返回的显示格式字符串

值类型	格式化类型	显示格式字符串
DATE	EUR	%d.%m.%Y
DATE	INTERVAL	%Y%m%d
DATE	ISO	%Y-%m-%d
DATE	JIS	%Y-%m-%d
DATE	USA	%m.%d.%Y
TIME	EUR	%H.%i.%s
TIME	INTERVAL	%H%i%s
TIME	ISO	%H:%i:%s
TIME	JIS	%H:%i:%s
TIME	USA	%h:%i:%s %p
DATETIME	EUR	%Y-%m-%d %H.%i.%s
DATETIME	INTERVAL	%Y%m%d%H%i%s
DATETIME	ISO	%Y-%m-%d %H:%i:%s
DATETIME	JIS	%Y-%m-%d %H:%i:%s
DATETIME	USA	%Y-%m-%d %H.%i.%s

【例4.83】使用GET_FORMAT()函数显示不同格式化类型下的格式字符串，SQL语句如下：

```
mysql> SELECT GET_FORMAT(DATE,'EUR'), GET_FORMAT(DATE,'USA');
+------------------------+------------------------+
| GET_FORMAT(DATE,'EUR') | GET_FORMAT(DATE,'USA') |
+------------------------+------------------------+
| %d.%m.%Y               | %m.%d.%Y               |
+------------------------+------------------------+
```

可以看到，不同类型的格式化字符串并不相同。

【例4.84】在DATE_FORMAT()函数中，使用GET_FORMAT()函数返回的显示格式字符串来显示指定的日期值，SQL语句如下：

```
mysql> SELECT DATE_FORMAT('2000-10-05 22:23:00', GET_FORMAT(DATE,'USA') );
+-------------------------------------------------------------+
| DATE_FORMAT('2000-10-05 22:23:00', GET_FORMAT(DATE,'USA') ) |
+-------------------------------------------------------------+
| 10.05.2000                                                  |
+-------------------------------------------------------------+
```

GET_FORMAT(DATE,'USA')返回的显示格式字符串为%m.%d.%Y，对照表4.3中的显示格式（%m以数字形式显示月份，%d以数字形式显示日期，%Y以4位数字形式显示年），因此结果为10.05.2000。

4.5 条件判断函数

条件判断函数也称为控制流程函数，根据满足的不同条件，执行相应的流程。MySQL中进行条件判断的函数有IF()、IFNULL()和CASE()。本节将分别介绍各个函数的用法。

4.5.1 IF()函数

IF()函数的语法格式如下：

```
IF(expr, v1, v2)
```

如果表达式expr是TRUE(expr <> 0 and expr <> NULL)，则返回值为v1；否则返回值为v2。IF()的返回值为数字或字符串，具体取决于其所在的语境。

【例4.85】使用IF()函数进行条件判断，SQL语句如下：

```
mysql> SELECT IF(1>2,2,3), IF(1<2,'yes ','no'),
IF(STRCMP('test','test1'),'no','yes');
+-----------+------------------+-----------------------------------+
|IF(1>2,2,3)|IF(1<2,'yes ','no')|IF(STRCMP('test','test1'),'no','yes')|
+-----------+------------------+-----------------------------------+
|         3 | yes              | no                                |
+-----------+------------------+-----------------------------------+
```

1>2的结果为FALSE，因此IF(1>2,2,3)返回第二个表达式的值；1<2的结果为TRUE，因此IF(1<2,'yes ', 'no')返回第一个表达式的值；"test"小于"test1"，结果为true，因此IF(STRCMP('test','test1'),'no','yes')返回第一个表达式的值。

> 提示 如果v1和v2中只有一个明确是NULL，则IF()函数的结果类型为非NULL表达式的结果类型。

4.5.2 IFNULL()函数

IFNULL()函数的语法格式如下：

```
IFNULL(v1,v2)
```

如果v1不为NULL，则IFNULL()的返回值为v1；否则返回值为v2。IFNULL()的返回值是数字或者字符串，具体取决于其所在的语境。

【例4.86】使用IFNULL()函数进行条件判断，SQL语句如下：

```
mysql> SELECT IFNULL(1,2), IFNULL(NULL,10), IFNULL(1/0, 'wrong');
+-------------+-----------------+----------------------+
| IFNULL(1,2) | IFNULL(NULL,10) | IFNULL(1/0, 'wrong') |
```

```
+--------------+----------------+-----------------------+
|       1      |       10       | wrong                 |
+--------------+----------------+-----------------------+
```

虽然IFNULL(1,2)的第二个值也不为空，但返回结果依然是第一个值；IFNULL(NULL,10)的第一个值为空，因此返回10；"1/0"的结果为空，因此IFNULL(1/0, 'wrong')返回字符串"wrong"。

4.5.3 CASE()函数

CASE()函数有两种语法格式，第一种语法格式如下：

```
CASE expr WHEN v1 THEN r1 [WHEN v2 THEN r2]...[ELSE rn+1] END
```

如果expr值等于某个vn，则返回对应THEN后面的结果；如果expr值与所有值都不相等，则返回ELSE后面的rn+1。

【例4.87】使用CASE expr WHEN语句执行分支操作，SQL语句如下：

```
mysql> SELECT CASE 2 WHEN 1 THEN 'one' WHEN 2 THEN 'two' ELSE 'more' END;
+------------------------------------------------------------+
| CASE 2 WHEN 1 THEN 'one' WHEN 2 THEN 'two' ELSE 'more' END |
+------------------------------------------------------------+
| two                                                        |
+------------------------------------------------------------+
```

expr值为2，与第二条分支语句WHEN后面的值相等，因此返回结果为"two"。

CASE的第二种语法格式如下：

```
CASE  WHEN v1 THEN r1 [WHEN v2 THEN r2]... ELSE rn+1] END
```

某个vn值为TRUE时，返回对应THEN后面的结果；如果所有值都不为TRUE，则返回ELSE后的rn+1。

【例4.88】使用CASE WHEN语句执行分支操作，SQL语句如下：

```
mysql> SELECT CASE WHEN 1<0 THEN 'true' ELSE 'false' END;
+--------------------------------------------+
| CASE WHEN 1<0 THEN 'true' ELSE 'false' END |
+--------------------------------------------+
| false                                      |
+--------------------------------------------+
```

1<0的结果为FALSE，因此函数返回值为ELSE后面的"false"。

> **提示** 一个CASE表达式的默认返回值类型是任何返回值的相容集合类型，但具体情况视其所在语境而定。如果用在字符串语境中，则返回结果为字符串；如果用在数字语境中，则返回结果为十进制值、实数值或整数值。

4.6 系统信息函数

MySQL中的系统信息有数据库的版本号、当前用户名和连接数、系统字符集、最后一个自动生成的ID值等。本节将介绍常用的获取系统信息的函数。

4.6.1 获取 MySQL 版本号、连接数和数据库名的函数

VERSION()返回指示MySQL服务器版本的字符串。这个字符串使用utf8字符集。

【例4.89】查看当前MySQL版本号，SQL语句如下：

```
mysql> SELECT VERSION();
+-----------+
| VERSION() |
+-----------+
| 9.0.1     |
+-----------+
```

CONNECTION_ID()返回MySQL服务器当前连接的次数，每个连接都有各自唯一的ID。

【例4.90】查看当前用户的连接数，SQL语句如下：

```
mysql> SELECT CONNECTION_ID();
+-----------------+
| CONNECTION_ID() |
+-----------------+
|              23 |
+-----------------+
```

这里返回23，返回值根据登录的次数会有所不同。

SHOW PROCESSLIST命令的输出结果显示有哪些线程正在运行，不仅可以查看当前所有的连接数，还可以查看当前的连接状态，帮助识别出有问题的查询语句等。

如果是root账号，则能看到所有用户的当前连接；如果是其他普通账号，则只能看到自己占用的连接。SHOW PROCESSLIST只列出前100条数据，如果想全部列出，可以使用SHOW FULL PROCESSLIST命令。

【例4.91】使用SHOW PROCESSLIST命令输出当前用户的连接信息，SQL语句如下：

```
MySQL> SHOW PROCESSLIST;
+--+---------------+--------+----+-------+-----+----------+-----------+
|Id|User           |Host    |db  |Command|Time |State     |Info       |
+--+---------------+--------+----+-------+-----+----------+-----------+
|4 |event_scheduler|localhost|NULL|Daemon|15274|Waiting on empty queue|NULL|
|23|root           |localhost:58788|NULL|Query|0|starting |SHOW PROCESSLIST|
+--+---------------+--------+----+-------+-----+----------+-----------+
```

各个列说明如下：

- ID列：用户登录MySQL时，系统分配的是"connection id"。
- User列：显示当前用户。如果不是root，那么这个命令只显示用户权限范围内的SQL语句。
- Host列：显示这个语句是从哪个IP的哪个端口上发出，可以用来追踪出现问题语句的用户。
- db列：显示这个进程目前连接的是哪个数据库。
- Command列：显示当前连接执行的命令，一般取值为休眠（Sleep）、查询（Query）、连接（Connect）。
- Time列：显示这个状态持续的时间，单位是秒。
- State列：显示使用当前连接的SQL语句的状态，是很重要的列。State只是语句执行中的某一个状态。一个SQL语句，以查询为例，可能需要经过Copying to tmp table、Sorting result、Sending data等状态才可以完成。后续会介绍所有状态。
- Info列：显示这个SQL语句，是判断问题语句的一个重要依据。

使用另一个命令行登录MySQL，此时将会有2个连接。在第二个登录的命令行下再次输入SHOW PROCESSLIST，结果如下：

```
mysql> SHOW PROCESSLIST;
+----+------+---------+--------+---------+--------+---------+---------+
| Id | User | Host    | db     | Command | Time   | State   | Info    |
+----+------+---------+--------+---------+--------+---------+---------+
| 1  | root |localhost:3602|NULL| Sleep  | 38     |         | NULL    |
| 2  | root |localhost:3272|NULL| Query  | 0      |NULL|show processlist|
+----+------+---------+--------+---------+--------+---------+---------+
```

可以看到，当前活动用户为已登录的连接Id为2的用户，正在执行的Command（操作命令）是Query（查询），使用的查询命令为SHOW PROCESSLIST；而连接Id为1的用户目前没有对数据进行操作，即为Sleep操作，而且已经经过了38秒。

DATABASE()和SCHEMA()函数返回使用utf8字符集的默认（当前）数据库名。

【例4.92】查看当前使用的数据库，SQL语句如下：

```
mysql> SELECT DATABASE(),SCHEMA();
+----------------+----------------+
| DATABASE()     | SCHEMA()       |
+----------------+----------------+
| test_db        | test_db        |
+----------------+----------------+
```

可以看到，两个函数的作用相同。

4.6.2　获取用户名的函数

USER()、CURRENT_USER、CURRENT_USER()、SYSTEM_USER()和SESSION_USER()这几个函数返回当前被MySQL服务器验证的用户名和主机名组合。这个值用于确定当前登录用户在MySQL中的存取权限。一般情况下，这几个函数的返回值相同。

【例4.93】获取当前登录用户名称，SQL语句如下：

```
mysql> SELECT USER(), CURRENT_USER(), SYSTEM_USER();
+----------------+----------------+----------------+
| USER()         | CURRENT_USER() | SYSTEM_USER()  |
+----------------+----------------+----------------+
| root@localhost | root@localhost | root@localhost |
+----------------+----------------+----------------+
```

返回结果指示了当前账户连接服务器时的用户名及所连接的客户主机，root为当前登录的用户名，localhost为登录的主机名。

4.6.3 获取字符串的字符集和排序方式的函数

CHARSET(str)返回字符串str使用字符集。

【例4.94】使用CHARSET()函数返回字符串使用的字符集，SQL语句如下：

```
mysql> SELECT CHARSET('abc'), CHARSET(CONVERT('abc' USING latin1)),
CHARSET(VERSION());
+---------------+-------------------------------------+-------------------+
|CHARSET('abc')|CHARSET(CONVERT('abc' USING latin1))| CHARSET(VERSION()) |
+---------------+-------------------------------------+-------------------+
| gbk           | latin1                              | utf8mb3           |
+---------------+-------------------------------------+-------------------+
```

CHARSET('abc')返回系统默认的字符集gbk；CHARSET(CONVERT('abc' USING latin1))返回的字符集为latin1；VERSION()返回的字符串使用utf8字符集，因此CHARSET(VERSION())返回结果为utf8。

COLLATION(str)返回字符串str的字符排列方式。

【例4.95】使用COLLATION()函数返回字符串排列方式，SQL语句如下：

```
mysql> SELECT COLLATION('abc'),COLLATION(CONVERT('abc' USING utf8));
+------------------+--------------------------------------+
| COLLATION('abc') | COLLATION(CONVERT('abc' USING utf8)) |
+------------------+--------------------------------------+
| gbk_chinese_ci   | utf8_general_ci                      |
+------------------+--------------------------------------+
```

可以看到，使用不同字符集时字符串的排列方式不同。

4.6.4 获取最后一个自动生成的 ID 值的函数

LAST_INSERT_ID()函数返回最后生成的AUTO_INCREMENT值。

【例4.96】使用SELECT LAST_INSERT_ID查看最后一个自动生成的列值。

1）一次插入一条记录

首先创建表worker，其Id字段带有AUTO_INCREMENT约束，SQL语句如下：

```
mysql> CREATE TABLE worker (Id INT AUTO_INCREMENT NOT NULL PRIMARY KEY, Name VARCHAR(30));
```

然后向表worker中插入两条记录：

```
mysql> INSERT INTO worker VALUES(NULL, 'jimy');
mysql> INSERT INTO worker VALUES(NULL, 'Tom');
mysql> SELECT * FROM worker;
+----+------+
| Id | Name |
+----+------+
|  1 | jimy |
|  2 | Tom  |
+----+------+
```

查看已经插入的数据可以发现，最后一条记录的Id字段值为2。

使用LAST_INSERT_ID()查看最后自动生成的Id值：

```
mysql> SELECT LAST_INSERT_ID();
+------------------+
| LAST_INSERT_ID() |
+------------------+
|                2 |
+------------------+
```

可以看到，一次插入一条记录时，返回值为最后一条记录的Id值。

2）一次插入多条记录

向表中同时插入多条记录，SQL语句如下：

```
mysql> INSERT INTO worker VALUES (NULL,'Kevin'),(NULL,'Michal'),(NULL,'Nick');
```

查询已经插入的记录：

```
mysql> SELECT * FROM worker;
+----+--------+
| Id | Name   |
+----+--------+
|  1 | jimy   |
|  2 | Tom    |
|  3 | Kevin  |
|  4 | Michal |
|  5 | Nick   |
+----+--------+
```

可以看到最后一条记录的Id字段值为5。

使用LAST_INSERT_ID()查看最后自动生成的Id值：

```
mysql> SELECT LAST_INSERT_ID();
+------------------+
| LAST_INSERT_ID() |
+------------------+
```

```
|                  3 |
+--------------------+
```

结果显示，LAST_INSERT_ID值是3而不是5，这是为什么呢？在向数据表中插入一条新记录时，LAST_INSERT_ID()返回带有AUTO_INCREMENT约束的字段最新生成的值2。继续向表中同时添加3条记录，读者可能以为这时LAST_INSERT_ID值为5，可显示结果却为3，这是因为当使用一条INSERT语句插入多行时，LAST_INSERT_ID()只返回插入第一行数据时产生的值，在这里为第3条记录。之所以会这样，是因为这使依靠其他服务器复制同样的INSERT语句变得简单。

> **提示** LAST_INSERT_ID与数据表无关，如果向表a插入数据后再向表b插入数据，那么LAST_INSERT_ID返回表b中的Id值。

4.7 加密函数

加密函数主要用来对数据进行加密和解密处理，以保证某些重要数据不被别人获取。这些函数在保证数据库安全时非常有用。本节将介绍各种加密函数的作用和使用方法。

4.7.1 加密函数 MD5(str)

MD5(str)为字符串计算出一个MD5 128比特校验和。该校验和以32位十六进制数字的二进制字符串形式返回，若参数为NULL，则会返回NULL。

【例4.97】使用MD5()函数加密字符串，SQL语句如下：

```
mysql> SELECT MD5 ('mypwd');
+----------------------------------+
| MD5 ('mypwd')                    |
+----------------------------------+
| 318bcb4be908d0da6448a0db76908d78 |
+----------------------------------+
```

可以看到，"mypwd"经MD5加密后的结果为318bcb4be908d0da6448a0db76908d78。

4.7.2 加密函数 SHA(str)

SHA(str)函数用于计算给定字符串str的加密哈希值。当参数为NULL时，SHA(str)函数将返回NULL。与MD5相比，SHA加密算法提供了更强的安全性能，因为它生成的哈希值更长，碰撞的可能性更低。

【例4.98】使用SHA()函数加密字符串，SQL语句如下：

```
mysql> SELECT SHA('tom123456');
+------------------------------------------+
```

```
| SHA('tom123456')                         |
+------------------------------------------+
| 8218b487f490cb484f45c31403eb1f597a2b531a |
+------------------------------------------+
```

4.7.3 加密函数 SHA2(str, hash_length)

SHA2(str, hash_length)函数用于对给定的字符串str进行加密,其加密算法为SHA-2。这个函数的第二个参数hash_length用于指定加密后哈希值的长度。hash_length支持的值为224、256、384、512和0。其中,0等同于256。

【例4.99】使用SHA2()加密字符串,SQL语句如下:

```
mysql> SELECT SHA2('tom123456',0) A,sha2('tom123456',256) B\G
*************************** 1. row ***************************
A: 9242a986a9edbd14a60450e9284a372efeff7e9f6209f675fdc4457f55de5e27
B: 9242a986a9edbd14a60450e9284a372efeff7e9f6209f675fdc4457f55de5e27
```

可以看到,hash_length的值为256和0时,结果都是一样的。

4.8 其他函数

本节将要介绍的函数不能笼统地分为哪一类,但是这些函数也非常有用,例如重复指定操作函数、改变字符集函数、IP地址与数字转换函数等。本节将介绍这些函数的作用和使用方法。

4.8.1 格式化函数 FORMAT(x,n)

FORMAT(x,n)将数字x格式化,并以四舍五入的方式保留小数点后n位,结果以字符串的形式返回。若n为0,则返回的结果函数中不含小数部分。

【例4.100】使用FORMAT函数格式化数字,并保留指定数值的小数位数,SQL语句如下:

```
MySQL> SELECT FORMAT(12332.123456, 4), FORMAT(12332.1,4), FORMAT(12332.2,0);
+-------------------------+-------------------+-------------------+
| FORMAT(12332.123456, 4) | FORMAT(12332.1,4) | FORMAT(12332.2,0) |
+-------------------------+-------------------+-------------------+
| 12,332.1235             | 12,332.1000       | 12,332            |
+-------------------------+-------------------+-------------------+
```

由结果可以看到,FORMAT(12332.123456, 4)保留4位小数值,并进行四舍五入,结果为12332.1235;FORMAT(12332.1,4)保留4位小数值,位数不够的用0补齐,结果为12332.1000;FORMAT(12332.2,0)不保留小数部分,返回结果为整数12332。

4.8.2 不同进制的数字进行转换的函数

CONV(N, from_base, to_base)将参数N由from_base进制转换为to_base进制。如有任意一个参数为NULL，则返回值为NULL。参数N被理解为一个整数，但也可以指定为一个字符串。最小基数为2，最大基数为36。

【例4.101】 使用CONV()函数在不同进制之间转换，SQL语句如下：

```
mysql> SELECT CONV('a',16,2), CONV(15,10,2), CONV(15,10,8), CONV(15,10,16);
+----------------+---------------+---------------+----------------+
| CONV('a',16,2) | CONV(15,10,2) | CONV(15,10,8) | CONV(15,10,16) |
+----------------+---------------+---------------+----------------+
| 1010           | 1111          | 17            | F              |
+----------------+---------------+---------------+----------------+
```

CONV('a',16,2)将十六进制的a转换为二进制数值，十六进制的a表示十进制的数值10，而十进制的数值10正好等于二进制的数值1010；CONV(15,10,2)将十进制的数值15转换为二进制值，结果为1111；CONV(15,10,8)将十进制的数值15转换为八进制值，结果为17；CONV(15,10,16)将十进制的数值15转换为十六进制值，结果为F。

进制说明：

- 二进制：采用0和1两个数字来表示，它以2为基数，逢二进一。
- 八进制：采用0、1、2、3、4、5、6、7八个数字来表示，逢八进一，以数字0开头。
- 十进制：采用0~9共10个数字来表示，逢十进一。
- 十六进制：由0~9、A~F组成，以数字0x开头。它与十进制的对应关系是：十六进制的0~9对应十进制的0~9，十六进制的A~F对应十进制的10~15。

4.8.3 IP地址与数字相互转换的函数

INET_ATON(expr)函数用于将给定的IP地址字符串转换为整数值。这个函数可以处理4位或8位的IP地址。

【例4.102】 使用INET_ATON()函数将字符串网络点地址转换为数值网络地址，SQL语句如下：

```
mysql> SELECT INET_ATON('209.207.224.40');
+-----------------------------+
| INET_ATON('209.207.224.40') |
+-----------------------------+
|                  3520061480 |
+-----------------------------+
```

INET_NTOA(expr)函数用于将给定的数值网络地址（4位或8位）转换为点地址表示形式的字符串。

【例4.103】使用INET_NTOA()函数将数值网络地址转换为字符串网络点地址，SQL语句如下：

```
mysql> SELECT INET_NTOA(3520061480);
+-----------------------+
| INET_NTOA(3520061480) |
+-----------------------+
| 209.207.224.40        |
+-----------------------+
```

可以看到，INET_NTOA()和INET_ATON()互为反函数。

4.8.4 加锁函数和解锁函数

GET_LOCK(str,timeout)设法使用字符串str给定的名字得到一个锁，超时为timeout秒。若成功得到锁，则返回1；若操作超时，则返回0；若发生错误，则返回NULL。假如有一个用GET_LOCK()得到的锁，当执行RELEASE_LOCK()或连接断开（正常或非正常）时，这个锁就会被解除。

RELEASE_LOCK(str)解开被GET_LOCK()获取的、用字符串str命名的锁。若锁被解开，则返回1；若该线程尚未创建锁，则返回0（此时锁没有被解开）；若命名的锁不存在，则返回NULL。若该锁从未被GET_LOCK()调用获取，或锁已经被提前解开，则该锁不存在。

IS_FREE_LOCK(str)检查名为str的锁是否可以使用（换言之，是否被封锁）。若锁可以使用，则返回1（没有人在用这个锁）；若这个锁正在被使用，则返回0；若出现错误（诸如不正确的参数），则返回NULL。

IS_USED_LOCK(str)检查名为str的锁是否正在被使用（换言之，是否被封锁）。若被封锁，则返回使用该锁的客户端的连接标识符（Connection ID）；否则，返回NULL。

【例4.104】使用加锁、解锁函数，SQL语句如下：

```
mysql> SELECT GET_LOCK('lock1',10) AS GetLock, IS_USED_LOCK('lock1') AS ISUsedLock,
IS_FREE_LOCK('lock1') AS ISFreeLock, RELEASE_LOCK('lock1') AS ReleaseLock;
+---------+------------+------------+-------------+
| GetLock | ISUsedLock | ISFreeLock | ReleaseLock |
+---------+------------+------------+-------------+
|       1 |         23 |          0 |           1 |
+---------+------------+------------+-------------+
```

GET_LOCK('lock1',10)的返回值为1，说明成功得到了一个名称为"lock1"的锁，持续时间为10秒。IS_USED_LOCK('lock1')的返回值为当前连接ID，表示名称为"lock1"的锁正在被使用。IS_FREE_LOCK('lock1')的返回值为0，说明名称为"lock1"的锁正在被使用。RELEASE_LOCK('lock1'的)返回值为1，说明解锁成功。

4.8.5 重复执行指定操作的函数

BENCHMARK(count,expr)函数重复执行count次表达式expr。它可以用于计算MySQL处理表达式的速度。结果值通常为0（0只是表示处理过程很快，并不是没有花费时间）。另一个作用是它可以在MySQL客户端内部报告语句执行的时间。

【例4.105】 使用BENCHMARK()重复执行指定函数。

（1）使用SHA()函数加密密码，SQL语句如下：

```
mysql> SELECT SHA('newpwd');
+------------------------------------------+
| SHA('newpwd')                            |
+------------------------------------------+
| c7f005b657906521157aa3fc261afe886d51f792 |
+------------------------------------------+
1 row in set (0.00 sec)
```

可以看到，执行SHA()函数花费的时间为0.00sec。

（2）使用BENCHMARK函数重复执行SHA()操作500000次，SQL语句如下：

```
mysql> SELECT BENCHMARK(500000, SHA('newpwd'));
+----------------------------------+
| BENCHMARK(500000, SHA('newpwd')) |
+----------------------------------+
|                                0 |
+----------------------------------+
1 row in set (0.29 sec)
```

可以看到，使用BENCHMARK执行500000次花费的时间为0.29 sec，明显比执行一次的时间延长了。

> **提示** BENCHMARK报告的时间是在客户端经过的时间，而不是在服务器端的CPU中的时间，每次执行后报告的时间并不一定是相同的。读者可以多次执行该语句，查看结果。

4.8.6 改变字符集的函数

CONVERT(... USING ...)函数被用来在不同的字符集之间转换数据。

【例4.106】 使用CONVERT()函数改变字符串的默认字符集，SQL语句如下：

```
MySQL> SELECT CHARSET('string'),CHARSET(CONVERT('string' USING latin1));
+-------------------+----------------------------------------+
| CHARSET('string') | CHARSET(CONVERT('string' USING latin1)) |
+-------------------+----------------------------------------+
| gbk               | latin1                                 |
+-------------------+----------------------------------------+
```

通过CONVERT()将字符串"string"的默认字符集gbk改为latin1。

4.8.7 改变数据类型的函数

CAST(x , AS type)和CONVERT(x, type)函数将一个类型的值转换为另一个类型的值，可转换的type有BINARY、CHAR(n)、DATE、TIME、DATETIME、DECIMAL、SIGNED、UNSIGNED。

【例4.107】 使用CAST()和CONVERT()函数进行数据类型的转换，SQL语句如下：

```
mysql> SELECT CAST(100 AS CHAR(2)), CON     VERT('2018-10-01 12:12:12',TIME);
+----------------------+------------------------------------+
| CAST(100 AS CHAR(2)) | CONVERT('2018-10-01 12:12:12',TIME) |
+----------------------+------------------------------------+
| 10                   | 12:12:12                           |
+----------------------+------------------------------------+
```

可以看到，CAST(100 AS CHAR(2))将整数数据100转换为带有两个显示宽度的字符串类型，结果为"10"；CONVERT('2018-10-01 12:12:12',TIME)将DATETIME类型的值转换为TIME类型的值，结果为"12:12:12"。

4.9 窗 口 函 数

在MySQL 8.0版本之前，没有排名函数，当需要在查询中实现排名时，必须手写@变量，比较麻烦。

从MySQL 8.0版本开始支持窗口函数，用它可以实现很多新的查询方式。窗口函数类似于SUM()、COUNT()那样的集合函数，但它并不会将多行查询结果合并为一行，而是将结果放回多行当中。也就是说，窗口函数是不需要GROUP BY的。

下面通过案例来讲述通过窗口函数实现排名效果的方法。

首先创建公司部门表branch，包含部门的名称（name）和部门人数（brcount）两个字段，创建语句如下：

```
CREATE TABLE branch
(
name          char(255) NOT NULL,
brcount       INT(11) NOT NULL
);

INSERT INTO branch(name,brcount)
      VALUES('branch1',5),
            ('branch2',10),
            ('branch3',8),
            ('branch4',20),
            ('branch5',9);
```

查询数据表branch中的数据：

```
mysql> SELECT * FROM branch;
+---------+---------+
| name    | brcount |
+---------+---------+
| branch1 |       5 |
| branch2 |      10 |
| branch3 |       8 |
| branch4 |      20 |
| branch5 |       9 |
+---------+---------+
```

对公司部门人数从小到大进行排序，可以利用窗口函数来实现：

```
mysql> SELECT *, rank() OVER w1 AS `rank` FROM branch  window w1 AS (ORDER BY brcount);
+---------+---------+------+
| name    | brcount | rank |
+---------+---------+------+
| branch1 |       5 |    1 |
| branch3 |       8 |    2 |
| branch5 |       9 |    3 |
| branch2 |      10 |    4 |
| branch4 |      20 |    5 |
+---------+---------+------+
```

这里创建了名称为"w1"的窗口函数，规定对brcount字段进行排序，然后在SELECT子句中对窗口函数w1执行rank()方法，将结果输出为rank字段。

需要注意，这里的window w1是可选的。例如，在每一行中加入员工的总数，可以这样操作：

```
mysql> SELECT *, SUM(brcount) over() as total_count FROM branch;
+---------+---------+-------------+
| name    | brcount | total_count |
+---------+---------+-------------+
| branch1 |       5 |          52 |
| branch2 |      10 |          52 |
| branch3 |       8 |          52 |
| branch4 |      20 |          52 |
| branch5 |       9 |          52 |
+---------+---------+-------------+
```

可以一次性查询出每个部门的员工人数占总人数的百分比，查询结果如下：

```
mysql> SELECT *,(brcount)/(sum(brcount) over()) AS rate FROM branch;
+---------+---------+--------+
| name    | brcount | rate   |
+---------+---------+--------+
| branch1 |       5 | 0.0962 |
| branch2 |      10 | 0.1923 |
| branch3 |       8 | 0.1538 |
| branch4 |      20 | 0.3846 |
| branch5 |       9 | 0.1731 |
+---------+---------+--------+
```

第 5 章
查询数据

数据库管理系统的一个重要功能就是数据查询。数据查询不只是简单地返回数据库中存储的数据,还应根据需要对数据进行筛选以及确定数据以什么样的格式显示。MySQL提供了强大、灵活的SELECT语句来实现这些操作。本章将介绍如何使用SELECT语句进行数据查询。

5.1 基本查询语句

MySQL从数据表中查询数据的基本语句为SELECT语句。SELECT语句的基本格式如下:

```
SELECT
    {* | <字段列表>}
    [
        FROM <表1>,<表2>...
        [WHERE <表达式>
        [GROUP BY <group by definition>]
        [HAVING <expression> [{<operator> <expression>}...]]
        [ORDER BY <order by definition>]
        [LIMIT [<offset>,] <row count>]
    ]
SELECT [字段1,字段2,...,字段n]
FROM [表或视图]
WHERE [查询条件];
```

其中,各条子句的含义如下:

- {*|<字段列表>}:包含星号通配符和字段列表,表示查询的字段。其中,字段列表至少包含一个字段名称。如果要查询多个字段,多个字段之间用英文逗号隔开,最后一个字段后不加逗号。
- FROM <表1>,<表2>...:表1和表2表示查询数据的来源,可以是单个或者多个表。

- WHERE<表达式>：该子句是可选项，如果选择该项，将限定查询行必须满足的查询条件。
- GROUP BY <字段>：该子句告诉MySQL如何显示查询出来的数据，并按照指定的字段分组。
- [HAVING <查询条件>]：用来过滤数据。和WHERE不同的是，WHERE在数据分组前过滤，而HAVING在数据分组后进行过滤。
- [ORDER BY <字段>]：该子句告诉MySQL按什么样的顺序显示查询出来的数据，可以使用的排序方法有升序（ASC）、降序（DESC）。
- [LIMIT [<offset>,] <row count>]：该子句告诉MySQL每次显示查询出来的数据条数。

SELECT [字段1,字段2,...,字段n]的可选参数比较多，读者可能无法一下子完全理解。不要紧，接下来将从最简单的查询语句开始，通过一步一步地深入学习，读者会对各个参数的作用有清晰的认识。

下面以一个例子说明如何使用SELECT从单张表中获取数据。

首先定义数据表，SQL语句如下：

```
CREATE TABLE fruits
(
f_id      char(10)       NOT NULL,
s_id      INT            NOT NULL,
f_name    char(255)      NOT NULL,
f_price   decimal(8,2)   NOT NULL,
PRIMARY KEY(f_id)
);
```

为了演示如何使用SELECT语句，需要插入如下数据：

```
mysql> INSERT INTO fruits (f_id, s_id, f_name, f_price)
    VALUES('a1', 101,'apple',5.2),
    ('b1',101,'blackberry', 10.2),
    ('bs1',102,'orange', 11.2),
    ('bs2',105,'melon',8.2),
    ('t1',102,'banana', 10.3),
    ('t2',102,'grape', 5.3),
    ('o2',103,'coconut', 9.2),
    ('c0',101,'cherry', 3.2),
    ('a2',103, 'apricot',2.2),
    ('l2',104,'lemon', 6.4),
    ('b2',104,'berry', 7.6),
    ('m1',106,'mango', 15.7),
    ('m2',105,'xbabay', 2.6),
    ('t4',107,'xbababa', 3.6),
    ('m3',105,'xxtt', 11.6),
    ('b5',107,'xxxx', 3.6);
```

使用SELECT语句查询f_id字段的数据：

```
mysql> SELECT f_id, f_name FROM fruits;
+------+-------------+
| f_id | f_name      |
+------+-------------+
```

```
| a1   | apple      |
| a2   | apricot    |
| b1   | blackberry |
| b2   | berry      |
| b5   | xxxx       |
| bs1  | orange     |
| bs2  | melon      |
| c0   | cherry     |
| l2   | lemon      |
| m1   | mango      |
| m2   | xbabay     |
| m3   | xxtt       |
| o2   | coconut    |
| t1   | banana     |
| t2   | grape      |
| t4   | xbababa    |
+------+------------+
```

该语句的执行过程是,SELECT语句决定了要查询的列值,在这里查询f_id和f_name两个字段的值;FROM子句指定了数据的来源,这里为数据表fruits,因此返回结果为表fruits中f_id和f_name两个字段下的所有数据。其显示顺序为添加到表中的顺序。

5.2 单表查询

单表查询是指从一张表的数据中查询所需的数据。本节将介绍单表查询中的各种基本的查询方式,主要有查询所有字段、查询指定字段、查询指定记录、查询空值、多条件的查询、对查询结果进行排序等。

5.2.1 查询所有字段

1. 在SELECT语句中使用星号(*)通配符查询所有字段

SELECT查询记录最简单的形式是从一张表中检索所有记录,实现的方法是使用星号(*)通配符来表示所有列的名称。语法格式如下:

SELECT * FROM 表名;

【例5.1】从表fruits中检索所有字段的数据,SQL语句如下:

```
mysql> SELECT * FROM fruits;
+------+------+------------+---------+
| f_id | s_id | f_name     | f_price |
+------+------+------------+---------+
| a1   | 101  | apple      |    5.20 |
| a2   | 103  | apricot    |    2.20 |
| b1   | 101  | blackberry |   10.20 |
| b2   | 104  | berry      |    7.60 |
```

```
| b5   | 107 | xxxx    |  3.60 |
| bs1  | 102 | orange  | 11.20 |
| bs2  | 105 | melon   |  8.20 |
| c0   | 101 | cherry  |  3.20 |
| l2   | 104 | lemon   |  6.40 |
| m1   | 106 | mango   | 15.70 |
| m2   | 105 | xbabay  |  2.60 |
| m3   | 105 | xxtt    | 11.60 |
| o2   | 103 | coconut |  9.20 |
| t1   | 102 | banana  | 10.30 |
| t2   | 102 | grape   |  5.30 |
| t4   | 107 | xbababa |  3.60 |
+------+-----+---------+-------+
```

可以看到，使用星号（*）通配符时，将返回所有列，列按照表定义时的顺序显示。

2. 在SELECT语句中指定所有字段

下面介绍另外一种查询所有字段值的方法。根据SELECT语句的格式，SELECT关键字后面的字段名为将要查找的数据，因此可以将表中所有字段的名称跟在SELECT子句后面，如果忘记了字段名称，可以使用DESC命令查看表的结构。但有时，表中的字段可能比较多，不一定能记得所有字段的名称，该方法就会很不方便，因此不建议使用。例如，查询表fruits中的所有数据，SQL语句也可以书写如下：

```
SELECT f_id, s_id ,f_name, f_price FROM fruits;
```

查询结果与【例5.1】相同。

> **提示** 一般情况下，除非需要使用表中所有的字段数据，最好不要使用通配符'*'。使用通配符虽然可以节省输入查询语句的时间，但是获取不需要的列数据通常会降低查询和所使用的应用程序的效率。通配符的优势是，当不知道所需要的列的名称时，可以通过它来获取它们。

5.2.2 查询指定字段

1. 查询单个字段

查询表中的某一个字段，语法格式为：

```
SELECT 列名 FROM 表名;
```

【例5.2】查询表fruits中f_name列的所有水果名称，SQL语句如下：

```
SELECT f_name FROM fruits;
```

该语句使用SELECT声明从表fruits中获取f_name字段下的所有水果名称，指定字段的名称紧跟在SELECT关键字之后。查询结果如下：

```
mysql> SELECT f_name FROM fruits;
+------------+
| f_name     |
+------------+
| apple      |
| apricot    |
| blackberry |
| berry      |
| xxxx       |
| orange     |
| melon      |
| cherry     |
| lemon      |
| mango      |
| xbabay     |
| xxtt       |
| coconut    |
| banana     |
| grape      |
| xbababa    |
+------------+
```

输出结果显示了表fruits中f_name字段下的所有数据。

2. 查询多个字段

使用SELECT声明，可以获取多个字段下的数据，只需要在关键字SELECT后面指定要查找的字段的名称，不同字段名称之间用逗号（，）分隔开，最后一个字段后面不需要加逗号。语法格式如下：

```
SELECT 字段名1,字段名2,...,字段名n FROM 表名;
```

【例5.3】从表fruits中获取f_name和f_price两列，SQL语句如下：

```
SELECT f_name, f_price FROM fruits;
```

该语句使用SELECT声明从表fruits中获取f_name和f_price两个字段下的所有水果名称和价格，两个字段之间用逗号分隔开。查询结果如下：

```
mysql> SELECT f_name, f_price FROM fruits;
+------------+---------+
| f_name     | f_price |
+------------+---------+
| apple      |    5.20 |
| apricot    |    2.20 |
| blackberry |   10.20 |
| berry      |    7.60 |
| xxxx       |    3.60 |
| orange     |   11.20 |
| melon      |    8.20 |
| cherry     |    3.20 |
| lemon      |    6.40 |
```

```
| mango       |   15.70 |
| xbabay      |    2.60 |
| xxtt        |   11.60 |
| coconut     |    9.20 |
| banana      |   10.30 |
| grape       |    5.30 |
| xbababa     |    3.60 |
+-------------+---------+
```

输出结果显示了表fruits中f_name和f_price两个字段下的所有数据。

5.2.3 查询指定记录

数据库中包含大量的数据，根据特殊要求，可能只需要查询表中的指定数据，即对数据进行过滤。在SELECT语句中，通过WHERE子句可以对数据进行过滤，语法格式为：

```
SELECT 字段名1,字段名2,...,字段名n
FROM 表名
WHERE 查询条件
```

在WHERE子句中，MySQL提供了一系列的条件判断符，如表5.1所示。

表5.1　WHERE 条件判断符

判 断 符	说　　明
=	相等
<> , !=	不相等
<	小于
<=	小于或等于
>	大于
>=	大于或等于
BETWEEN	位于两值之间

【例5.4】查询价格为10.2元的水果的名称，SQL语句如下：

```
SELECT f_name, f_price FROM fruits WHERE f_price = 10.2;
```

该语句使用SELECT声明从表fruits中获取价格等于10.2元的水果的数据，查询结果如下：

```
+------------+---------+
| f_name     | f_price |
+------------+---------+
| blackberry |   10.20 |
+------------+---------+
```

从查询结果中可以看到，价格是10.2元的水果的名称是blackberry，其他的均不满足查询条件。本例采用了简单的相等过滤，查询指定列f_price中是否有值10.20。相等过滤还可以用来比较字符串，下面给出一个例子。

【例5.5】查找名称为"apple"的水果的价格,SQL语句如下:

```
SELECT f_name, f_price FROM fruits WHERE f_name = 'apple';
```

该语句使用SELECT声明从表fruits中获取名称为"apple"的水果的价格,查询结果如下:

```
+--------+---------+
| f_name | f_price |
+--------+---------+
| apple  |  5.20   |
+--------+---------+
```

从查询结果中可以看到,只有名称为"apple"的行被返回,其他的均不满足查询条件。

【例5.6】查询价格小于10元的水果的名称,SQL语句如下:

```
SELECT f_name, f_price FROM fruits WHERE f_price < 10;
```

该语句使用SELECT声明从表fruits中获取价格低于10元的水果名称,即f_price小于10的水果信息被返回,查询结果如下:

```
+---------+---------+
| f_name  | f_price |
+---------+---------+
| apple   |  5.20   |
| apricot |  2.20   |
| berry   |  7.60   |
| xxxx    |  3.60   |
| melon   |  8.20   |
| cherry  |  3.20   |
| lemon   |  6.40   |
| xbabay  |  2.60   |
| coconut |  9.20   |
| grape   |  5.30   |
| xbababa |  3.60   |
+---------+---------+
```

可以看到,查询结果中所有记录的f_price字段的值均小于10.00元,而大于或等于10.00元的记录没有被返回。

5.2.4 带 IN 关键字的查询

IN操作符用来查询满足指定范围内的条件的记录。使用IN操作符,将所有检索条件用圆括号括起来,检索条件之间用逗号分隔开,只要满足条件范围内的一个值即为匹配项。

【例5.7】查询s_id(水果供应商)为101和102的记录,SQL语句如下:

```
SELECT s_id,f_name, f_price FROM fruits WHERE s_id IN (101,102) ORDER BY f_name;
```

查询结果如下:

```
+------+------------+---------+
| s_id | f_name     | f_price |
+------+------------+---------+
| 101  | apple      |    5.20 |
| 102  | banana     |   10.30 |
| 101  | blackberry |   10.20 |
| 101  | cherry     |    3.20 |
| 102  | grape      |    5.30 |
| 102  | orange     |   11.20 |
+------+------------+---------+
```

相反，可以使用关键字NOT来检索不在条件范围内的记录。

【例5.8】查询所有s_id不等于101也不等于102的记录，SQL语句如下：

```
SELECT s_id,f_name, f_price FROM fruits WHERE s_id NOT IN (101,102) ORDER BY f_name;
```

查询结果如下：

```
+------+---------+---------+
| s_id | f_name  | f_price |
+------+---------+---------+
| 103  | apricot |    2.20 |
| 104  | berry   |    7.60 |
| 103  | coconut |    9.20 |
| 104  | lemon   |    6.40 |
| 106  | mango   |   15.70 |
| 105  | melon   |    8.20 |
| 107  | xbababa |    3.60 |
| 105  | xbabay  |    2.60 |
| 105  | xxtt    |   11.60 |
| 107  | xxxx    |    3.60 |
+------+---------+---------+
```

可以看到，该语句在IN关键字前面加上了NOT关键字，这使得查询的结果与【例5.7】结果正好相反，【例5.7】检索了s_id等于101和102的记录，而这里所要求查询的记录中s_id字段值不等于这两个值中的任何一个。

5.2.5 带 BETWEEN AND 的范围查询

BETWEEN AND用来查询某个范围内的值，该操作符需要两个参数，即范围的开始值和结束值。如果字段值满足指定范围的查询条件，则这些记录被返回。

【例5.9】查询价格在2.00元到10.20元之间的水果名称及其价格，SQL语句如下：

```
SELECT f_name, f_price FROM fruits WHERE f_price BETWEEN 2.00 AND 10.20;
```

查询结果如下：

```
+------------+---------+
| f_name     | f_price |
+------------+---------+
```

```
| apple       |   5.20 |
| apricot     |   2.20 |
| blackberry  |  10.20 |
| berry       |   7.60 |
| xxxx        |   3.60 |
| melon       |   8.20 |
| cherry      |   3.20 |
| lemon       |   6.40 |
| xbabay      |   2.60 |
| coconut     |   9.20 |
| grape       |   5.30 |
| xbababa     |   3.60 |
+-------------+--------+
```

可以看到，返回结果包含了价格从2.00元到10.20元之间的字段值，并且端点值10.20也包括在返回结果中，即BETWEEN AND匹配范围中的所有值，包括开始值和结束值。

BETWEEN AND操作符前可以加关键字NOT，表示指定范围之外的值，如果字段值不满足指定范围内的查询条件，则这些记录被返回。

【例5.10】查询价格在2.00元到10.20元之外的水果名称及其价格，SQL语句如下：

```
SELECT f_name, f_price FROM fruits WHERE f_price NOT BETWEEN 2.00 AND 10.20;
```

查询结果如下：

```
+--------+---------+
| f_name | f_price |
+--------+---------+
| orange |   11.20 |
| mango  |   15.70 |
| xxtt   |   11.60 |
| banana |   10.30 |
+--------+---------+
```

由结果可以看到，返回的记录只有f_price字段值大于10.20的。其实，f_price字段值小于2.00的记录也满足查询条件。因此，如果表中有f_price字段值小于2.00的记录，也应当作为查询结果。

5.2.6 带LIKE的字符匹配查询

在前面的检索操作中，讲述了如何查询多个字段的记录，如何进行比较查询或者是查询一个条件范围内的记录，如果要查找所有包含字符"ge"的水果名称，该如何查找呢？简单的比较操作在这里已经行不通了，需要使用通配符进行匹配查找，通过创建查找模式对表中的数据进行比较。执行这个任务的关键字是LIKE。

通配符是一种在SQL的WHERE条件子句中拥有特殊意思的字符。SQL语句中支持多种通配符，可以和LIKE一起使用的通配符有"%"和"_"。

1. 百分号通配符（%）匹配任意长度的字符，甚至包括零字符

【例5.11】查找所有以 "b" 字母开头的水果，SQL语句如下：

```
SELECT f_id, f_name FROM fruits WHERE f_name LIKE 'b%';
```

查询结果如下：

```
+------+------------+
| f_id | f_name     |
+------+------------+
| b1   | blackberry |
| b2   | berry      |
| t1   | banana     |
+------+------------+
```

该语句的查询结果返回所有以 "b" 开头的f_id和f_name。"%"告诉MySQL，返回所有以字母 "b" 开头的记录，而不管 "b" 后面有多少个字符。

在搜索匹配时，通配符 "%" 可以放在不同位置，如【例5.12】所示。

【例5.12】在表fruits中，查询f_name中包含字母 "g" 的记录，SQL语句如下：

```
SELECT f_id, f_name FROM fruits WHERE f_name LIKE '%g%';
```

查询结果如下：

```
+------+--------+
| f_id | f_name |
+------+--------+
| bs1  | orange |
| m1   | mango  |
| t2   | grape  |
+------+--------+
```

该语句查询字符串中包含字母 "g" 的水果名称，只要名字中有字母 "g"，不管前面或后面有多少个字符，都满足查询的条件。

【例5.13】在表fruits中，查询f_name中以 "b" 开头并以 "y" 结尾的水果的名称，SQL语句如下：

```
SELECT f_name FROM fruits WHERE f_name LIKE 'b%y';
```

查询结果如下：

```
+------------+
| f_name     |
+------------+
| blackberry |
| berry      |
+------------+
```

通过以上查询结果可以看到，"%" 用于匹配在指定位置的任意数目的字符。

2. 下画线通配符（_）一次只能匹配一个任意字符

下画线通配符（_）的用法和"%"相同，区别是"%"可以匹配多个字符，而"_"只能匹配任意单个字符。如果要匹配多个字符，则需要使用相同个数的"_"，并且各个"_"之间没有空格。

【例5.14】 在表fruits中，查询以字母"y"结尾且"y"前面只有4个字母的记录，SQL语句如下：

```
SELECT f_id, f_name FROM fruits WHERE f_name LIKE '_ _ _ _y';
```

查询结果如下：

```
+------+--------+
| f_id | f_name |
+------+--------+
| b2   | berry  |
+------+--------+
```

从结果中可以看到，以"y"结尾且前面只有4个字母的记录只有一条。其他记录的f_name字段也有以"y"结尾的，但其总的字符串长度不为5，因此不在返回结果中。

5.2.7 查询空值

创建数据表的时候，设计者可以指定某列中是否包含空值（NULL）。空值不同于0，也不同于空字符串。空值一般表示数据未知、不适用或将在以后添加。在SELECT语句中，使用IS NULL子句可以查询某字段内容为空的记录。

下面在数据库中创建数据表customers，该表中包含了本章中需要用到的数据，SQL语句如下：

```
CREATE TABLE customers
(
c_id         int           NOT NULL AUTO_INCREMENT,
c_name       char(50)      NOT NULL,
c_address    char(50)      NULL,
c_city       char(50)      NULL,
c_zip        char(10)      NULL,
c_contact    char(50)      NULL,
c_email      char(255)     NULL,
PRIMARY KEY (c_id)
);
```

为了演示，需要插入数据，SQL语句如下：

```
INSERT INTO customers(c_id, c_name, c_address, c_city, c_zip, c_contact, c_email)
VALUES(10001, 'RedHook', '200 Street ', 'Tianjin', '300000', 'LiMing',
'LMing@163.com'),
(10002, 'Stars', '333 Fromage Lane', 'Dalian', '116000', 'Zhangbo',
'Jerry@hotmail.com'),
(10003, 'Netbhood', '1 Sunny Place', 'Qingdao', '266000', 'LuoCong', NULL),
```

```
(10004, 'JOTO', '829 Riverside Drive', 'Haikou', '570000', 'YangShan',
'sam@hotmail.com');
```

【例5.15】 查询表customers中c_email为空的记录的c_id、c_name和c_email字段值，SQL语句如下：

```
SELECT c_id, c_name,c_email FROM customers WHERE c_email IS NULL;
```

查询结果如下：

```
+-------+----------+---------+
| c_id  | c_name   | c_email |
+-------+----------+---------+
| 10003 | Netbhood | NULL    |
+-------+----------+---------+
```

可以看到，表customers中有一条字段c_email的值为NULL的记录。

与IS NULL相反的是NOT NULL，该关键字查找字段不为空的记录。

【例5.16】 查询表customers中c_email不为空的记录的c_id、c_name和c_email字段值，SQL语句如下：

```
SELECT c_id, c_name,c_email FROM customers WHERE c_email IS NOT NULL;
```

查询结果如下：

```
+-------+---------+-------------------+
| c_id  | c_name  | c_email           |
+-------+---------+-------------------+
| 10001 | RedHook | LMing@163.com     |
| 10002 | Stars   | Jerry@hotmail.com |
| 10004 | JOTO    | sam@hotmail.com   |
+-------+---------+-------------------+
```

可以看到，查询出来的记录的c_email字段都不为空值。

5.2.8 带AND的多条件查询

使用SELECT查询时，可以增加查询的限制条件，这样可以使查询的结果更加精确。MySQL在WHERE子句中使用AND操作符限定只有满足所有查询条件的记录才会被返回。可以使用AND连接两个甚至多个查询条件，多个条件表达式之间用AND分开。

【例5.17】 在表fruits中查询s_id = 101并且f_price≥5元的水果id、价格和名称，SQL语句如下：

```
SELECT f_id, f_price, f_name FROM fruits WHERE s_id = '101' AND f_price >=5;
```

查询结果如下：

```
+------+---------+------------+
| f_id | f_price | f_name     |
+------+---------+------------+
```

```
| a1    |   5.20   | apple      |
| b1    |  10.20   | blackberry |
+-------+----------+------------+
```

上面的语句检索了s_id=101的水果供应商提供的所有价格大于或等于5元的水果的名称和价格。WHERE子句中的条件分为两部分，AND关键字指示MySQL返回所有同时满足两个条件的行。s_id=101的水果供应商提供的价格小于5元的水果，或者s_id不等于101的水果（不管其价格为多少），均不是查询的结果。

> **提示** 上述例子的WHERE子句中只包含了一个AND语句，把两个过滤条件组合在一起。实际上可以添加多个AND过滤条件，增加条件的同时需增加一个AND关键字。

【例5.18】在表fruits中查询s_id = 101或者102，并且f_price≥5、f_name='apple' 的水果价格和名称，SQL语句如下：

```
SELECT f_id, f_price, f_name FROM fruits WHERE s_id IN('101', '102') AND f_price >= 5 AND f_name = 'apple';
```

查询结果如下：

```
+------+---------+--------+
| f_id | f_price | f_name |
+------+---------+--------+
| a1   |   5.20  | apple  |
+------+---------+--------+
```

可以看到，符合查询条件的记录只有一条。

5.2.9 带 OR 的多条件查询

与AND相反，在WHERE声明中使用OR操作符，表示只要满足其中一个条件的记录即可返回。OR也可以连接两个甚至多个查询条件，多个条件表达式之间用OR分开。

【例5.19】查询s_id=101或者s_id=102的水果供应商提供的水果的f_price和f_name，SQL语句如下：

```
SELECT s_id,f_name, f_price FROM fruits WHERE s_id = 101 OR s_id = 102;
```

查询结果如下：

```
+------+------------+---------+
| s_id | f_name     | f_price |
+------+------------+---------+
|  101 | apple      |    5.20 |
|  101 | blackberry |   10.20 |
|  102 | orange     |   11.20 |
|  101 | cherry     |    3.20 |
|  102 | banana     |   10.30 |
|  102 | grape      |    5.30 |
+------+------------+---------+
```

结果显示了s_id=101和s_id=102的水果供应商提供的水果的名称和价格。OR操作符告诉MySQL，检索的时候只需要满足其中的一个条件，不需要全部满足。如果这里使用AND，则将检索不到符合条件的数据。

在这里，也可以使用IN操作符实现与OR相同的功能，如【例5.20】所示。

【例5.20】查询s_id=101或者s_id=102的水果供应商提供的水果的f_price和f_name，SQL语句如下：

```
SELECT s_id,f_name, f_price FROM fruits WHERE s_id IN(101,102);
```

查询结果如下：

```
+------+------------+---------+
| s_id | f_name     | f_price |
+------+------------+---------+
|  101 | apple      |    5.20 |
|  101 | blackberry |   10.20 |
|  102 | orange     |   11.20 |
|  101 | cherry     |    3.20 |
|  102 | banana     |   10.30 |
|  102 | grape      |    5.30 |
+------+------------+---------+
```

在这里可以看到，OR操作符和IN操作符使用后的结果是一样的，它们可以实现相同的功能，但是使用IN操作符使得检索语句更加简洁明了，并且IN的执行速度要快于OR。更重要的是，使用IN操作符可以执行更加复杂的嵌套查询（后面章节将会讲述）。

> 提示　OR可以和AND一起使用，但是在使用时要注意两者的优先级，由于AND的优先级高于OR，因此先对AND两边的操作数进行操作，再与OR中的操作数结合。

5.2.10　查询结果不重复

从前面的例子中可以看到，SELECT查询返回所有匹配的行。例如，查询表fruits中所有的s_id，其结果为：

```
+------+
| s_id |
+------+
|  101 |
|  103 |
|  101 |
|  104 |
|  107 |
|  102 |
|  105 |
|  101 |
|  104 |
|  106 |
```

```
| 105  |
| 105  |
| 103  |
| 102  |
| 102  |
| 107  |
+------+
```

查询结果返回了16条记录，其中有一些重复的s_id值。有时出于对数据分析的需求，需要消除重复的记录，该如何操作呢？在SELECT语句中，可以使用DISTINCT关键字指示MySQL消除重复的记录，语法格式如下：

```
SELECT DISTINCT 字段名 FROM 表名;
```

【例5.21】查询表fruits中的s_id字段，返回s_id字段值且不得重复，SQL语句如下：

```
SELECT DISTINCT s_id FROM fruits;
```

查询结果如下：

```
+------+
| s_id |
+------+
| 101  |
| 103  |
| 104  |
| 107  |
| 102  |
| 105  |
| 106  |
+------+
```

可以看到，这次查询结果只返回了7条记录的s_id值，且不再有重复的值。SELECT DISTINCT s_id告诉MySQL只返回不同的s_id值。

5.2.11 对查询结果排序

从前面的查询结果中可以发现，有些字段的值是没有任何顺序的，MySQL可以通过在SELECT语句中使用ORDER BY子句对查询的结果进行排序。

1. 单列排序

例如，查询f_name字段，查询结果如下：

```
mysql> SELECT f_name FROM fruits;
+------------+
| f_name     |
+------------+
| apple      |
| apricot    |
| blackberry |
| berry      |
```

```
| xxxx         |
| orange       |
| melon        |
| cherry       |
| lemon        |
| mango        |
| xbabay       |
| xxtt         |
| coconut      |
| banana       |
| grape        |
| xbababa      |
+--------------+
```

可以看到，查询的数据并没有以一种特定的顺序显示。如果没有对查询结果进行排序，就将根据它们插入数据表中的顺序来显示。

下面使用ORDER BY子句对指定的列数据进行排序。

【例5.22】查询表fruits中的f_name字段，并对其进行排序，SQL语句如下：

```
mysql> SELECT f_name FROM fruits ORDER BY f_name;
+--------------+
| f_name       |
+--------------+
| apple        |
| apricot      |
| banana       |
| berry        |
| blackberry   |
| cherry       |
| coconut      |
| grape        |
| lemon        |
| mango        |
| melon        |
| orange       |
| xbababa      |
| xbabay       |
| xxtt         |
| xxxx         |
+--------------+
```

该语句查询的结果和前面的相同，不同的是，通过指定ORDER BY子句，MySQL对name列的数据按字母表的顺序进行了升序排列。

2. 多列排序

有时，需要根据多列值进行排序。比如，如果要显示一个学生列表，可能会有多个学生的姓氏是相同的，因此还需要根据学生的名进行排序。对多列数据进行排序时，要将需要排序的列用逗号隔开。

【例5.23】 查询表fruits中的f_name和f_price字段，先按f_name排序，再按f_price排序，SQL语句如下：

```
SELECT f_name, f_price FROM fruits ORDER BY f_name, f_price;
```

查询结果如下：

```
+------------+---------+
| f_name     | f_price |
+------------+---------+
| apple      |    5.20 |
| apricot    |    2.20 |
| banana     |   10.30 |
| berry      |    7.60 |
| blackberry |   10.20 |
| cherry     |    3.20 |
| coconut    |    9.20 |
| grape      |    5.30 |
| lemon      |    6.40 |
| mango      |   15.70 |
| melon      |    8.20 |
| orange     |   11.20 |
| xbababa    |    3.60 |
| xbabay     |    2.60 |
| xxtt       |   11.60 |
| xxxx       |    3.60 |
+------------+---------+
```

> **提示** 在对多列进行排序的时候，排序的第一列必须有相同的列值，才会对第二列进行排序。如果第一列数据中所有值都是唯一的，将不再对第二列进行排序。

3. 指定排序方向

默认情况下，查询数据按字母升序进行排序（A~Z），但数据的排序并不仅限于此，还可以使用ORDER BY对查询结果进行降序排序（Z~A）。这可以通过关键字DESC实现，【例5.24】将说明如何进行降序排列。

【例5.24】 查询表fruits中的f_name和f_price字段，对结果按f_price降序排序，SQL语句如下：

```
SELECT f_name, f_price FROM fruits ORDER BY f_price DESC;
```

查询结果如下：

```
+------------+---------+
| f_name     | f_price |
+------------+---------+
| mango      |   15.70 |
| xxtt       |   11.60 |
| orange     |   11.20 |
| banana     |   10.30 |
| blackberry |   10.20 |
```

```
| coconut   |    9.20 |
| melon     |    8.20 |
| berry     |    7.60 |
| lemon     |    6.40 |
| grape     |    5.30 |
| apple     |    5.20 |
| xxxx      |    3.60 |
| xbababa   |    3.60 |
| cherry    |    3.20 |
| xbabay    |    2.60 |
| apricot   |    2.20 |
+-----------+---------+
```

提示 与DESC相反的是ASC（升序），它将字段列中的数据按字母表顺序升序排列。实际上，在排序的时候ASC是默认的排序方式，所以加不加ASC关键字都可以。

也可以对多列进行不同的顺序排序，如【例5.25】所示。

【例5.25】查询表fruits，先按f_price降序排列，再按f_name升序排列，SQL语句如下：

```
SELECT f_price, f_name FROM fruits ORDER BY f_price DESC, f_name;
```

查询结果如下：

```
+---------+------------+
| f_price | f_name     |
+---------+------------+
|   15.70 | mango      |
|   11.60 | xxtt       |
|   11.20 | orange     |
|   10.30 | banana     |
|   10.20 | blackberry |
|    9.20 | coconut    |
|    8.20 | melon      |
|    7.60 | berry      |
|    6.40 | lemon      |
|    5.30 | grape      |
|    5.20 | apple      |
|    3.60 | xbababa    |
|    3.60 | xxxx       |
|    3.20 | cherry     |
|    2.60 | xbabay     |
|    2.20 | apricot    |
+---------+------------+
```

提示 DESC关键字只对其前面的列进行降序排列，在这里只对f_price排序，而并没有对f_name进行排序。因此，f_price按降序排列，而f_name仍按升序排列。如果要对多列都进行降序排列，则必须在每一列的列名后面加DESC关键字。

5.2.12 分组查询

分组查询是对数据按照某个或多个字段进行分组。MySQL中使用GROUP BY关键字对数据进行分组，语法格式如下：

```
[GROUP BY 字段] [HAVING <条件表达式>]
```

"字段"为进行分组时所依据的列名称；"HAVING <条件表达式>"指定显示满足表达式限定条件的结果。

1. 创建分组

GROUP BY关键字通常和集合函数一起使用，比如MAX()、MIN()、COUNT()、SUM()、AVG()。例如，要返回每个水果供应商提供的水果种类，就要在分组过程中用到COUNT()函数，把数据分为多个逻辑组，并对每个组进行集合计算。

【例5.26】根据s_id对表fruits中的数据进行分组，SQL语句如下：

```sql
SELECT s_id, COUNT(*) AS Total FROM fruits GROUP BY s_id;
```

查询结果如下：

```
+------+-------+
| s_id | Total |
+------+-------+
|  101 |     3 |
|  103 |     2 |
|  104 |     2 |
|  107 |     2 |
|  102 |     3 |
|  105 |     3 |
|  106 |     1 |
+------+-------+
```

s_id表示供应商的ID，Total字段使用COUNT()函数计算得出，GROUP BY子句按照s_id升序排序并对数据分组，可以看到s_id为101、102、105的供应商都提供3种水果，s_id为103、104、107的供应商都提供2种水果，s_id为106的供应商只提供1种水果。

如果要查看每个供应商提供的水果的种类，该怎么办呢？在MySQL中，可以在GROUP BY子句中使用GROUP_CONCAT()函数，将每个分组中的各个字段的值显示出来。

【例5.27】根据s_id对表fruits中的数据进行分组，将每个供应商提供的水果名称显示出来，SQL语句如下：

```sql
SELECT s_id, GROUP_CONCAT(f_name) AS Names FROM fruits GROUP BY s_id;
```

查询结果如下：

```
+------+-------------------------+
| s_id | Names                   |
+------+-------------------------+
```

```
| 101 | apple,blackberry,cherry |
| 102 | orange,banana,grape     |
| 103 | apricot,coconut         |
| 104 | berry,lemon             |
| 105 | melon,xbabay,xxtt       |
| 106 | mango                   |
| 107 | xxxx,xbababa            |
+-----+-------------------------+
```

由结果可以看到，GROUP_CONCAT()函数将每个分组中的名称显示出来了，其名称的个数与COUNT()函数计算出来的相同。

2. 使用HAVING过滤分组

GROUP BY可以和HAVING一起限定显示记录所需满足的条件，只有满足条件的分组才会被显示。

【例5.28】根据s_id对表fruits中的数据进行分组，并显示水果种类大于1的分组信息，SQL语句如下：

```
SELECT s_id, GROUP_CONCAT(f_name) AS Names FROM fruits GROUP BY s_id HAVING COUNT(f_name) > 1;
```

查询结果如下：

```
+-----+-------------------------+
| s_id| Names                   |
+-----+-------------------------+
| 101 | apple,blackberry,cherry |
| 102 | orange,banana,grape     |
| 103 | apricot,coconut         |
| 104 | berry,lemon             |
| 105 | melon,xbabay,xxtt       |
| 107 | xxxx,xbababa            |
+-----+-------------------------+
```

s_id为101、102、103、104、105、107的供应商提供的水果种类大于1，满足HAVING子句条件，因此出现在返回结果中；而s_id为106的供应商提供的水果种类等于1，不满足限定条件，因此不在返回结果中。

> **提示** HAVING关键字与WHERE关键字都是用来过滤数据的，两者有什么区别呢？最重要的区别是，HAVING在数据分组之后进行过滤来选择分组，而WHERE在分组之前选择记录。另外，WHERE排除的记录不再包括在分组中。

3. 在GROUP BY子句中使用WITH ROLLUP

使用WITH ROLLUP关键字之后，可以在所有查询出的分组记录之后增加一条记录，该记录计算查询出的所有记录的总和，即统计记录数量。

【例5.29】 根据s_id对表fruits中的数据进行分组，并显示记录数量，SQL语句如下：

```
SELECT s_id, COUNT(*) AS Total FROM fruits GROUP BY s_id WITH ROLLUP;
```

查询结果如下：

```
+------+-------+
| s_id | Total |
+------+-------+
| 101  |   3   |
| 102  |   3   |
| 103  |   2   |
| 104  |   2   |
| 105  |   3   |
| 106  |   1   |
| 107  |   2   |
| NULL |  16   |
+------+-------+
```

由结果可以看到，通过GROUP BY分组之后，在显示结果的最后面新添加了一行，该行Total列的值正好是上面所有数值之和。

4. 多字段分组

使用GROUP BY可以对多个字段进行分组，GROUP BY关键字后面跟需要分组的字段。MySQL根据多字段的值来进行层次分组，分组层次从左到右，即先按第1个字段分组，然后在第1个字段值相同的记录中根据第2个字段的值进行分组，以此类推。

【例5.30】 根据s_id和f_name字段对表fruits中的数据进行分组，SQL语句如下：

```
mysql> SELECT * FROM fruits GROUP BY s_id,f_name;
```

查询结果如下：

```
+------+------+------------+---------+
| f_id | s_id | f_name     | f_price |
+------+------+------------+---------+
| a1   | 101  | apple      |   5.20  |
| a2   | 103  | apricot    |   2.20  |
| b1   | 101  | blackberry |  10.20  |
| b2   | 104  | berry      |   7.60  |
| b5   | 107  | xxxx       |   3.60  |
| bs1  | 102  | orange     |  11.20  |
| bs2  | 105  | melon      |   8.20  |
| c0   | 101  | cherry     |   3.20  |
| l2   | 104  | lemon      |   6.40  |
| m1   | 106  | mango      |  15.70  |
| m2   | 105  | xbabay     |   2.60  |
| m3   | 105  | xxtt       |  11.60  |
| o2   | 103  | coconut    |   9.20  |
| t1   | 102  | banana     |  10.30  |
| t2   | 102  | grape      |   5.30  |
```

```
| t4   | 107  | xbababa     |    3.60 |
+------+------+-------------+---------+
```

由结果可以看到,查询记录先按照s_id进行分组,再对f_name字段按不同的取值进行分组。

> **注意** 如果上面的例子报错为SQL模式问题,可以在my.ini文件中的sql_mode中删除ONLY_FULL_GROP_BY,然后重启MySQL服务器即可解决问题。

5. GROUP BY和ORDER BY一起使用

某些情况下需要对分组进行排序。在前面的介绍中,ORDER BY用来对查询的记录排序,它和GROUP BY一起使用就可以完成对分组的排序。

为了演示效果,首先创建数据表,SQL语句如下:

```
CREATE TABLE orderitems
(
o_num       int           NOT NULL,
o_item      int           NOT NULL,
f_id        char(10)      NOT NULL,
quantity    int           NOT NULL,
item_price  decimal(8,2)  NOT NULL,
PRIMARY KEY (o_num,o_item)
) ;
```

然后插入演示数据,SQL语句如下:

```
INSERT INTO orderitems(o_num, o_item, f_id, quantity, item_price)
VALUES(30001, 1, 'a1', 10, 5.2),
(30001, 2, 'b2', 3, 7.6),
(30001, 3, 'bs1', 5, 11.2),
(30001, 4, 'bs2', 15, 9.2),
(30002, 1, 'b3', 2, 20.0),
(30003, 1, 'c0', 100, 10),
(30004, 1, 'o2', 50, 2.50),
(30005, 1, 'c0', 5, 10),
(30005, 2, 'b1', 10, 8.99),
(30005, 3, 'a2', 10, 2.2),
(30005, 4, 'm1', 5, 14.99);
```

【例5.31】 查询订单价格大于100的订单号和总订单价格,SQL语句如下:

```
SELECT o_num, SUM(quantity * item_price) AS orderTotal FROM orderitems
GROUP BY o_num HAVING SUM(quantity*item_price) >= 100;
```

查询结果如下:

```
+-------+------------+
| o_num | orderTotal |
+-------+------------+
| 30001 |     268.80 |
| 30003 |    1000.00 |
| 30004 |     125.00 |
```

```
| 30005 |    236.85 |
+-------+-----------+
```

从返回的结果中可以看到，orderTotal列的总订单价格并没有按照一定顺序显示。接下来使用ORDER BY关键字按总订单价格升序显示结果，SQL语句如下：

```
SELECT o_num, SUM(quantity * item_price) AS orderTotal FROM orderitems
GROUP BY o_num HAVING SUM(quantity*item_price) >= 100 ORDER BY orderTotal;
```

查询结果如下：

```
+-------+------------+
| o_num | orderTotal |
+-------+------------+
| 30004 |     125.00 |
| 30005 |     236.85 |
| 30001 |     268.80 |
| 30003 |    1000.00 |
+-------+------------+
```

由结果可以看到，GROUP BY子句按订单号对数据进行分组，SUM()函数便可以返回总的订单价格，HAVING子句对分组数据进行过滤，只返回总价格大于100的订单，最后使用ORDER BY子句升序输出。

> **提示** 当使用ROLLUP时，不能同时使用ORDER BY子句进行排序，即ROLLUP和ORDER BY是互相排斥的。

5.2.13 使用 LIMIT 限制查询结果的数量

SELECT返回所有匹配的行，有可能是表中所有的行，若仅需要返回第一行或者前几行，可使用LIMIT关键字，基本语法格式如下：

```
LIMIT [位置偏移量,] 行数
```

"位置偏移量"指示MySQL从哪一行开始显示，是一个可选参数，如果不指定"位置偏移量"，将会从表中的第一条记录开始（第一条记录的位置偏移量是0，第二条记录的位置偏移量是1，以此类推）；"行数"指示返回的记录行数。

【例5.32】显示表fruits中的前4行记录，SQL语句如下：

```
SELECT * From fruits LIMIT 4;
```

查询结果如下：

```
+------+------+------------+---------+
| f_id | s_id | f_name     | f_price |
+------+------+------------+---------+
| a1   |  101 | apple      |    5.20 |
| a2   |  103 | apricot    |    2.20 |
| b1   |  101 | blackberry |   10.20 |
```

```
| b2   | 104  | berry      |    7.60 |
+------+------+------------+---------+
```

由于该语句没有指定返回记录的"位置偏移量"参数，因此显示结果从第一行开始。又因为"行数"参数为4，所以返回的结果为表中的前4行记录。

【例5.33】 在表fruits中，使用LIMIT子句，返回从第5条记录开始的行数长度为3的记录，SQL语句如下：

```
SELECT * From fruits LIMIT 4, 3;
```

查询结果如下：

```
+------+------+--------+---------+
| f_id | s_id | f_name | f_price |
+------+------+--------+---------+
| b5   | 107  | xxxx   |    3.60 |
| bs1  | 102  | orange |   11.20 |
| bs2  | 105  | melon  |    8.20 |
+------+------+--------+---------+
```

第一个数字"4"表示从第5行开始（位置偏移量从0开始，第5行的位置偏移量为4），第二个数字3表示返回3行记录。因此，该语句返回从第5条记录行开始之后的3行记录。

带一个参数的LIMIT指定从查询结果的首行开始，唯一的参数表示返回的行数，即"LIMIT n"与"LIMIT 0,n"等价。带两个参数的LIMIT可以返回从任何一个位置开始的指定长度的行数。

> **提示** MySQL 9.0中可以使用"LIMIT 4 OFFSET 3;"，意思是获取从第5行记录开始后面的3行记录，和"LIMIT 4,3;"返回的结果相同。

5.3 使用聚合函数查询

有时候并不需要返回表中的实际数据，而需要对数据进行总结。MySQL提供了一些聚合函数，可以对获取的数据进行分析和报告。这些函数的功能包括计算数据表中记录行数的总数、计算某个字段列下数据的总和，以及计算表中某个字段下的最大值、最小值或者平均值。这些聚合函数的名称和作用如表5.2所示。

表5.2 MySQL 聚合函数

函数	作用
COUNT()	返回某列的行数
SUM()	返回某列值的和
AVG()	返回某列的平均值
MAX()	返回某列的最大值
MIN()	返回某列的最小值

接下来，将详细介绍各个函数的使用方法。

5.3.1　COUNT()函数

COUNT()函数统计数据表中包含的记录行的总数，或者根据查询结果返回列中包含的数据行数。其使用方法有两种：

- COUNT(*)：计算表中的总行数，而不管某列是否有数值或者是否为NULL。
- COUNT(字段名)：计算指定列下的总行数，计算时将忽略NULL的行。

【例5.34】查询表customers中的总行数，SQL语句如下：

```
mysql> SELECT COUNT(*) AS cust_num FROM customers;
+----------+
| cust_num |
+----------+
|        4 |
+----------+
```

由查询结果可以看到，COUNT(*)返回表customers中记录的总行数，而不管其值是什么，返回的总数的名称为cust_num。

【例5.35】查询表customers中有电子邮箱的顾客的总数，SQL语句如下：

```
mysql> SELECT COUNT(c_email) AS email_num  FROM customers;
+-----------+
| email_num |
+-----------+
|         3 |
+-----------+
```

由查询结果可以看到，只有3个顾客有email，email为NULL的记录没有被COUNT()函数计算。

> **提示**　【例5.34】和【例5.35】中不同的数值说明了两种方式在计算总数的时候对待NULL的不同方式：指定列的值为空的行被COUNT()函数忽略；如果不指定列，而在COUNT()函数中使用星号（*），则所有记录都不会被忽略。

前面介绍分组查询的时候，介绍了如何用COUNT()函数与GROUP BY关键字一起来计算不同分组中的记录总数。下面再来看一个例子。

【例5.36】在表orderitems中，使用COUNT()函数统计不同订单号中订购的水果种类，SQL语句如下：

```
mysql> SELECT o_num, COUNT(f_id) FROM orderitems GROUP BY o_num;
+-------+-------------+
| o_num | COUNT(f_id) |
+-------+-------------+
| 30001 |           4 |
| 30002 |           1 |
```

```
| 30003 |            1 |
| 30004 |            1 |
| 30005 |            4 |
+-------+--------------+
```

从查询结果中可以看到，GROUP BY关键字先按照订单号进行分组，然后计算每个分组中的总记录数。

5.3.2 SUM()函数

SUM()是一个求和函数，返回指定列值的总和。

【例5.37】在表orderitems中，查询30005号订单一共购买的水果的总量，SQL语句如下：

```
mysql>SELECT SUM(quantity) AS items_total FROM orderitems WHERE o_num = 30005;
+-------------+
| items_total |
+-------------+
|          30 |
+-------------+
```

SUM(quantity)函数返回订单中所有水果数量之和，WHERE子句指定查询的订单号为30005。

SUM()可以与GROUP BY一起使用，来计算每个分组的总和。

【例5.38】在表orderitems中，使用SUM()函数统计不同订单号中订购的水果的总量，SQL语句如下：

```
mysql> SELECT o_num, SUM(quantity) AS items_total FROM orderitems GROUP BY o_num;
+-------+-------------+
| o_num | items_total |
+-------+-------------+
| 30001 |          33 |
| 30002 |           2 |
| 30003 |         100 |
| 30004 |          50 |
| 30005 |          30 |
+-------+-------------+
```

由查询结果可以看到，GROUP BY按照订单号o_num进行分组，SUM()函数计算每个分组中订购的水果的总量。

> 提示 SUM()函数在计算时会忽略列值为NULL的行。

5.3.3 AVG()函数

AVG()函数通过计算返回的行数和每一行数据的和，求得指定列数据的平均值。

【例5.39】在表fruits中，查询s_id=103的水果供应商的水果价格平均值，SQL语句如下：

```
mysql> SELECT AVG(f_price) AS avg_price FROM fruits WHERE s_id = 103;
+-----------+
| avg_price |
+-----------+
|  5.700000 |
+-----------+
```

本例中，查询语句增加了一个WHERE子句，并且添加了查询过滤条件，只查询s_id = 103的记录中的f_price。因此，通过AVG()函数计算的结果只是指定的供应商水果的价格平均值，而不是市场上所有水果价格的平均值。

AVG()可以与GROUP BY一起使用，来计算每个分组的平均值。

【例5.40】在表fruits中，查询每一个水果供应商的水果价格的平均值，SQL语句如下：

```
mysql> SELECT s_id,AVG(f_price) AS avg_price FROM fruits GROUP BY s_id;
+------+-----------+
| s_id | avg_price |
+------+-----------+
|  101 |  6.200000 |
|  103 |  5.700000 |
|  104 |  7.000000 |
|  107 |  3.600000 |
|  102 |  8.933333 |
|  105 |  7.466667 |
|  106 | 15.700000 |
+------+-----------+
```

GROUP BY关键字根据s_id字段对记录进行分组，然后计算出每个分组的平均值。这种分组求平均值的方法非常有用，例如，求不同班级学生成绩的平均值、求不同部门员工的平均工资、求各地的年平均气温等。

> 提示　在使用AVG()函数时，其参数为要计算平均值的列名称。如果要得到多个列的平均值，则需要在每一列上使用AVG()函数。

5.3.4 MAX()函数

MAX()返回指定列中的最大值。

【例5.41】在表fruits中，查找市场上水果价格的最高值，SQL语句如下：

```
mysql>SELECT MAX(f_price) AS max_price FROM fruits;
+-----------+
| max_price |
+-----------+
|   15.70   |
+-----------+
```

由结果可以看到，MAX()函数查询出f_price字段的最大值为15.70。

MAX()也可以和GROUP BY关键字一起使用，来计算每个分组中的最大值。

【例5.42】在表fruits中，查找不同水果供应商提供的水果的最高价格，SQL语句如下：

```
mysql> SELECT s_id, MAX(f_price) AS max_price FROM fruits GROUP BY s_id;
+------+-----------+
| s_id | max_price |
+------+-----------+
| 101  |     10.20 |
| 103  |      9.20 |
| 104  |      7.60 |
| 107  |      3.60 |
| 102  |     11.20 |
| 105  |     11.60 |
| 106  |     15.70 |
+------+-----------+
```

由结果可以看到，GROUP BY关键字根据s_id字段对记录进行分组，然后计算出每个分组中的最大值。

MAX()函数不仅适用于数值类型，也可用于字符类型。

【例5.43】在表fruits中查找f_name的最大值，SQL语句如下：

```
mysql> SELECT MAX(f_name) FROM fruits;
+-------------+
| MAX(f_name) |
+-------------+
| xxxx        |
+-------------+
```

由结果可以看到，MAX()函数可以对字母进行大小判断，并返回最大的字符或者字符串。

> **提示** MAX()函数除了用来找出最大的列值或日期值之外，还可以返回任意列中的最大值，包括返回字符类型的最大值。在对字符类型数据进行比较时，按照字符的ASCII码值大小进行比较，对于a~z，a的ASCII码值最小，z的最大。在比较时，先比较第一个字母，如果相等，继续比较下一个字符，直到两个字符不相等或者字符结束为止。例如，"b"与"t"比较时，"t"为最大值；"bcd"与"bca"比较时，"bcd"为最大值。

5.3.5 MIN()函数

MIN()返回指定列中的最小值。

【例5.44】在表fruits中，查找市场上水果价格的最低值，SQL语句如下：

```
mysql>SELECT MIN(f_price) AS min_price FROM fruits;
+-----------+
| min_price |
+-----------+
```

```
|    2.20   |
+-----------+
```

由结果可以看到，MIN()函数查询出f_price字段的最小值为2.20。

MIN()也可以和GROUP BY关键字一起使用，用来计算每个分组中的最小值。

【例5.45】 在表fruits中，查找不同水果供应商提供的水果的最低价格，SQL语句如下：

```
mysql> SELECT s_id, MIN(f_price) AS min_price  FROM fruits GROUP BY s_id;
+------+-----------+
| s_id | min_price |
+------+-----------+
|  101 |      3.20 |
|  103 |      2.20 |
|  104 |      6.40 |
|  107 |      3.60 |
|  102 |      5.30 |
|  105 |      2.60 |
|  106 |     15.70 |
+------+-----------+
```

由结果可以看到，GROUP BY关键字根据s_id字段对记录进行分组，然后计算出每个分组中的最小值。MIN()函数与MAX()函数类似，不仅适用于数值类型，也可用于字符类型。

5.4 连接查询

连接是关系数据库的主要特点。连接查询是关系数据库中最主要的查询，主要包括内连接（Inner Join）、外连接（Out Join）等。通过连接运算符可以实现多表查询。在关系数据库管理系统中，建立表时各数据之间的关系不必确定，常把一个实体的所有信息存放在一张表中。当查询数据时，通过连接操作查询出存放在多张表中的不同实体的信息。当两张或多张表中存在相同意义的字段时，便可以通过这些字段对不同的表进行连接查询。本节将介绍多表之间的内连接查询、外连接查询以及复合条件连接查询。

5.4.1 内连接查询

内连接使用比较运算符进行表间某（些）列数据的比较操作，并列出这些表中与连接条件相匹配的数据行，并组合成新的记录。也就是说，在内连接查询中，只有满足条件的记录才能出现在结果关系中。

为了演示的需要，首先创建数据表suppliers，SQL语句如下：

```
CREATE TABLE suppliers
(
s_id      int       NOT NULL AUTO_INCREMENT,
s_name    char(50) NOT NULL,
s_city    char(50) NULL,
```

```
s_zip      char(10) NULL,
s_call     CHAR(50) NOT NULL,
PRIMARY KEY (s_id)
);
```

插入需要演示的数据，SQL 语句如下：

```
INSERT INTO suppliers(s_id, s_name,s_city,  s_zip, s_call)
VALUES(101,'FastFruit Inc.','Tianjin','300000','48075'),
(102,'LT Supplies','Chongqing','400000','44333'),
(103,'ACME','Shanghai','200000','90046'),
(104,'FNK Inc.','Zhongshan','528437','11111'),
(105,'Good Set','Taiyuang','030000', '22222'),
(106,'Just Eat Ours','Beijing','010', '45678'),
(107,'DK Inc.','Zhengzhou','450000', '33332');
```

【例5.46】在表 fruits 和表 suppliers 之间使用内连接查询。

首先查看两张表的结构：

```
mysql> DESC fruits;
+---------+--------------+------+-----+---------+-------+
| Field   | Type         | Null | Key | Default | Extra |
+---------+--------------+------+-----+---------+-------+
| f_id    | char(10)     | NO   | PRI | NULL    |       |
| s_id    | int          | NO   |     | NULL    |       |
| f_name  | char(255)    | NO   |     | NULL    |       |
| f_price | decimal(8,2) | NO   |     | NULL    |       |
+---------+--------------+------+-----+---------+-------+

mysql> DESC suppliers;
+--------+----------+------+-----+--------+----------------+
| Field  | Type     | Null | Key | Default| Extra          |
+--------+----------+------+-----+--------+----------------+
| s_id   | int      | NO   | PRI | NULL   | auto_increment |
| s_name | char(50) | NO   |     | NULL   |                |
| s_city | char(50) | YES  |     | NULL   |                |
| s_zip  | char(10) | YES  |     | NULL   |                |
| s_call | char(50) | NO   |     | NULL   |                |
+--------+----------+------+-----+--------+----------------+
```

由结果可以看到，表 fruits 和表 suppliers 中都有相同数据类型的字段 s_id，两张表通过 s_id 字段建立联系。

接下来从表 fruits 中查询 f_name、f_price 字段，从表 suppliers 中查询 s_id、s_name 字段，SQL 语句如下：

```
mysql> SELECT suppliers.s_id, s_name,f_name, f_price  FROM fruits ,suppliers WHERE
fruits.s_id = suppliers.s_id;
+------+----------------+------------+---------+
| s_id | s_name         | f_name     | f_price |
+------+----------------+------------+---------+
| 101  | FastFruit Inc. | apple      |   5.20  |
```

```
| 103 | ACME           | apricot    |  2.20 |
| 101 | FastFruit Inc. | blackberry | 10.20 |
| 104 | FNK Inc.       | berry      |  7.60 |
| 107 | DK Inc.        | xxxx       |  3.60 |
| 102 | LT Supplies    | orange     | 11.20 |
| 105 | Good Set       | melon      |  8.20 |
| 101 | FastFruit Inc. | cherry     |  3.20 |
| 104 | FNK Inc.       | lemon      |  6.40 |
| 106 | Just Eat Ours  | mango      | 15.70 |
| 105 | Good Set       | xbabay     |  2.60 |
| 105 | Good Set       | xxtt       | 11.60 |
| 103 | ACME           | coconut    |  9.20 |
| 102 | LT Supplies    | banana     | 10.30 |
| 102 | LT Supplies    | grape      |  5.30 |
| 107 | DK Inc.        | xbababa    |  3.60 |
+------+----------------+------------+--------+
```

本例的SELECT语句与前面所介绍的一个最大的差别是：SELECT后面指定的列分别属于两张不同的表，f_name和f_price字段在表fruits中，而s_name字段在表suppliers中，s_id字段在表fruits和表suppliers中都存在；同时FROM子句列出了表fruits和suppliers。WHERE子句在这里作为过滤条件，表明只有两张表中的s_id字段值相等的时候，才符合连接查询的条件。从返回的结果可以看到，显示的记录是由两张表中的不同列值组成的新记录。

> **提示** 由于表fruits和表suppliers中有相同的字段s_id，因此在比较的时候需要完全限定表名（格式为"表名.列名"）。如果只给出s_id，MySQL不知道指的是哪一个表的字段，将会返回错误信息。

下面的内连接查询语句返回与【例5.46】完全相同的结果。

【例5.47】在表fruits和表suppliers之间，使用INNER JOIN语法进行内连接查询，SQL语句如下：

```
mysql> SELECT suppliers.s_id, s_name,f_name, f_price FROM fruits INNER JOIN
suppliers ON fruits.s_id = suppliers.s_id;
+------+----------------+------------+--------+
| s_id | s_name         | f_name     | f_price|
+------+----------------+------------+--------+
| 101  | FastFruit Inc. | apple      |  5.20  |
| 103  | ACME           | apricot    |  2.20  |
| 101  | FastFruit Inc. | blackberry | 10.20  |
| 104  | FNK Inc.       | berry      |  7.60  |
| 107  | DK Inc.        | xxxx       |  3.60  |
| 102  | LT Supplies    | orange     | 11.20  |
| 105  | Good Set       | melon      |  8.20  |
| 101  | FastFruit Inc. | cherry     |  3.20  |
| 104  | FNK Inc.       | lemon      |  6.40  |
| 106  | Just Eat Ours  | mango      | 15.70  |
| 105  | Good Set       | xbabay     |  2.60  |
| 105  | Good Set       | xxtt       | 11.60  |
```

```
| 103  | ACME          | coconut    |  9.20 |
| 102  | LT Supplies   | banana     | 10.30 |
| 102  | LT Supplies   | grape      |  5.30 |
| 107  | DK Inc.       | xbababa    |  3.60 |
+------+---------------+------------+-------+
```

在这里的查询语句中,两张表之间的关系通过INNER JOIN指定。使用这种语法的时候,连接的条件使用ON子句而不是WHERE,ON和WHERE后面指定的条件相同。

如果一个连接查询涉及的两张表是同一张,这种查询称为自连接查询。自连接是一种特殊的内连接,它是指相互连接的表在物理上为同一张表,但可以在逻辑上分为两张表。

【例5.48】 查询f_id= 'a1'的水果供应商提供的水果种类,SQL语句如下:

```
mysql> SELECT f1.f_id, f1.f_name  FROM fruits AS f1, fruits AS f2 WHERE f1.s_id
= f2.s_id AND f2.f_id = 'a1';
+------+------------+
| f_id | f_name     |
+------+------------+
| a1   | apple      |
| b1   | blackberry |
| c0   | cherry     |
+------+------------+
```

此处查询的两张表是相同的表,为了防止产生二义性,对表使用了别名。表fruits第一次出现的别名为f1,第二次出现的别名为f2,在使用SELECT语句返回列时,明确指出返回以f1为前缀的列的全名;WHERE连接两张表,并按照第二张表的f_id对数据进行过滤,返回所需数据。

5.4.2 外连接查询

外连接查询将查询多张表中相关联的行。内连接查询,返回的查询结果集合中仅是符合查询条件和连接条件的行。有时候需要包含没有关联的行中的数据,即在返回的查询结果集合中,不仅包含符合连接条件的行,还包含左表(左外连接或左连接)和右表(右外连接或右连接)。外连接分为左外连接(或左连接)和右外连接(或右连接):

- LEFT JOIN(左连接):返回左表中的所有记录和右表中与左表中的行匹配的记录。
- RIGHT JOIN(右连接):返回右表中的所有记录和左表中与右表中的行匹配的记录。

1. LEFT JOIN(左连接)

左连接的结果包括LEFT OUTER子句中指定的左表的所有行,而不仅仅是连接列所匹配的行。如果左表的某行在右表中没有匹配行,则在相关联的结果行中,右表的所有选择的表列均为空值。

为了演示的需要,首先创建表orders,SQL语句如下:

```
CREATE TABLE orders
(
o_num    int     NOT NULL AUTO_INCREMENT,
```

```
o_date datetime NOT NULL,
c_id    int      NOT NULL,
PRIMARY KEY (o_num)
) ;
```

然后插入需要演示的数据，SQL语句如下：

```
INSERT INTO orders(o_num, o_date, c_id)
VALUES(30001, '2008-09-01', 10001),
(30002, '2008-09-12', 10003),
(30003, '2008-09-30', 10004),
(30004, '2008-10-03', 10005),
(30005, '2008-10-08', 10001);
```

【例5.49】在表customers和表orders中，查询所有客户，包括没有订单的客户，SQL语句如下：

```
mysql> SELECT customers.c_id, orders.o_num  FROM customers LEFT OUTER JOIN orders
ON customers.c_id = orders.c_id;
+-------+-------+
| c_id  | o_num |
+-------+-------+
| 10001 | 30005 |
| 10001 | 30001 |
| 10002 |  NULL |
| 10003 | 30002 |
| 10004 | 30003 |
+-------+-------+
```

结果显示了5条记录，c_id等于10002的客户目前并没有下订单，所以对应的表orders中并没有该客户的订单信息，因此该条记录只取出了表customers中相应的值，而从表orders中取出的值为NULL。

2. RIGHT JOIN（右连接）

右连接是左连接的反向连接，将返回右表中的所有行和左表中与右表中的行匹配的行。如果右表的某行在左表中没有匹配行，左表将返回空值。

【例5.50】在表customers和表orders中，查询所有订单，包括没有客户的订单，SQL语句如下：

```
mysql> SELECT customers.c_id, orders.o_num   FROM customers RIGHT OUTER JOIN orders
ON customers.c_id = orders.c_id;
+-------+-------+
| c_id  | o_num |
+-------+-------+
| 10001 | 30001 |
| 10003 | 30002 |
| 10004 | 30003 |
|  NULL | 30004 |
| 10001 | 30005 |
+-------+-------+
```

结果显示了5条记录，订单号等于30004的订单的客户可能由于某种原因取消了该订单，对应的表customers中并没有该客户的信息，因此该条记录只取出了表orders中相应的值，而从表customers中取出的值为NULL。

5.4.3 复合条件连接查询

复合条件连接查询是在连接查询的过程中，通过添加过滤条件来限制查询的结果，使查询的结果更加准确。

【例5.51】在表customers和表orders中，使用INNER JOIN语法查询表customers中c_id为10001的客户的订单信息，SQL语句如下：

```
mysql> SELECT customers.c_id, orders.o_num FROM customers INNER JOIN orders
  ON customers.c_id = orders.c_id AND customers.c_id = 10001;
+-------+-------+
| c_id  | o_num |
+-------+-------+
| 10001 | 30001 |
| 10001 | 30005 |
+-------+-------+
```

返回的结果中只有两条记录，这是因为在连接查询时，指定查询c_id为10001的订单信息，添加了过滤条件之后，返回的结果将会变少。

继续使用连接查询，并对查询的结果进行排序。

【例5.52】在表fruits和表suppliers之间，使用INNER JOIN语法进行内连接查询，并对查询结果排序，SQL语句如下：

```
mysql> SELECT suppliers.s_id, s_name,f_name, f_price FROM fruits INNER JOIN
suppliers ON fruits.s_id = suppliers.s_id ORDER BY fruits.s_id;
+------+----------------+------------+---------+
| s_id | s_name         | f_name     | f_price |
+------+----------------+------------+---------+
| 101  | FastFruit Inc. | apple      |    5.20 |
| 101  | FastFruit Inc. | blackberry |   10.20 |
| 101  | FastFruit Inc. | cherry     |    3.20 |
| 102  | LT Supplies    | orange     |   11.20 |
| 102  | LT Supplies    | banana     |   10.30 |
| 102  | LT Supplies    | grape      |    5.30 |
| 103  | ACME           | apricot    |    2.20 |
| 103  | ACME           | coconut    |    9.20 |
| 104  | FNK Inc.       | berry      |    7.60 |
| 104  | FNK Inc.       | lemon      |    6.40 |
| 105  | Good Set       | melon      |    8.20 |
| 105  | Good Set       | xbabay     |    2.60 |
| 105  | Good Set       | xxtt       |   11.60 |
| 106  | Just Eat Ours  | mango      |   15.70 |
| 107  | DK Inc.        | xxxx       |    3.60 |
```

```
| 107 | DK Inc.       | xbababa      |    3.60 |
+-----+---------------+--------------+---------+
```

由结果可以看到,内连接查询的结果按照suppliers.s_id字段进行了升序排序。

5.5 子查询

子查询指一个查询语句嵌套在另一个查询语句内部的查询,这个特性从MySQL 4.1开始引入。在SELECT子句中先计算子查询,子查询结果作为外层另一个查询的过滤条件,查询可以基于一张表或者多张表。子查询中常用的操作符有ANY(SOME)、ALL、IN、EXISTS。子查询可以添加到SELECT、UPDATE和DELETE语句中,而且可以进行多层嵌套。子查询中也可以使用比较运算符,如"<""<="">"">="和"!="等。本节将介绍如何在SELECT语句中嵌套子查询。

5.5.1 带 ANY、SOME 关键字的子查询

ANY和SOME是同义词,表示满足其中任意一个条件。它们允许创建一个表达式对子查询的返回值列表进行比较,只要满足内层子查询中的任何一个比较条件,就返回一个结果作为外层查询的条件。

下面定义两张表tbl1和tbl2:

```
CREATE table tbl1 ( num1 INT NOT NULL);
CREATE table tbl2 ( num2 INT NOT NULL);
```

分别向两张表中插入数据:

```
INSERT INTO tbl1 values(1), (5), (13), (27);
INSERT INTO tbl2 values(6), (14), (11), (20);
```

ANY关键字接在一个比较操作符的后面,表示若与子查询返回的任何值比较的结果为TRUE,则返回TRUE。

【例5.53】返回表tbl2中的num2列的所有值,然后将表tbl1中的num1列的值与num2列的值进行比较,只要num1列中的值大于num2列中的任何一个值,即为符合查询条件的结果。

```
mysql> SELECT num1 FROM tbl1 WHERE num1 > ANY (SELECT num2 FROM tbl2);
+------+
| num1 |
+------+
|   13 |
|   27 |
+------+
```

在子查询中,返回的是表tbl2的num2列的值,即(6,14,11,20),然后将表tbl1中的num1列的值与(6,14,11,20)进行比较,只要num1列中的值大于(6,14,11,20)中的任意一个,即为符合条件的结果,因此返回的结果为(13,27)。

5.5.2 带 ALL 关键字的子查询

ALL关键字与ANY和SOME不同，使用ALL时需要同时满足所有内层查询的条件。

ALL关键字接在一个比较操作符的后面，表示与子查询返回的所有值进行比较，若结果为TRUE，则返回TRUE。

下面修改【例5.53】，用ALL关键字替换ANY。

【例5.54】返回表tbl1中比表tbl2的num2列所有值都大的值，SQL语句如下：

```
mysql> SELECT num1 FROM tbl1 WHERE num1 > ALL (SELECT num2 FROM tbl2);
+------+
| num1 |
+------+
|   27 |
+------+
```

在子查询中，返回的是表tbl2的值，即num2列的所有（6,14,11,20），然后将表tbl1中的num1列的值与之进行比较，大于所有num2列值的num1值只有27，因此返回结果为27。

5.5.3 带 EXISTS 关键字的子查询

EXISTS关键字后面的参数是一个任意的子查询，系统对子查询进行运算以判断它是否返回行，如果至少返回一行，那么EXISTS返回的结果为TRUE，此时外层查询语句将进行查询；如果子查询没有返回任何行，那么EXISTS返回的结果是FALSE，此时外层查询语句将不进行查询。

【例5.55】查询表suppliers中是否存在s_id=107的水果供应商，如果存在，则查询表fruits中的记录，SQL语句如下：

```
mysql> SELECT * FROM fruits WHERE EXISTS (SELECT s_name FROM suppliers WHERE s_id = 107);
+------+------+------------+---------+
| f_id | s_id | f_name     | f_price |
+------+------+------------+---------+
| a1   | 101  | apple      |    5.20 |
| a2   | 103  | apricot    |    2.20 |
| b1   | 101  | blackberry |   10.20 |
| b2   | 104  | berry      |    7.60 |
| b5   | 107  | xxxx       |    3.60 |
| bs1  | 102  | orange     |   11.20 |
| bs2  | 105  | melon      |    8.20 |
| c0   | 101  | cherry     |    3.20 |
| l2   | 104  | lemon      |    6.40 |
| m1   | 106  | mango      |   15.70 |
| m2   | 105  | xbabay     |    2.60 |
| m3   | 105  | xxtt       |   11.60 |
| o2   | 103  | coconut    |    9.20 |
| t1   | 102  | banana     |   10.30 |
```

```
| t2   | 102  | grape       |    5.30 |
| t4   | 107  | xbababa     |    3.60 |
+------+------+-------------+---------+
```

由结果可以看到，内层查询结果表明表suppliers中存在s_id=107的记录，因此EXISTS表达式返回TRUE；外层查询语句接收TRUE之后对表fruits进行查询，返回所有的记录。

EXISTS关键字可以和条件表达式一起使用。

【例5.56】 查询表suppliers中是否存在s_id=107的水果供应商，如果存在，则查询表fruits中的f_price>10.20的记录，SQL语句如下：

```
mysql> SELECT * FROM fruits WHERE f_price>10.20 AND EXISTS (SELECT s_name FROM
suppliers WHERE s_id = 107);
+------+------+---------+---------+
| f_id | s_id | f_name  | f_price |
+------+------+---------+---------+
| bs1  | 102  | orange  |   11.20 |
| m1   | 106  | mango   |   15.70 |
| m3   | 105  | xxtt    |   11.60 |
| t1   | 102  | banana  |   10.30 |
+------+------+---------+---------+
```

由结果可以看到，内层查询结果表明表suppliers中存在s_id=107的记录，因此EXISTS表达式返回TRUE；外层查询语句接收TRUE之后根据查询条件f_price > 10.20对表fruits进行查询，返回结果为4条f_price>10.20的记录。

NOT EXISTS与EXISTS的使用方法相同，返回的结果相反。子查询如果至少返回一行，那么NOT EXISTS返回的结果为FALSE，此时外层查询语句将不进行查询；如果子查询没有返回任何行，那么NOT EXISTS返回的结果是TRUE，此时外层查询语句将进行查询。

【例5.57】 查询表suppliers中是否存在s_id=107的水果供应商，如果不存在，则查询表fruits中的记录，SQL语句如下：

```
mysql> SELECT * FROM fruits WHERE NOT EXISTS (SELECT s_name FROM suppliers WHERE
s_id = 107);
Empty set (0.00 sec)
```

查询语句SELECT s_name FROM suppliers WHERE s_id = 107对表suppliers进行查询并返回了一条记录，因此NOT EXISTS表达式返回FALSE，外层表达式接收FALSE，将不再查询表fruits中的记录。

> **提示** EXISTS和NOT EXISTS的结果只取决于是否会返回行，而不取决于这些行的内容，所以这个子查询输入列表通常是无关紧要的。

5.5.4 带IN关键字的子查询

IN关键字进行子查询时，内层查询语句仅仅返回一个数据列，这个数据列里的值将提供给外层查询语句进行比较操作。

【例5.58】 在表orderitems中查询f_id为c0的订单号,并根据订单号查询客户c_id,SQL语句如下:

```
mysql> SELECT c_id FROM orders WHERE o_num IN  (SELECT o_num  FROM orderitems WHERE f_id = 'c0');
+-------+
| c_id  |
+-------+
| 10004 |
| 10001 |
+-------+
```

查询结果中的c_id有两个值,分别为10001和10004。

上述查询过程可以分步执行,首先内层子查询查出表orderitems中符合条件的订单号,单独执行内查询,查询结果如下:

```
mysql> SELECT o_num  FROM orderitems WHERE f_id = 'c0';
+-------+
| o_num |
+-------+
| 30003 |
| 30005 |
+-------+
```

然后执行外层查询,在表orders查询订单号等于30003或30005的客户c_id。嵌套子查询语句还可以写为如下形式,实现相同的效果:

```
mysql> SELECT c_id FROM orders WHERE o_num IN (30003, 30005);
+-------+
| c_id  |
+-------+
| 10004 |
| 10001 |
+-------+
```

这个例子说明在处理SELECT语句的时候,MySQL实际上执行了两个操作过程,即先执行内层子查询,再执行外层查询,内层子查询的结果作为外层查询的比较条件。

SELECT语句中可以使用NOT IN关键字,其作用与IN正好相反。

【例5.59】 与【例5.58】类似,但是在SELECT语句中使用NOT IN关键字,SQL语句如下:

```
mysql> SELECT c_id FROM orders WHERE o_num NOT IN (SELECT o_num  FROM orderitems WHERE f_id = 'c0');
+-------+
| c_id  |
+-------+
| 10001 |
| 10003 |
| 10005 |
+-------+
```

这里返回的结果中有3条记录。由【例5.58】可知，子查询返回的订单值有两个，即30003和30005，但为什么这里还有值为10001的c_id呢？这是因为c_id等于10001的客户的订单不只一个，可以查看订单表orders中的记录。

```
mysql> SELECT * FROM orders;
+-------+---------------------+-------+
| o_num | o_date              | c_id  |
+-------+---------------------+-------+
| 30001 | 2008-09-01 00:00:00 | 10001 |
| 30002 | 2008-09-12 00:00:00 | 10003 |
| 30003 | 2008-09-30 00:00:00 | 10004 |
| 30004 | 2008-10-03 00:00:00 | 10005 |
| 30005 | 2008-10-08 00:00:00 | 10001 |
+-------+---------------------+-------+
```

可以看到，虽然排除了订单号为30003和30005的客户c_id，但是o_num为30001和30005的订单都是c_id为10001的客户的订单。因此，结果中只是排除了订单号，但是仍然有可能选择同一个客户。

> **提示** 子查询的功能也可以通过连接查询完成，但是子查询使得MySQL代码更容易阅读和编写。

5.5.5 带比较运算符的子查询

在前面介绍带ANY、ALL关键字的子查询时使用了">"比较运算符，子查询时还可以使用其他的比较运算符，如"<""<="" ="">="和"!="等。

【例5.60】在表suppliers中查询s_city等于"Tianjin"的水果供应商s_id，然后在表fruits中查询该供应商提供的所有的水果种类，SQL语句如下：

```
SELECT s_id, f_name FROM fruits WHERE s_id =(SELECT s1.s_id FROM suppliers AS s1
WHERE s1.s_city = 'Tianjin');
```

该嵌套查询首先在表suppliers中查找s_city等于Tianjin的水果供应商的s_id，单独执行子查询查看s_id的值：

```
mysql> SELECT s1.s_id FROM suppliers AS s1 WHERE s1.s_city = 'Tianjin';
+------+
| s_id |
+------+
| 101  |
+------+
```

然后在外层查询时，在表fruits中查找s_id等于101的水果供应商提供的水果的种类，查询结果如下：

```
mysql> SELECT s_id, f_name FROM fruits WHERE s_id = (SELECT s1.s_id FROM suppliers
AS s1 WHERE s1.s_city = 'Tianjin');
```

```
+------+------------+
| s_id | f_name     |
+------+------------+
| 101  | apple      |
| 101  | blackberry |
| 101  | cherry     |
+------+------------+
```

结果表明,"Tianjin"地区的水果供应商提供的水果种类有3种,分别为apple、blackberry与cherry。

【例5.61】在表suppliers中查询s_city等于"Tianjin"的水果供应商的s_id,然后在表fruits中查询所有非该供应商提供的水果的种类,SQL语句如下:

```
mysql> SELECT s_id, f_name FROM fruits WHERE s_id <>  (SELECT s1.s_id FROM suppliers AS s1 WHERE s1.s_city = 'Tianjin');
+------+---------+
| s_id | f_name  |
+------+---------+
| 103  | apricot |
| 104  | berry   |
| 107  | xxxx    |
| 102  | orange  |
| 105  | melon   |
| 104  | lemon   |
| 106  | mango   |
| 105  | xbabay  |
| 105  | xxtt    |
| 103  | coconut |
| 102  | banana  |
| 102  | grape   |
| 107  | xbababa |
+------+---------+
```

该嵌套查询执行过程与【例5.60】相同,但这里使用了"<>"运算符,因此返回的结果和【例5.60】的正好相反。

5.6 合并查询结果

利用UNION关键字,可以联合多条SELECT语句,并将它们的结果组合成单个结果集。合并时,两张表对应的列数和数据类型必须相同。各个SELECT语句之间使用UNION或UNION ALL关键字分隔。不带ALL关键字的UNION,执行时会删除重复的记录,所有返回行都是唯一的;使用关键字ALL的作用是不删除重复行,也不对结果进行自动排序。基本语法格式如下:

```
SELECT column,... FROM table1
UNION [ALL]
SELECT column,... FROM table2
```

【例5.62】查询所有价格小于9的水果的信息,查询s_id等于101和103的所有水果的信息,使用UNION连接查询结果,SQL语句如下:

```
SELECT s_id, f_name, f_price FROM fruits WHERE f_price < 9.0 UNION SELECT s_id,
f_name, f_price FROM fruits WHERE s_id IN(101,103);
```

查询结果如下:

```
+------+------------+---------+
| s_id | f_name     | f_price |
+------+------------+---------+
| 101  | apple      |    5.20 |
| 103  | apricot    |    2.20 |
| 104  | berry      |    7.60 |
| 107  | xxxx       |    3.60 |
| 105  | melon      |    8.20 |
| 101  | cherry     |    3.20 |
| 104  | lemon      |    6.40 |
| 105  | xbabay     |    2.60 |
| 102  | grape      |    5.30 |
| 107  | xbababa    |    3.60 |
| 101  | blackberry |   10.20 |
| 103  | coconut    |    9.20 |
+------+------------+---------+
```

如前所述,UNION将多个SELECT语句的结果组合成一个结果集合。可以分开查看每个SELECT语句的结果:

```
mysql> SELECT s_id, f_name, f_price FROM fruits WHERE f_price < 9.0;
+------+---------+---------+
| s_id | f_name  | f_price |
+------+---------+---------+
| 101  | apple   |    5.20 |
| 103  | apricot |    2.20 |
| 104  | berry   |    7.60 |
| 107  | xxxx    |    3.60 |
| 105  | melon   |    8.20 |
| 101  | cherry  |    3.20 |
| 104  | lemon   |    6.40 |
| 105  | xbabay  |    2.60 |
| 102  | grape   |    5.30 |
| 107  | xbababa |    3.60 |
+------+---------+---------+

mysql> SELECT s_id, f_name, f_price FROM fruits WHERE s_id IN(101,103);
+------+------------+---------+
| s_id | f_name     | f_price |
+------+------------+---------+
| 101  | apple      |    5.20 |
| 103  | apricot    |    2.20 |
| 101  | blackberry |   10.20 |
```

```
| 101  | cherry       |    3.20 |
| 103  | coconut      |    9.20 |
+------+--------------+---------+
```

由分开查询的结果可以看到,第一条SELECT语句查询价格小于9的水果,第二条SELECT语句查询s_id为101和103的水果供应商提供的水果。使用UNION将两条SELECT语句分隔开,执行完毕之后把输出结果组合成单个的结果集,并删除重复的记录。

使用UNION ALL会包含重复的行。在【例5.62】中,分开查询时,两个返回结果中有相同的记录,UNION从查询结果集中自动去除了重复的行。如果要返回所有匹配行,而不进行删除,可以使用UNION ALL。

【例5.63】查询所有价格小于9的水果的信息,查询s_id等于101和103的所有水果的信息,使用UNION ALL连接查询结果,SQL语句如下:

```
SELECT s_id, f_name, f_price FROM fruits WHERE f_price < 9.0 UNION ALL
SELECT s_id, f_name, f_price FROM fruits WHERE s_id IN(101,103);
```

查询结果如下:

```
+------+------------+---------+
| s_id | f_name     | f_price |
+------+------------+---------+
| 101  | apple      |    5.20 |
| 103  | apricot    |    2.20 |
| 104  | berry      |    7.60 |
| 107  | xxxx       |    3.60 |
| 105  | melon      |    8.20 |
| 101  | cherry     |    3.20 |
| 104  | lemon      |    6.40 |
| 105  | xbabay     |    2.60 |
| 102  | grape      |    5.30 |
| 107  | xbababa    |    3.60 |
| 101  | apple      |    5.20 |
| 103  | apricot    |    2.20 |
| 101  | blackberry |   10.20 |
| 101  | cherry     |    3.20 |
| 103  | coconut    |    9.20 |
+------+------------+---------+
```

由结果可以看到,这里总的记录数等于两条SELECT语句返回的记录数之和,连接查询结果并没有去除重复的行。

提示 使用UNION ALL时不会删除重复行,并且执行时需要的资源少。因此,当知道有重复行但是想保留这些行、确定查询结果中不会有重复数据或者不需要去掉重复数据的时候,应当使用UNION ALL以提高查询效率。

5.7　为表和字段取别名

在前面介绍分组查询、聚合函数查询和嵌套子查询时，有的地方使用了AS关键字为查询结果中的某一列指定一个特定的名字；在介绍内连接查询时，则对相同的表fruits分别指定了两个不同的名字。因此，我们可以为字段或者表取一个别名，在查询时，使用别名替代其指定的内容。本节将介绍如何为字段和表创建别名以及如何使用别名。

5.7.1　为表取别名

当表名字很长或者要执行一些特殊查询时，为了方便操作，可以为表指定别名，用这个别名替代表原来的名称。为表取别名的基本语法格式如下：

```
表名 [AS] 表别名
```

"表名"为数据库中存储的数据表的名称，"表别名"为查询时指定的表的新名称，AS关键字为可选参数。

【例5.64】为表orders取别名"o"，查询o_num为30001的订单的下单日期，SQL语句如下：

```
SELECT * FROM orders AS o WHERE o.o_num = 30001;
```

这里"orders AS o"代码表示为表orders取别名为"o"，指定过滤条件时直接使用o代替orders，查询结果如下：

```
+-------+---------------------+-------+
| o_num | o_date              | c_id  |
+-------+---------------------+-------+
| 30001 | 2008-09-01 00:00:00 | 10001 |
+-------+---------------------+-------+
```

【例5.65】为表customers和表orders分别取别名，并进行连接查询，SQL语句如下：

```
mysql> SELECT c.c_id, o.o_num FROM customers AS c LEFT OUTER JOIN orders AS o ON c.c_id = o.c_id;
+-------+-------+
| c_id  | o_num |
+-------+-------+
| 10001 | 30001 |
| 10001 | 30005 |
| 10002 | NULL  |
| 10003 | 30002 |
| 10004 | 30003 |
+-------+-------+
```

由结果可以看到，MySQL可以同时为多张表取别名，而且表别名可以放在不同的位置，如WHERE子句、SELECT列表、ON子句以及ORDER BY子句等。

在前面介绍内连接查询时，指出自连接是一种特殊的内连接，在连接查询中的两张表是同一张表，其查询语句如下：

```
mysql> SELECT f1.f_id, f1.f_name FROM fruits AS f1, fruits AS f2 WHERE f1.s_id
= f2.s_id AND f2.f_id = 'a1';
+------+------------+
| f_id | f_name     |
+------+------------+
| a1   | apple      |
| b1   | blackberry |
| c0   | cherry     |
+------+------------+
```

在这里，如果不使用表别名，MySQL将不知道引用的是哪张fruits表实例。这是表别名一个非常有用的地方。

> 提示　在为表取别名时，要保证不能与数据库中其他表的名称发生冲突。

5.7.2　为字段取别名

在使用SELECT语句显示查询结果时，MySQL会显示每个SELECT后面指定的输出列。在有些情况下，显示的列的名称会很长或者名称不够直观，此时可以指定列别名，替换字段或表达式。为字段取别名的基本语法格式为：

```
列名 [AS] 列别名
```

"列名"为表中字段定义的名称，"列别名"为字段的新名称，AS关键字为可选参数。

【例5.66】查询表fruits，为f_name取别名fruit_name，为f_price取别名fruit_price，为表fruits取别名f1，查询表中f_price < 8的水果的名称，SQL语句如下：

```
mysql> SELECT f1.f_name AS fruit_name, f1.f_price AS fruit_price FROM fruits AS
f1 WHERE f1.f_price < 8;
+------------+-------------+
| fruit_name | fruit_price |
+------------+-------------+
| apple      |        5.20 |
| apricot    |        2.20 |
| berry      |        7.60 |
| xxxx       |        3.60 |
| cherry     |        3.20 |
| lemon      |        6.40 |
| xbabay     |        2.60 |
| grape      |        5.30 |
| xbababa    |        3.60 |
+------------+-------------+
```

也可以为SELECT子句中的计算字段取别名。例如，对使用COUNT聚合函数或者CONCAT等系统函数执行的结果字段取别名。

【例5.67】 查询表suppliers中的字段s_name和s_city，使用CONCAT函数连接这两个字段值，并取列别名为suppliers_title。

先不对连接后的值取别名，SQL语句如下：

```
mysql> SELECT CONCAT(TRIM(s_name) , ' (', TRIM(s_city), ')') FROM suppliers
 ORDER BY s_name;
+------------------------------------------------+
| CONCAT(TRIM(s_name) , ' (',  TRIM(s_city), ')') |
+------------------------------------------------+
| ACME (Shanghai)                                |
| DK Inc. (Zhengzhou)                            |
| FastFruit Inc. (Tianjin)                       |
| FNK Inc. (Zhongshan)                           |
| Good Set (Taiyuan)                             |
| Just Eat Ours (Beijing)                        |
| LT Supplies (Chongqing)                        |
+------------------------------------------------+
```

由结果可以看到，显示结果的列名称为SELECT子句后面的计算字段。实际上计算之后的列是没有名字的，这样的结果不容易让人理解，如果为字段取一个别名，将会使结果清晰，SQL语句如下：

```
mysql> SELECT CONCAT(TRIM(s_name) , ' (', TRIM(s_city), ')') AS suppliers_title
FROM suppliers ORDER BY s_name;
+--------------------------+
| suppliers_title          |
+--------------------------+
| ACME (Shanghai)          |
| DK Inc. (Zhengzhou)      |
| FastFruit Inc. (Tianjin) |
| FNK Inc. (Zhongshan)     |
| Good Set (Taiyuan)       |
| Just Eat Ours (Beijing)  |
| LT Supplies (Chongqing)  |
+--------------------------+
```

SELECT子句中增加了AS suppliers_title，它指示MySQL为计算字段创建一个别名suppliers_title，显示结果为指定的列别名，这样就增强了查询结果的可读性。

> **提示** 表别名只在执行查询的时候使用，并不在返回结果中显示；而列别名定义之后，将返回给客户端显示，显示字段列的别名。

5.8 使用正则表达式查询

正则表达式通常用来检索或替换那些符合某个模式的文本内容，根据指定的匹配模式匹配文本中符合要求的特殊字符串。例如，从一个文本文件中提取电话号码，查找一篇文章中重复的单词或者替换用户输入的某些敏感词语等,这些地方都可以使用正则表达式。正则表达式强大而且灵活,可以应用于非常复杂的查询。

MySQL中使用REGEXP关键字指定正则表达式的字符匹配模式。表5.3列出了REGEXP操作符中常用的字符匹配。

表5.3 正则表达式常用字符匹配

选项	说明	例子	匹配值示例
^	匹配文本的开始字符	'^b'匹配以字母 b 开头的字符串	book,big,banana,bike
$	匹配文本的结束字符	'st$'匹配以 st 结尾的字符串	test,resist,persist
.	匹配任何单个字符	'b.t'匹配任何有字符 b 或 t 的字符串	bit,bat,but,bite
*	匹配 0 个或多个在它前面的字符	'f*n'匹配字符n前面有任意字符 f 的字符串	fn,fan,faan,fabcn
+	匹配前面的字符 1 次或多次	'ba+ '匹配以 b 开头后面紧跟至少一个 a 的字符串	ba,bay,bare,battle
<字符串>	匹配包含指定的字符串的文本	'fa'匹配包含 fa 的字符串	fan,afa,faad
[字符集合]	匹配字符集合中的任何一个字符	'[xz]' 匹配包含 x 或者 z 的字符串	dizzy,zebra,x-ray,extra
[^]	匹配不在括号中的任何字符	'[^abc]'匹配任何不包含 a、b 或 c 的字符串	desk,fox,f8ke
字符串{n,}	匹配前面的字符串至少 n 次	b{2}匹配 2 个或更多的 b	bbb,bbbb,bbbbbbb
字符串{n,m}	匹配前面的字符串至少 n 次,至多 m 次。如果 n 为 0,此参数为可选参数	b{2,4}匹配含最少 2 个、最多 4 个 b 的字符串	bb,bbb,bbbb

下文将详细介绍如何在MySQL中使用正则表达式。

5.8.1 查询以特定字符或字符串开头的记录

字符"^"匹配以特定字符或者字符串开头的文本。

【例5.68】在表fruits中,查询f_name字段中以字母"b"开头的记录,SQL语句如下:

```
mysql> SELECT * FROM fruits WHERE f_name REGEXP '^b';
+------+------+------------+---------+
| f_id | s_id | f_name     | f_price |
+------+------+------------+---------+
| b1   | 101  | blackberry | 10.20   |
| b2   | 104  | berry      | 7.60    |
| t1   | 102  | banana     | 10.30   |
+------+------+------------+---------+
```

表fruits中有3条记录的f_name字段值是以字母b开头的,因此返回结果中有3条记录。

【例5.69】在表fruits中，查询f_name字段中以"be"开头的记录，SQL语句如下：

```
mysql> SELECT * FROM fruits WHERE f_name REGEXP '^be';
+------+------+--------+---------+
| f_id | s_id | f_name | f_price |
+------+------+--------+---------+
| b2   | 104  | berry  |   7.60  |
+------+------+--------+---------+
```

只有berry是以"be"开头的，因此查询结果中只有1条记录。

5.8.2 查询以特定字符或字符串结尾的记录

字符"$"匹配以特定字符或者字符串结尾的文本。

【例5.70】在表fruits中，查询f_name字段中以字母"y"结尾的记录，SQL语句如下：

```
mysql> SELECT * FROM fruits WHERE f_name REGEXP 'y$';
+------+------+------------+---------+
| f_id | s_id | f_name     | f_price |
+------+------+------------+---------+
| b1   | 101  | blackberry |  10.20  |
| b2   | 104  | berry      |   7.60  |
| c0   | 101  | cherry     |   3.20  |
| m2   | 105  | xbabay     |   2.60  |
+------+------+------------+---------+
```

表fruits中有4条记录的f_name字段值是以字母"y"结尾的，因此返回结果中有4条记录。

【例5.71】在表fruits中，查询f_name字段中以字符串"rry"结尾的记录，SQL语句如下：

```
mysql> SELECT * FROM fruits WHERE f_name REGEXP 'rry$';
+------+------+------------+---------+
| f_id | s_id | f_name     | f_price |
+------+------+------------+---------+
| b1   | 101  | blackberry |  10.20  |
| b2   | 104  | berry      |   7.60  |
| c0   | 101  | cherry     |   3.20  |
+------+------+------------+---------+
```

表fruits中有3条记录的f_name字段值是以字符串"rry"结尾的，因此返回结果中有3条记录。

5.8.3 用符号"."来替代字符串中的任意一个字符

字符"."匹配任意一个字符。

【例5.72】在表fruits中，查询f_name字段中包含字母"a"与"g"且两个字母之间只有一个字母的记录，SQL语句如下：

```
mysql> SELECT * FROM fruits WHERE f_name REGEXP 'a.g';
+------+------+--------+---------+
| f_id | s_id | f_name | f_price |
+------+------+--------+---------+
| bs1  | 102  | orange |  11.20  |
| m1   | 106  | mango  |  15.70  |
+------+------+--------+---------+
```

在查询语句中，"a.g"指定匹配字符中要有字母a和g，且两个字母之间包含单个字符，并不限定匹配的字符的位置和所在查询字符串的总长度，因此orange和mango都符合匹配条件。

5.8.4 使用"*"和"+"来匹配多个字符

星号（*）匹配前面的字符任意多次，包括0次。加号（+）匹配前面的字符至少一次。

【例5.73】在表fruits中，查询f_name字段中以字母"b"开头且"b"后面出现紧跟字母"a"的记录，SQL语句如下：

```
mysql> SELECT * FROM fruits WHERE f_name REGEXP '^ba*';
+------+------+------------+---------+
| f_id | s_id | f_name     | f_price |
+------+------+------------+---------+
| b1   | 101  | blackberry |  10.20  |
| b2   | 104  | berry      |   7.60  |
| t1   | 102  | banana     |  10.30  |
+------+------+------------+---------+
```

星号（*）可以匹配任意多个字符，因此blackberry和berry中字母"b"后面虽然没有出现字母"a"，但是也满足匹配条件。

【例5.74】在表fruits中，查询f_name字段中以字母"b"开头且"b"后面字母"a"至少出现一次的记录，SQL语句如下：

```
mysql> SELECT * FROM fruits WHERE f_name REGEXP '^ba+';
+------+------+--------+---------+
| f_id | s_id | f_name | f_price |
+------+------+--------+---------+
| t1   | 102  | banana |  10.30  |
+------+------+--------+---------+
```

"a+"匹配字母"a"至少一次，只有banana满足匹配条件。

5.8.5 匹配指定字符串

正则表达式可以匹配指定字符串，只要这个字符串在查询文本中即可，如果要匹配多个字符串，则多个字符串之间使用"|"隔开。

【例5.75】在表fruits中，查询f_name字段中包含字符串"on"的记录，SQL语句如下：

```
mysql> SELECT * FROM fruits WHERE f_name REGEXP 'on';
```

```
+------+------+---------+---------+
| f_id | s_id | f_name  | f_price |
+------+------+---------+---------+
| bs2  | 105  | melon   |    8.20 |
| l2   | 104  | lemon   |    6.40 |
| o2   | 103  | coconut |    9.20 |
+------+------+---------+---------+
```

可以看到，f_name字段的melon、lemon和coconut 3个值中都包含有字符串"on"，满足匹配条件。

【例5.76】 在表fruits中，查询f_name字段中包含字符串"on"或者"ap"的记录，SQL语句如下：

```
mysql> SELECT * FROM fruits WHERE f_name REGEXP 'on|ap';
+------+------+---------+---------+
| f_id | s_id | f_name  | f_price |
+------+------+---------+---------+
| a1   | 101  | apple   |    5.20 |
| a2   | 103  | apricot |    2.20 |
| bs2  | 105  | melon   |    8.20 |
| l2   | 104  | lemon   |    6.40 |
| o2   | 103  | coconut |    9.20 |
| t2   | 102  | grape   |    5.30 |
+------+------+---------+---------+
```

可以看到，f_name字段的melon、lemon和coconut 3个值中都包含有字符串"on"，apple和apricot中包含字符串"ap"，满足匹配条件。

之前介绍过，LIKE运算符也可以匹配指定的字符串，但与REGEXP不同，LIKE匹配的字符串如果在文本中间出现，则找不到它，相应的行也不会返回。REGEXP在文本内进行匹配，如果被匹配的字符串在文本中出现，REGEXP将会找到它，相应的行也会被返回。对比结果如【例5.77】所示。

【例5.77】 在表fruits中，使用LIKE运算符查询f_name字段中值为"on"的记录，SQL语句如下：

```
mysql> SELECT * FROM fruits WHERE f_name LIKE 'on';
Empty set (0.00 sec)
```

f_name字段中没有值为"on"的记录，返回结果为空。读者可以体会一下两者的区别。

5.8.6　匹配指定字符中的任意一个

方括号（[]）指定一个字符集合，只要匹配其中任何一个字符，即为所查找的文本。

【例5.78】 在表fruits中，查找f_name字段中包含字母"o"或者"t"的记录，SQL语句如下：

```
mysql> SELECT * FROM fruits WHERE f_name REGEXP '[ot]';
```

```
+------+------+---------+---------+
| f_id | s_id | f_name  | f_price |
+------+------+---------+---------+
| a2   | 103  | apricot |    2.20 |
| bs1  | 102  | orange  |   11.20 |
| bs2  | 105  | melon   |    8.20 |
| l2   | 104  | lemon   |    6.40 |
| m1   | 106  | mango   |   15.70 |
| m3   | 105  | xxtt    |   11.60 |
| o2   | 103  | coconut |    9.20 |
+------+------+---------+---------+
```

由查询结果可以看到，所有返回的记录的f_name字段的值中都包含有字母o或者t，或者两个都有。

方括号（[]）还可以指定数值集合。

【例5.79】在表fruits中，查询s_id字段中包含4、5或者6的记录，SQL语句如下：

```
mysql> SELECT * FROM fruits WHERE s_id REGEXP '[456]';
+------+------+---------+---------+
| f_id | s_id | f_name  | f_price |
+------+------+---------+---------+
| b2   | 104  | berry   |    7.60 |
| bs2  | 105  | melon   |    8.20 |
| l2   | 104  | lemon   |    6.40 |
| m1   | 106  | mango   |   15.70 |
| m2   | 105  | xbabay  |    2.60 |
| m3   | 105  | xxtt    |   11.60 |
+------+------+---------+---------+
```

在查询结果中，s_id字段值中只要有4、5、6这3个数字中的1个，即满足匹配条件。

匹配集合"[456]"也可以写成"[4-6]"，即指定集合区间。例如，"[a-z]"表示集合区间为a~z的字母，"[0-9]"表示集合区间为所有数字。

5.8.7 匹配指定字符以外的字符

"[^字符集合]"匹配不在指定集合中的任何字符。

【例5.80】在表fruits中，查询f_id字段中包含字母a~e和数字1~2以外字符的记录，SQL语句如下：

```
mysql> SELECT * FROM fruits WHERE f_id REGEXP '[^a-e1-2]';
+------+------+---------+---------+
| f_id | s_id | f_name  | f_price |
+------+------+---------+---------+
| b5   | 107  | xxxx    |    3.60 |
| bs1  | 102  | orange  |   11.20 |
| bs2  | 105  | melon   |    8.20 |
| c0   | 101  | cherry  |    3.20 |
| l2   | 104  | lemon   |    6.40 |
```

```
| m1    | 106  | mango    |  15.70 |
| m2    | 105  | xbabay   |   2.60 |
| m3    | 105  | xxtt     |  11.60 |
| o2    | 103  | coconut  |   9.20 |
| t1    | 102  | banana   |  10.30 |
| t2    | 102  | grape    |   5.30 |
| t4    | 107  | xbababa  |   3.60 |
+-------+------+----------+--------+
```

返回记录中的f_id字段值中包含指定字母和数字以外的值，如s、m、o、t等，这些字母均不在a~e与1~2中，满足匹配条件。

5.8.8 使用{n,}或者{n,m}来指定字符串连续出现的次数

"字符串{n,}"表示至少匹配n次前面的字符；"字符串{n,m}"表示匹配前面的字符串不少于n次，不多于m次。例如，"a{2,}"表示字母"a"至少连续出现2次，也可以大于2次；"a{2,4}"表示字母"a"最少连续出现2次，最多不能超过4次。

【例5.81】在表fruits中，查询f_name字段中字母"x"至少出现2次的记录，SQL语句如下：

```
mysql> SELECT * FROM fruits WHERE f_name REGEXP 'x{2,}';
+------+------+--------+---------+
| f_id | s_id | f_name | f_price |
+------+------+--------+---------+
| b5   | 107  | xxxx   |   3.60  |
| m3   | 105  | xxtt   |  11.60  |
+------+------+--------+---------+
```

可以看到，f_name字段的"xxxx"包含了4个字母"x"，"xxtt"包含两个字母"x"，均满足匹配条件。

【例5.82】在表fruits中，查询f_name字段中字符串"ba"最少出现1次、最多出现3次的记录，SQL语句如下：

```
mysql> SELECT * FROM fruits WHERE f_name REGEXP 'ba{1,3}';
+------+------+---------+---------+
| f_id | s_id | f_name  | f_price |
+------+------+---------+---------+
| m2   | 105  | xbabay  |   2.60  |
| t1   | 102  | banana  |  10.30  |
| t4   | 107  | xbababa |   3.60  |
+------+------+---------+---------+
```

可以看到，"ba"在f_name字段的xbabay值中出现了2次，在banana中出现了1次，在xbababa中出现了3次，均满足匹配条件。

5.9 通用表表达式

通用表表达式简称为CTE（Common Table Expressions）。CTE是命名的临时结果集，作用范围是当前语句。CTE可以理解为一个可以复用的子查询，当然跟子查询还是有点区别的，CTE可以引用其他CTE，但子查询不能引用其他子查询。

CTE的语法格式如下：

```
with_clause:
   WITH [RECURSIVE]
       cte_name [(col_name [, col_name] ...)] AS (subquery)
       [, cte_name [(col_name [, col_name] ...)] AS (subquery)] ...
```

使用WITH语句创建CTE的情况如下：

（1）在SELECT、UPDATE、DELETE语句的开头：

```
WITH ... SELECT ...
WITH ... UPDATE ...
WITH ... DELETE ...
```

（2）在子查询的开头：

```
SELECT ... WHERE id IN (WITH ... SELECT ...) ...
SELECT * FROM (WITH ... SELECT ...) AS dt ...
```

（3）在包含 SELECT声明的语句之前：

```
INSERT ... WITH ... SELECT ...
REPLACE ... WITH ... SELECT ...
CREATE TABLE ... WITH ... SELECT ...
CREATE VIEW ... WITH ... SELECT ...
DECLARE CURSOR ... WITH ... SELECT ...
EXPLAIN ... WITH ... SELECT ...
```

下面通过案例来讲述通用表表达式的使用方法。

创建商品表goods，该数据表包含上下级关系的数据，具体字段包含商品编号（id）、商品名称（name）、上级商品的编号（gid）。创建语句如下：

```
CREATE TABLE goods(
id int(11),
name varchar(30),
gid int(11),
PRIMARY KEY (`id`));
```

插入演示数据，SQL语句如下：

```
INSERT INTO goods (id, name, gid) VALUES (1, '商品',0);
INSERT INTO goods (id, name, gid) VALUES (2, '水果',1);
```

```
INSERT INTO goods (id, name, gid) VALUES (3, '蔬菜',1);
INSERT INTO goods (id, name, gid) VALUES (4, '苹果',2);
INSERT INTO goods (id, name, gid) VALUES (5, '香蕉',2);
INSERT INTO goods (id, name, gid) VALUES (6, '菠菜',3);
INSERT INTO goods (id, name, gid) VALUES (7, '萝卜',3);
```

下面开始查询每个商品对应的上级商品名称。

首先使用子查询的方式，SQL语句如下：

```
mysql> SELECT g.*, (SELECT name FROM goods where id = g.gid) as pname FROM goods AS g;
+----+------+-----+-------+
| id | name | gid | pname |
+----+------+-----+-------+
|  1 | 商品 |   0 | NULL  |
|  2 | 水果 |   1 | 商品  |
|  3 | 蔬菜 |   1 | 商品  |
|  4 | 苹果 |   2 | 水果  |
|  5 | 香蕉 |   2 | 水果  |
|  6 | 菠菜 |   3 | 蔬菜  |
|  7 | 萝卜 |   3 | 蔬菜  |
+----+------+-----+-------+
```

接着使用CTE的方式完成上述功能，SQL语句如下：

```
mysql> WITH cte as (SELECT * FROM goods) SELECT g.*, (SELECT cte.name FROM cte WHERE cte.id = g.gid) AS gname FROM goods AS g;
+----+------+-----+-------+
| id | name | gid | gname |
+----+------+-----+-------+
|  1 | 商品 |   0 | NULL  |
|  2 | 水果 |   1 | 商品  |
|  3 | 蔬菜 |   1 | 商品  |
|  4 | 苹果 |   2 | 水果  |
|  5 | 香蕉 |   2 | 水果  |
|  6 | 菠菜 |   3 | 蔬菜  |
|  7 | 萝卜 |   3 | 蔬菜  |
+----+------+-----+-------+
```

从结果可以看出，CTE是一个可以重复使用的结果集。相比于子查询，CTE的效率会更高，因为非递归的CTE只会查询一次并可以重复使用。

CTE可以引用其他CTE的结果。例如，在下面的语句中，cte2就引用了cte1中的结果。

```
mysql> with cte1 as (select * from goods), cte2 as ( select g.*, cte1.name as gname from goods as g left join cte1 on g.gid = cte1.id) select * from cte2;
+----+------+-----+-------+
| id | name | gid | gname |
+----+------+-----+-------+
|  1 | 商品 |   0 | NULL  |
|  2 | 水果 |   1 | 商品  |
|  3 | 蔬菜 |   1 | 商品  |
```

```
|  4 | 苹果  |    2 | 水果   |
|  5 | 香蕉  |    2 | 水果   |
|  6 | 菠菜  |    3 | 蔬菜   |
|  7 | 萝卜  |    3 | 蔬菜   |
+----+-------+------+--------+
```

还有一种特殊的CTE，就是递归CTE，其子查询会引用自身。WITH子句必须以WITH RECURSIVE开头。

CTE递归子查询包括两部分：SEED查询和RECURSIVE查询，中间由UNION [ALL]或UNION DISTINCT分隔。SEED查询会被执行一次，以创建初始数据子集。RECURSIVE查询会被重复执行以返回数据子集，直到获得完整结果集。当迭代不会生成任何新行时，递归会停止。可以参看下面的案例：

```
mysql> WITH RECURSIVE cte(n) AS (SELECT 1 UNION ALL SELECT n + 1 FROM cte WHERE n < 8) SELECT * FROM cte;
+------+
| n    |
+------+
|    1 |
|    2 |
|    3 |
|    4 |
|    5 |
|    6 |
|    7 |
|    8 |
+------+
```

上面的语句会递归显示8行，每行分别显示数字1~8。

递归的过程如下：

（1）首先执行SELECT 1得到结果 1，即当前n的值为1。

（2）接着执行SELECT N + 1 FROM cte WHERE n < 8，因为当前n为1，所以WHERE条件成立，生成新行，SELECT n + 1得到结果2，即当前n的值为2。

（3）继续执行SELECT n + 1 FROM cte WHERE n < 8，因为当前n为2，所以WHERE条件成立，生成新行，SELECT n + 1得到结果3，即当前n的值为3。

（4）一直递归下去。

（5）直到当n为8时，where条件不成立，无法生成新行，递归停止。

下面使用递归CTE来查询每个商品到顶级商品的层次，SQL语句如下：

```
mysql>with recursive cte as ( select id, name, cast('0' as char(255)) as path from goods where gid = 0 union all select goods.id,goods.name, concat(cte.path, ',', cte.id) as path from goods inner join cte on goods.gid = cte.id) select * from cte;
+-----+------+------+
| id  | name | path |
+-----+------+------+
|  1  | 商品 | 0    |
```

```
|   2 | 水果| 0,1   |
|   3 | 蔬菜| 0,1   |
|   4 | 苹果| 0,1,2 |
|   5 | 香蕉| 0,1,2 |
|   6 | 菠菜| 0,1,3 |
|   7 | 萝卜| 0,1,3 |
+-----+----+-------+
```

查询一个指定商品的所有父级商品，SQL语句如下：

```
mysql> with recursive cte as (select id, name, gid from goods where id = 7
    union all select goods.id, goods.name, goods.gid from goods inner join cte on cte.gid
= goods.id) select * from cte;
+-----+-----+-----+
| id  | name| gid |
+-----+-----+-----+
|   7 | 萝卜|  3  |
|   3 | 蔬菜|  1  |
|   1 | 商品|  0  |
+-----+-----+-----+
```

第 6 章
插入、更新与删除数据

存储在系统中的数据是数据库管理系统的核心,数据库被设计用来管理数据的存储和访问,以及维护数据的完整性。MySQL中提供了功能丰富的数据库管理语句,包括有效地向数据库中插入数据的INSERT语句、更新数据的UPDATE语句以及当数据不再使用时删除数据的DELETE语句。本章将详细介绍如何在MySQL中使用这些语句操作数据。

6.1 插入数据

在使用数据库之前,数据库中必须有数据。在MySQL中,使用INSERT语句向数据库表中插入新的数据。插入数据的方式有为表的所有字段插入数据、为表的指定字段插入数据、同时插入多条数据、将查询结果插入表中。下面将分别介绍这些内容。

6.1.1 为表的所有字段插入数据

使用基本的INSERT语句插入数据时,要求指定表名称和要插入的值,基本语法格式如下:

```
INSERT INTO table_name (column_list) VALUES (value_list);
```

"table_name"指定要插入数据的表名,"column_list"指定要插入数据的哪些列,value_list指定每个列应插入的数据。注意,使用该语句时,字段列和数据值的数量必须相同。

本章将使用样例表person进行演示,表person的创建语句如下:

```
CREATE TABLE person
(
id      INT UNSIGNED NOT NULL AUTO_INCREMENT,
name    CHAR(40) NOT NULL DEFAULT '',
age     INT NOT NULL DEFAULT 0,
info    CHAR(50) NULL,
```

```
PRIMARY KEY (id)
);
```

向表中所有字段插入数据的方法有两种：一种是指定所有字段名，另一种是完全不指定字段名。

【例6.1】在表person中，插入一条新记录，id值为1，name值为Green，age值为21，info值为Lawyer。

执行插入操作之前，使用SELECT语句查看表中的数据：

```
mysql> SELECT * FROM person;
Empty set (0.00 sec)
```

结果显示当前表为空，没有数据。接下来执行插入操作：

```
mysql> INSERT INTO person (id ,name, age , info) VALUES (1,'Green', 21, 'Lawyer');
```

语句执行完毕后，查看插入结果：

```
mysql> SELECT * FROM person;
+----+--------+-----+--------+
| id | name   | age | info   |
+----+--------+-----+--------+
|  1 | Green  |  21 | Lawyer |
+----+--------+-----+--------+
```

可以看到插入记录成功。在插入记录时，指定了表person的所有字段，因此将为每一个字段插入新的值。

INSERT语句后面的列名称顺序可以不是表person定义时的顺序，即插入数据时，不需要按照表定义的顺序插入，只需保证值的顺序与列字段的顺序相同即可，如【例6.2】所示。

【例6.2】在表person中，插入一条新记录，id值为2，name值为Suse，age值为22，info值为dancer，SQL语句如下：

```
mysql> INSERT INTO person (age ,name, id , info) VALUES (22, 'Suse', 2, 'dancer');
```

语句执行完毕后，查看插入结果：

```
mysql> SELECT * FROM person;
+----+--------+-----+--------+
| id | name   | age | info   |
+----+--------+-----+--------+
|  1 | Green  |  21 | Lawyer |
|  2 | Suse   |  22 | dancer |
+----+--------+-----+--------+
```

由结果可以看到，INSERT语句成功插入了一条记录。

使用INSERT插入数据时，允许列名称列表column_list为空，此时，在值列表中需要为表的每一个字段指定值，并且值的顺序必须和数据表中字段定义时的顺序相同，如【例6.3】所示。

【例6.3】在表person中，插入一条新记录，id值为3，name值为Mary，age值为24，info值为Musician，SQL语句如下：

```
mysql> INSERT INTO person VALUES (3,'Mary', 24, 'Musician');
```

语句执行完毕后,查看插入结果:

```
mysql> SELECT * FROM person;
+----+--------+------+------------+
| id | name   | age  | info       |
+----+--------+------+------------+
|  1 | Green  |   21 | Lawyer     |
|  2 | Suse   |   22 | dancer     |
|  3 | Mary   |   24 | Musician   |
+----+--------+------+------------+
```

可以看到插入记录成功。数据库中增加了一条id为3的记录,其他字段值为指定的插入值。本例的INSERT语句中没有指定插入字段列表,只有一个值列表。在这种情况下,值列表为每一个字段列指定插入值,并且这些值的顺序必须和表person中字段定义的顺序相同。

> **提示** 在MySQL中,当使用INSERT语句插入数据时,可以忽略插入数据的列名称。但是,如果这样做,那么VALUES关键字后面的值不仅要求完整,而且顺序必须与表定义时列的顺序相同。如果表的结构发生改变,例如对列进行增加、删除或改变位置操作,这将会影响到用这种方式插入数据时的顺序。因此,如果表结构发生变化,那么在使用这种方法插入数据时,可能需要调整VALUES关键字后面的值的顺序。为了避免这个问题,建议在插入数据时指定列名称。这样,即使表结构发生改变,也不会影响到插入数据的顺序。

6.1.2 为表的指定字段插入数据

为表的指定字段插入数据,就是在INSERT语句中只向部分字段插入值,而其他字段的值为表定义时的默认值。

【例6.4】在表person中,插入一条新记录,name值为Willam,age值为20,info值为sports man,SQL语句如下:

```
mysql> INSERT INTO person (name, age,info) VALUES('Willam', 20, 'sports man');
```

提示信息表示插入一条记录成功。

使用SELECT查询表中的记录,查询结果如下:

```
mysql> SELECT * FROM person;
+----+--------+------+------------+
| id | name   | age  | info       |
+----+--------+------+------------+
|  1 | Green  |   21 | Lawyer     |
|  2 | Suse   |   22 | dancer     |
|  3 | Mary   |   24 | Musician   |
|  4 | Willam |   20 | sports man |
+----+--------+------+------------+
```

可以看到插入记录成功。该id字段自动添加了一个整数值4。在这里，id字段为表的主键，不能为空，系统会自动为该字段插入自增的序列值。

在插入数据时，如果某些字段没有指定插入值，MySQL将插入该字段定义时的默认值。下面的例子说明在没有指定列字段时，插入默认值。

【例6.5】在表person中，插入一条新记录，name值为laura，age值为25，SQL语句如下：

```
mysql> INSERT INTO person (name, age ) VALUES ('Laura', 25);
```

语句执行完毕后，查看插入结果：

```
mysql> SELECT * FROM person;
+----+--------+-----+------------+
| id | name   | age | info       |
+----+--------+-----+------------+
|  1 | Green  |  21 | Lawyer     |
|  2 | Suse   |  22 | dancer     |
|  3 | Mary   |  24 | Musician   |
|  4 | Willam |  20 | sports man |
|  5 | Laura  |  25 | NULL       |
+----+--------+-----+------------+
```

可以看到，在本例插入语句中，没有指定info字段值，info字段在定义时默认值为NULL，因此系统自动为该字段插入NULL。

> 提示 要保证每个插入值的类型和对应列的数据类型匹配，如果类型不同，将无法插入，并且MySQL会产生错误。

6.1.3 同时插入多条数据

INSERT语句可以同时向数据表中插入多条数据，插入时指定多个值列表，每个值列表之间用逗号分隔开，基本语法格式如下：

```
INSERT INTO table_name (column_list)
VALUES (value_list1), (value_list2),...,(value_listn);
```

"value_list1,value_list2,...,value_listn"表示第1,2,...,n个插入数据的字段的值列表。

【例6.6】在表person中，在name、age和info字段指定插入值，同时插入3条新记录，SQL语句如下：

```
INSERT INTO person(name, age, info) VALUES ('Evans',27, 'secretary'),
('Dale',22, 'cook'),('Edison',28, 'singer');
```

语句执行完毕后，查看插入结果：

```
mysql> SELECT * FROM person;
+----+--------+-----+------------+
| id | name   | age | info       |
+----+--------+-----+------------+
|  1 | Green  |  21 | Lawyer     |
|  2 | Suse   |  22 | dancer     |
```

```
|  3 | Mary   | 24 | Musician    |
|  4 | Willam | 20 | sports man  |
|  5 | Laura  | 25 | NULL        |
|  6 | Evans  | 27 | secretary   |
|  7 | Dale   | 22 | cook        |
|  8 | Edison | 28 | singer      |
+----+--------+-----+------------+
```

由结果可以看到，INSERT语句执行后，表person中添加了3条记录，其name和age字段分别为指定的值，id字段为MySQL添加的默认的自增值。

使用INSERT同时插入多条数据时，MySQL会返回一些在执行单行插入时没有给出的额外信息，这些信息的含义如下：

- Records：表明插入的数据条数。
- Duplicates：表明插入时被忽略的数据，原因可能是这些数据中包含了重复的主键值。
- Warnings：表明有问题的数据值，例如发生数据类型转换。

【例6.7】 在表person中，不指定插入列表，同时插入2条新记录，SQL语句如下：

```
INSERT INTO person VALUES (9,'Harry',21, 'magician'), (NULL,'Harriet',19,
'pianist');
```

语句执行完毕后，查看插入结果：

```
mysql> SELECT * FROM person;
+----+---------+-----+------------+
| id | name    | age | info       |
+----+---------+-----+------------+
|  1 | Green   | 21  | Lawyer     |
|  2 | Suse    | 22  | dancer     |
|  3 | Mary    | 24  | Musician   |
|  4 | Willam  | 20  | sports man |
|  5 | Laura   | 25  | NULL       |
|  6 | Evans   | 27  | secretary  |
|  7 | Dale    | 22  | cook       |
|  8 | Edison  | 28  | singer     |
|  9 | Harry   | 21  | magician   |
| 10 | Harriet | 19  | pianist    |
+----+---------+-----+------------+
```

由结果可以看到，INSERT语句执行后，表person中添加了2条记录。与前面介绍单个INSERT语法不同，person表名后面没有指定插入字段列表，因此，对于VALUES关键字后面的多个值列表，都要为每一条记录的每一个字段列指定插入值，并且这些值的顺序必须和表person中字段定义的顺序相同。带有AUTO_INCREMENT属性的id字段插入NULL，系统会自动为该字段插入唯一的自增编号。

提示 一个同时插入多行记录的INSERT语句等同于多个单行插入的INSERT语句，但是前者的处理效率更高。因为MySQL执行单条INSERT语句插入多行数据比使用多条INSERT语句快，所以在插入多行记录时最好选择使用单条INSERT语句的方式。

6.1.4 将查询结果插入表中

INSERT语句不仅可以用来在向数据表插入数据时指定插入数据的列值，还可以将SELECT语句查询的结果插入表中。如果想要从另外一张表中合并个人信息到表person，不需要把每一条数据的值一个一个地输入，只需要使用由一条INSERT语句和一条SELECT语句组成的组合语句，即可快速地从一张或多张表中向另一张表中插入多行。基本语法格式如下：

```
INSERT INTO table_name1 (column_list1)
SELECT (column_list2) FROM table_name2 WHERE (condition)
```

"table_name1"指定待插入数据的表；"column_list1"指定待插入表中要插入数据的列；"table_name2"指定待插入的数据是从哪张表中查询出来的；column_list2指定数据来源表的查询列，该列必须和column_list1列中的字段个数相同，数据类型相同；condition指定SELECT语句的查询条件。

【例6.8】从表person_old中查询所有的记录，并将其插入表person中。

首先，创建一个名为"person_old"的数据表，其表结构与表person的相同，SQL语句如下：

```
CREATE TABLE person_old
(
id      INT UNSIGNED NOT NULL AUTO_INCREMENT,
name    CHAR(40) NOT NULL DEFAULT '',
age     INT NOT NULL DEFAULT 0,
info    CHAR(50) NULL,
PRIMARY KEY (id)
);
```

向表person_old中添加两条记录，SQL语句如下：

```
mysql> INSERT INTO person_old VALUES (11,'Harry',20,'student'), (12,'Beckham',31,'police');
mysql> SELECT * FROM person_old;
+----+---------+-----+---------+
| id | name    | age | info    |
+----+---------+-----+---------+
| 11 | Harry   | 20  | student |
| 12 | Beckham | 31  | police  |
+----+---------+-----+---------+
```

由结果可以看到，插入记录成功，现在表person_old中有两条记录。

接下来将表person_old中的所有记录插入表person中，SQL语句如下：

```
INSERT INTO person(id, name, age, info) SELECT id, name, age, info FROM person_old;
```

语句执行完毕后，查看插入结果：

```
mysql> SELECT * FROM person;
+----+---------+-----+-------------+
| id | name    | age | info        |
```

```
+----+---------+-----+------------+
|  1 | Green   | 21  | Lawyer     |
|  2 | Suse    | 22  | dancer     |
|  3 | Mary    | 24  | Musician   |
|  4 | Willam  | 20  | sports man |
|  5 | Laura   | 25  | NULL       |
|  6 | Evans   | 27  | secretary  |
|  7 | Dale    | 22  | cook       |
|  8 | Edison  | 28  | singer     |
|  9 | Harry   | 21  | magician   |
| 10 | Harriet | 19  | pianist    |
| 11 | Harry   | 20  | student    |
| 12 | Beckham | 31  | police     |
+----+---------+-----+------------+
```

由结果可以看到，INSERT语句执行后，表person中多了两条记录，这两条记录和表person_old中的完全相同，说明数据转移成功。这里的id字段为自增的主键，在插入的时候要保证该字段值的唯一性，如果不能确定，可以在插入的时候忽略该字段，只插入其他字段的值。

> **提示** 这个例子中使用的表person_old和表person的定义相同。事实上，MySQL不关心SELECT返回的列名，它根据列的位置进行插入，SELECT中的第1列对应待插入表的第1列，第2列对应待插入表的第2列，以此类推。这种插入数据的用法，使得在不同结果的表之间也可以方便地转移数据。

6.2 更新数据

表中有数据之后，接下来可以对数据进行更新操作。MySQL中使用UPDATE语句更新表中的数据，可以更新特定的行或者同时更新所有的行。基本语法格式如下：

```
UPDATE table_name
SET column_name1 = value1,column_name2=value2,...,column_namen=valuen
WHERE (condition);
```

"column_name1,column_name2,...,column_namen"为指定更新的字段的名称；"value1, value2,...,valuen"为相对应的指定字段的更新值；condition指定更新的数据需要满足的条件。更新多列时，每个"列—值"对之间用逗号隔开，最后一列后面不需要逗号。

【例6.9】在表person中，更新id值为11的记录，将age字段值改为15，将name字段值改为LiMing，SQL语句如下：

```
UPDATE person SET age = 15, name='LiMing' WHERE id = 11;
```

执行更新操作前可以使用SELECT语句查看当前的数据，SQL语句如下：

```
mysql> SELECT * FROM person WHERE id=11;
+----+--------+-----+---------+
```

```
| id | name  | age | info    |
+----+-------+-----+---------+
| 11 | Harry | 20  | student |
+----+-------+-----+---------+
```

由结果可以看到，更新之前id等于11的记录的name字段值为harry、age字段值为20。
下面使用UPDATE语句更新数据，SQL语句如下：

```
mysql> UPDATE person SET age = 15, name='LiMing' WHERE id = 11;
```

语句执行完毕后，查看更新结果：

```
mysql> SELECT * FROM person WHERE id=11;
+----+--------+-----+---------+
| id | name   | age | info    |
+----+--------+-----+---------+
| 11 | LiMing | 15  | student |
+----+--------+-----+---------+
```

由结果可以看到，id等于11的数据中的name和age字段的值已经成功地被修改为指定值。

> **提示** 保证UPDATE以WHERE子句结束，通过WHERE子句指定被更新的数据所需要满足的条件。如果忽略WHERE子句，MySQL将更新表中所有的行。

【例6.10】在表person中，更新age值为19~22的记录，将info字段值都改为student，SQL语句如下：

```
UPDATE person SET info='student' WHERE id BETWEEN 19 AND 22;
```

执行更新操作前可以使用SELECT语句查看当前的数据，SQL语句如下：

```
mysql> SELECT * FROM person WHERE age BETWEEN 19 AND 22;
+----+---------+-----+-------------+
| id | name    | age | info        |
+----+---------+-----+-------------+
| 1  | Green   | 21  | Lawyer      |
| 2  | Suse    | 22  | dancer      |
| 4  | Willam  | 20  | sports man  |
| 7  | Dale    | 22  | cook        |
| 9  | Harry   | 21  | magician    |
| 10 | Harriet | 19  | pianist     |
+----+---------+-----+-------------+
```

可以看到，age字段值为19~22的记录的info字段值各不相同。
下面使用UPDATE语句更新数据，SQL语句如下：

```
mysql> UPDATE person SET info='student' WHERE age BETWEEN 19 AND 22;
```

语句执行完毕后，查看更新结果：

```
mysql> SELECT * FROM person WHERE age BETWEEN 19 AND 22;
+----+---------+-----+---------+
```

```
| id | name    | age | info    |
+----+---------+-----+---------+
|  1 | Green   |  21 | student |
|  2 | Suse    |  22 | student |
|  4 | Willam  |  20 | student |
|  7 | Dale    |  22 | student |
|  9 | Harry   |  21 | student |
| 10 | Harriet |  19 | student |
+----+---------+-----+---------+
```

由结果可以看到，UPDATE执行后，成功将表中符合条件的6条记录的info字段值都改为student。

6.3 删除数据

从数据表中删除数据需使用DELETE语句。DELETE语句允许WHERE子句指定删除条件。DELETE语句基本语法格式如下：

```
DELETE FROM table_name [WHERE <condition>];
```

"table_name"指定要执行删除操作的表；"[WHERE <condition>]"为可选参数，指定删除条件，如果没有WHERE子句，DELETE语句将删除表中的所有数据。

【例6.11】 在表person中，删除id等于11的记录。

执行删除操作前，使用SELECT语句查看当前id=11的记录，SQL语句如下：

```
mysql> SELECT * FROM person WHERE id=11;
+----+--------+-----+---------+
| id | name   | age | info    |
+----+--------+-----+---------+
| 11 | LiMing |  15 | student |
+----+--------+-----+---------+
```

可以看到，现在表中有id=11的记录。

下面使用DELETE语句删除该记录，SQL语句如下：

```
mysql> DELETE FROM person WHERE id = 11;
Query OK, 1 row affected (0.02 sec)
```

语句执行完毕后，查看删除结果：

```
mysql> SELECT * FROM person WHERE id=11;
Empty set (0.00 sec)
```

查询结果为空，说明删除操作成功。

【例6.12】 在表person中，使用DELETE语句同时删除多条记录。在【例6.10】中，使用UPDATE语句将age字段值为19~22的记录的info字段值修改为student，这里删除这些记录。

执行删除操作前，使用SELECT语句查看当前的数据，SQL语句如下：

```
mysql> SELECT * FROM person WHERE age BETWEEN 19 AND 22;
+----+---------+-----+---------+
| id | name    | age | info    |
+----+---------+-----+---------+
|  1 | Green   |  20 | student |
|  2 | Suse    |  21 | student |
|  4 | Willam  |  22 | student |
|  7 | Dale    |  22 | student |
|  9 | Harry   |  21 | student |
| 10 | Harriet |  19 | student |
+----+---------+-----+---------+
```

可以看到，age字段值为19~22的记录存在表中。

下面使用DELETE删除这些记录，SQL语句如下：

```
mysql> DELETE FROM person WHERE age BETWEEN 19 AND 22;
Query OK, 6 rows affected (0.00 sec)
```

语句执行完毕，查看删除结果：

```
mysql> SELECT * FROM person WHERE age BETWEEN 19 AND 22;
Empty set (0.00 sec)
```

查询结果为空，删除多条记录成功。

【例6.13】 删除表person中所有记录。

执行删除操作前，使用SELECT语句查看当前的数据，SQL语句如下：

```
mysql> SELECT * FROM person;
+----+---------+-----+-----------+
| id | name    | age | info      |
+----+---------+-----+-----------+
|  3 | Mary    |  24 | Musician  |
|  5 | Laura   |  25 | NULL      |
|  6 | Evans   |  27 | secretary |
| 12 | Beckham |  31 | police    |
+----+---------+-----+-----------+
```

结果显示表person中还有4条记录。执行DELETE语句删除这4条记录，SQL语句如下：

```
mysql> DELETE FROM person;
Query OK, 4 rows affected (0.00 sec)
```

语句执行完毕后，查看删除结果：

```
mysql> SELECT * FROM person;
Empty set (0.00 sec)
```

查询结果为空，说明删除表中所有记录成功，现在表person中已经没有任何数据记录了。

> **提示** 如果想删除表中的所有记录，还可以使用TRUNCATE TABLE语句。TRUNCATE将直接删除原来的表，并重新创建一张表，其语法结构为TRUNCATE TABLE table_name。TRUNCATE直接删除表而不是删除记录，因此执行速度比DELETE快。

6.4 为表增加计算列

什么叫计算列呢？简单来说就是某一列的值是通过别的列计算得来的。例如，a列值为1、b列值为2，c列不需要手动插入，定义a + b的结果为c的值，那么c就是计算列，是通过别的列计算得来的。

增加计算列的语法格式如下：

```
col_name data_type [GENERATED ALWAYS] AS (expression)
      [VIRTUAL | STORED] [UNIQUE [KEY]] [COMMENT comment]
      [NOT NULL | NULL] [[PRIMARY] KEY]
```

在MySQL 9.0中，CREAE TABLE和ALTER TABLE都支持增加计算列。下面以CREAE TABLE为例进行讲解。

【例6.14】定义数据表tb1，然后定义字段id、字段a、字段b和字段c，其中字段c为计算列，用于计算a+b的值。

首先创建测试表tb1，SQL语句如下：

```
CREATE TABLE tb1(
id int(9) NOT NULL AUTO_INCREMENT,
a int(9) DEFAULT NULL,
b int(9) DEFAULT NULL,
c int(9) GENERATED ALWAYS AS ((a + b)) VIRTUAL,
PRIMARY KEY (`id`)
);
```

插入演示数据，SQL语句如下：

```
insert into tb1(a,b) values (100,200);
```

查询数据表tb1中的数据，结果如下：

```
mysql> SELECT * FROM tb1;
+----+------+------+------+
| id | a    | b    | c    |
+----+------+------+------+
|  1 | 100  | 200  | 300  |
+----+------+------+------+
```

更新数据表tb1中的数据，SQL语句如下：

```
mysql> UPDATE tb1 SET a=500;
Query OK, 1 row affected (0.01 sec)
Rows matched: 1  Changed: 1  Warnings: 0
```

再次查看数据表tb1中的数据，结果如下：

```
mysql> SELECT * FROM tb1;
+----+------+------+------+
```

```
| id | a    | b    | c    |
+----+------+------+------+
| 1  | 500  | 200  | 700  |
+----+------+------+------+
```

从结果中可以看出，字段c中的数据始终是字段a和字段b的和，随着字段a和字段b中的数据变化，自动重新计算a+b的值。

6.5 DDL的原子化

从MySQL 9.0支持DDL（数据定义语言）的原子化。原子化DDL是指一个DDL操作是不可分割的，要么全成功要么全失败。它确保了在执行DDL操作时，如果其中一个操作失败，那么所有其他操作都将被回滚。这在处理复杂的数据库架构和高度可用的系统时尤为重要。DDL操作回滚日志写入数据字典表mysql.innodb_ddl_log（该表是隐藏的表，无法通过SHOW TABLES看到）中，用于回滚操作。通过设置参数，可将DDL操作日志打印输出到MySQL错误日志中。

下面通过案例来对比不同版本中DDL操作的区别。

分别在MySQL 5.7版本和MySQL 9.0版本中创建数据库和数据表，SQL语句如下：

```
CREATE DATABASE mytest;
USE mytest;

CREATE TABLE bk1
(
bookid            INT NOT NULL,
bookname          VARCHAR(255)
);
mysql> SHOW TABLES;
+------------------+
| Tables_in_mytest |
+------------------+
| bk1              |
+------------------+
```

（1）在MySQL 5.7版本中，测试步骤如下：

删除数据表bk1和数据表bk2，结果如下：

```
mysql> DROP TABLE BK1,BK2;
ERROR 1051 (42S02): Unknown table 'mytest.bk2'
```

再次查询数据库中的数据表名称，结果如下：

```
mysql> SHOW TABLES;
Empty set (0.00 sec)
```

从结果可以看出，虽然执行删除操作时报错了，但是仍然删除了数据表bk1。

（2）在MySQL 9.0版本中，测试步骤如下：

删除数据表bk1和数据表bk2，结果如下：

```
mysql> DROP TABLE bk1,bk2;
ERROR 1051 (42S02): Unknown table 'mytest.bk2'
```

再次查询数据库中的数据表名称，结果如下：

```
mysql> SHOW TABLES;
+------------------+
| Tables_in_mytest |
+------------------+
| bk1              |
+------------------+
```

从结果可以看出，数据表bk1并没有被删除，因为执行删除操作时报错了，删除操作被回滚了。

第 7 章
索引的设计和使用

索引用于快速找出在某个列中有一特定值的行。不使用索引,MySQL必须从第1条记录开始读完整张表,直到找出相关的行。表越大,查询数据所花费的时间越多。如果表中查询的列有一个索引,MySQL就能通过索引快速到达某个位置去搜寻数据文件,而不必查看所有数据。本章将介绍与索引相关的内容,包括索引的含义和特点、索引的分类、索引的设计原则以及如何创建和删除索引。

7.1 索引简介

索引是对数据库表中一列或多列的值进行排序的一种结构,使用索引可提高数据库中特定数据的查询效率。本节将介绍索引的含义、分类和设计原则。

7.1.1 索引的含义和特点

索引是一个单独的、存储在磁盘上的数据库结构,包含了对数据表里所有记录的引用指针。使用索引可以快速找出在某个或多个列中有一特定值的行,所有MySQL列类型都可以被索引。对相关列使用索引是加快查询操作速度的最佳途径。

例如,数据库中有2万条记录,现在要执行一个查询"SELECT * FROM table where num=10000",如果没有索引,就必须遍历整张表,直到num等于10000的这一行被找到为止;如果在num列上创建索引,MySQL不需要任何扫描,直接在索引里面找10000,就可以得知这一行的位置。可见,索引的建立可以加快数据库的查询速度。

索引是在存储引擎中实现的,每种存储引擎的索引都不一定完全相同,并且每种存储引擎也不一定支持所有索引类型。根据存储引擎来定义每张表的最大索引数和最大索引长度。所有存储引擎支持每张表至少16个索引,总索引长度至少为256字节。大多数存储引擎有更高的限制。MySQL中索引的存储类型有两种,即BTREE和HASH,具体和表的存储引擎相关:

MyISAM和InnoDB存储引擎只支持BTREE索引；MEMORY/HEAP存储引擎可以支持HASH和BTREE索引。

索引的优点主要有以下4点：

（1）通过创建唯一索引，可以保证数据库表中每一行数据的唯一性。
（2）可以大大加快数据的查询速度，这也是创建索引的主要原因。
（3）在实现数据的参考完整性方面，可以加速表和表之间的连接。
（4）在使用分组和排序子句进行数据查询时，可以显著减少查询中分组和排序的时间。

增加索引也有许多不利的方面，主要表现在如下3个方面：

（1）创建索引和维护索引要耗费时间，并且随着数据量的增加，所耗费的时间也会增加。
（2）索引需要占磁盘空间：除了数据表占数据空间之外，每一个索引还要占一定的物理空间。如果有大量的索引，则索引文件可能比数据文件更快达到最大文件尺寸。
（3）当对表中的数据进行增加、删除和修改时，索引也要动态地维护，这样就降低了数据的维护速度。

7.1.2 索引的分类

MySQL的索引可以分为以下几类。

1. 普通索引和唯一索引

普通索引是MySQL中的基本索引类型，允许在定义索引的列中插入重复值和空值。

唯一索引要求索引列的值必须唯一，但允许有空值。如果是组合索引，则列值的组合必须唯一。主键索引是一种特殊的唯一索引，不允许有空值。

2. 单列索引和组合索引

单列索引即一个索引只包含单个列，一张表可以有多个单列索引。

组合索引是指在表的多个字段组合上创建的索引，只有在查询条件中使用了这些字段的左边字段时，索引才会被使用。使用组合索引时遵循"最左前缀"原则。

3. 全文索引

全文索引类型为FULLTEXT，在定义索引的列上支持值的全文查找，允许在这些索引列中插入重复值和空值。全文索引可以在CHAR、VARCHAR或者TEXT类型的列上创建。MySQL中只有MyISAM存储引擎支持全文索引。

4. 空间索引

空间索引是对空间数据类型的字段建立的索引，MySQL中的空间数据类型有4种，分别是GEOMETRY、POINT、LINESTRING和POLYGON。MySQL使用SPATIAL关键字进行扩展，使得能够用与创建正规索引类似的语法创建空间索引。创建空间索引的列，必须声明为NOT NULL，空间索引只能在存储引擎为MyISAM的表中创建。

7.1.3 索引的设计原则

索引设计不合理或者缺少索引都会对数据库和应用程序的性能造成障碍。高效的索引对于获得良好的性能非常重要。设计索引时，应该考虑以下准则：

（1）索引并非越多越好，一张表中如有大量的索引，不仅占用磁盘空间，还会影响INSERT、DELETE、UPDATE等语句的性能，因为表中的数据更改时，索引也会进行调整和更新。

（2）避免对经常更新的表进行过多索引，并且索引中的列要尽可能少。应该为经常用于查询的字段创建索引，但要避免添加不必要的字段。

（3）数据量小的表最好不要使用索引，当数据较少时，查询花费的时间可能比遍历索引的时间还要短，因此索引可能不会产生优化效果。

（4）在条件表达式中经常用到的不同值较多的列上建立索引，在不同值很少的列上不要建立索引。比如学生表的"性别"字段上只有"男"与"女"两个不同值，因此无须建立索引，如果建立索引，不但不会提高查询效率，反而会严重降低数据更新速度。

（5）当唯一性是某种数据本身的特征时，指定唯一索引。使用唯一索引时要确保定义的列的数据完整性，以提高查询速度。

（6）在频繁进行排序或分组（即进行GROUP BY或ORDER BY操作）的列上建立索引，如果待排序的列有多个，可以在这些列上建立组合索引。

7.2 创建索引

MySQL支持多种方法在单个或多个列上创建索引：在创建表的语句CREATE TABLE中指定索引列、使用ALTER TABLE语句在已经存在的表上创建索引，或者使用CREATE INDEX语句在已经存在的表上添加索引。本节将详细介绍这3种方法。

7.2.1 创建表的时候创建索引

使用CREATE TABLE语句创建表时，除了可以定义列的数据类型，还可以定义主键约束、外键约束或者唯一性约束。而不论定义哪种约束，在定义约束的同时相当于在指定列上创建了一个索引。创建表时创建索引的基本语法格式如下：

```
CREATE TABLE table_name [col_name data_type]
[UNIQUE|FULLTEXT|SPATIAL] [INDEX|KEY] [index_name] (col_name [length]) [ASC | DESC]
```

UNIQUE、FULLTEXT和SPATIAL为可选参数，分别表示唯一索引、全文索引和空间索引；INDEX与KEY为同义词，两者作用相同，用来指定创建索引；col_name为需要创建索引的字段列，该列必须从数据表中定义的多个列中选择；index_name指定索引的名称，为可选参数，如果不指定，MySQL默认col_name为索引值；length为可选参数，表示索引的长度，只有字符串类型的字段才能指定索引长度；ASC或DESC指定升序或者降序的索引值存储。

1. 创建普通索引

普通索引是最基本的索引类型，没有唯一性之类的限制，其作用只是加快对数据的访问速度。

【例7.1】 在表book中的year_publication字段上建立普通索引，SQL语句如下：

```sql
CREATE TABLE book
(
bookid              INT NOT NULL,
bookname            VARCHAR(255) NOT NULL,
authors             VARCHAR(255) NOT NULL,
info                VARCHAR(255) NULL,
comment             VARCHAR(255) NULL,
year_publication    YEAR NOT NULL,
INDEX(year_publication)
);
```

该语句执行完毕之后，使用SHOW CREATE TABLE查看表结构：

```
mysql> SHOW CREATE table book \G
*************************** 1. row ***************************
       Table: book
Create Table: CREATE TABLE `book` (
  `bookid` int NOT NULL,
  `bookname` varchar(255) NOT NULL,
  `authors` varchar(255) NOT NULL,
  `info` varchar(255) DEFAULT NULL,
  `comment` varchar(255) DEFAULT NULL,
  `year_publication` year(4) NOT NULL,
  KEY `year_publication` (`year_publication`)
) ENGINE=InnoDB DEFAULT CHARSET=utf8mb4 COLLATE=utf8mb4_0900_ai_ci
```

由结果可以看到，表book的year_publication字段上已成功建立了索引，其索引名称year_publication为MySQL自动添加的。

使用EXPLAIN语句查看索引是否正在使用：

```
mysql> EXPLAIN SELECT * FROM book WHERE year_publication=1990 \G
*************************** 1. row ***************************
           id: 1
  select_type: SIMPLE
        table: book
         type: ref
possible_keys: year_publication
          key: year_publication
      key_len: 1
          ref: const
         rows: 1
        Extra: Using index condition
```

EXPLAIN语句输出结果的各行解释如下:

(1) id行是查询结果的编号。

(2) select_type行指定所使用的SELECT查询类型,这里值为SIMPLE,表示简单的SELECT,不使用UNION或子查询。其他可能的取值有PRIMARY、UNION、SUBQUERY等。

(3) table行指定数据库读取的数据表的名字,它们按被读取的先后顺序排列。

(4) type行指定本数据表与其他数据表之间的关联关系,可能的取值有system、const、eq_ref、ref、range、index和 All。

(5) possible_keys行给出了MySQL在搜索数据记录时可选用的各个索引。

(6) key行是MySQL实际选用的索引。

(7) key_len行给出索引按字节计算的长度,key_len数值越小,表示查询速度越快。

(8) ref行给出了关联关系中另一个数据表里的数据列名。

(9) rows行是MySQL在执行这个查询时预计会从这个数据表里读出的数据行的个数。

(10) Extra行提供了与关联操作有关的信息。

可以看到,possible_keys和key的值都为year_publication,表明查询时使用了索引。

2. 创建唯一索引

创建唯一索引的主要原因是减少查询索引列操作的执行时间,尤其当数据表比较庞大时。它与前面的普通索引类似,不同之处在于:索引列的值必须唯一,但允许有空值。如果是组合索引,则列值的组合必须唯一。

【例7.2】创建一张表t1,在表中的id字段上使用UNIQUE关键字创建唯一索引,SQL语句如下:

```sql
CREATE TABLE t1
(
id    INT NOT NULL,
name CHAR(30) NOT NULL,
UNIQUE INDEX UniqIdx(id)
);
```

该语句执行完毕之后,使用SHOW CREATE TABLE查看表结构:

```
mysql> SHOW CREATE table t1 \G
*************************** 1. row ***************************
       Table: t1
Create Table: CREATE TABLE `t1` (
  `id` int NOT NULL,
  `name` char(30) NOT NULL,
  UNIQUE KEY `UniqIdx` (`id`)
) ENGINE=InnoDB DEFAULT CHARSET=utf8mb4 COLLATE=utf8mb4_0900_ai_ci
```

由结果可以看到,id字段上已经成功建立了一个名为UniqIdx的唯一索引。

3. 创建单列索引

单列索引是在数据表中的某一个字段上创建的索引，一张表中可以创建多个单列索引。前面两个例子中创建的索引都为单列索引。

【例7.3】 创建一张表t2，在表中的name字段上创建单列索引，SQL语句如下：

```
CREATE TABLE t2
(
id   INT NOT NULL,
name CHAR(50) NULL,
INDEX SingleIdx(name(20))
);
```

该语句执行完毕之后，使用SHOW CREATE TABLE查看表结构：

```
mysql> SHOW CREATE table t2 \G
*************************** 1. row ***************************
       Table: t2
Create Table: CREATE TABLE `t2` (
  `id` int NOT NULL,
  `name` char(50) DEFAULT NULL,
  KEY `SingleIdx` (`name`(20))
) ENGINE=InnoDB DEFAULT CHARSET=utf8mb4 COLLATE=utf8mb4_0900_ai_ci
```

由结果可以看到，id字段上已经成功建立了一个名为SingleIdx的单列索引，索引长度为20。

4. 创建组合索引

组合索引是在多个字段上创建一个索引。

【例7.4】 创建表t3，在表中的id、name和age字段上建立组合索引，SQL语句如下：

```
CREATE TABLE t3
(
id   INT NOT NULL,
name CHAR(30) NOT NULL,
age  INT NOT NULL,
info VARCHAR(255),
INDEX MultiIdx(id, name, age)
);
```

该语句执行完毕之后，使用SHOW CREATE TABLE查看表结构：

```
mysql> SHOW CREATE table t3 \G
*** 1. row ***
       Table: t3
CREATE Table: CREATE TABLE `t3` (
  `id` int NOT NULL,
  `name` char(30) NOT NULL,
  `age` int NOT NULL,
  `info` varchar(255) DEFAULT NULL,
```

```
    KEY `MultiIdx` (`id`,`name`,`age`)
) ENGINE=InnoDB DEFAULT CHARSET=utf8mb4 COLLATE=utf8mb4_0900_ai_ci
```

由结果可以看到，id、name和age字段上已经成功建立了一个名为MultiIdx的组合索引。

组合索引可以起到几个索引的作用，但并不是随便查询哪个字段都可以使用组合索引，而是遵从"最左前缀"原则：利用索引中最左边的列集来匹配行，这样的列集称为最左前缀。例如，这里的组合索引由id、name和age三个字段构成，索引行中按id、name、age的顺序存放，那么MySQL可以搜索（id, name, age）、（id, name）或者id字段组合。如果列不构成索引最左面的前缀，那么MySQL不能使用局部索引，如（age）或者（name,age）组合不能使用索引。

在表t3中，查询id和name字段，使用EXPLAIN语句查看索引的使用情况：

```
mysql> EXPLAIN SELECT * FROM t3 WHERE id=1 AND name='joe' \G
*** 1. row ***
           id: 1
  select_type: SIMPLE
        table: t3
         type: ref
possible_keys: MultiIdx
          key: MultiIdx
      key_len: 94
          ref: const,const
         rows: 1
        Extra: Using where
```

可以看到，查询id和name字段时，使用了名称为"MultiIdx"的索引。如果查询（name,age）组合或者单独查询name和age字段，则结果如下：

```
*** 1. row ***
           id: 1
  select_type: SIMPLE
        table: t3
         type: ALL
possible_keys: NULL
          key: NULL
      key_len: NULL
          ref: NULL
         rows: 1
        Extra: Using where
```

此时，possible_keys和key的值均为NULL，并没有使用在表t3中创建的索引进行查询。

5. 创建全文索引

FULLTEXT索引可以用于全文搜索。只有MyISAM存储引擎支持FULLTEXT索引，并且只为CHAR、VARCHAR和TEXT列创建全文索引。全文索引总是对整个列进行索引，不支持局部（前缀）索引。

【例7.5】创建表t4，在表中的info字段上建立全文索引，SQL语句如下：

```
CREATE TABLE t4
```

```
(
id    INT NOT NULL,
name CHAR(30) NOT NULL,
age  INT NOT NULL,
info VARCHAR(255),
FULLTEXT INDEX FullTxtIdx(info)
) ENGINE=MyISAM;
```

> **提示** 因为MySQL 9.0中默认存储引擎为InnoDB,所以创建表时需要修改表的存储引擎为MyISAM,否则创建全文索引会出错。

语句执行完毕之后,使用SHOW CREATE TABLE查看表结构:

```
mysql> SHOW CREATE table t4 \G
*************************** 1. row ***************************
       Table: t4
Create Table: CREATE TABLE `t4` (
  `id` int NOT NULL,
  `name` char(30) NOT NULL,
  `age` int NOT NULL,
  `info` varchar(255) DEFAULT NULL,
  FULLTEXT KEY `FullTxtIdx` (`info`)
) ENGINE=MyISAM DEFAULT CHARSET=utf8mb4 COLLATE=utf8mb4_0900_ai_ci
```

由结果可以看到,info字段上已经成功建立了一个名为"FullTxtIdx"的全文索引。全文索引非常适合于大型数据集,对于小的数据集,它的用处比较小。

6. 创建空间索引

空间索引必须在MyISAM类型的表中创建,且空间类型的字段必须非空。

【例7.6】 创建表t5,在空间类型为GEOMETRY的字段上创建空间索引,SQL语句如下:

```
CREATE TABLE t5
(
g GEOMETRY NOT NULL,
SPATIAL INDEX spatIdx(g)
)ENGINE=MyISAM;
```

该语句执行完毕之后,使用SHOW CREATE TABLE查看表结构:

```
mysql> SHOW CREATE table t5 \G
*** 1. row ***
       Table: t5
CREATE Table: CREATE TABLE `t5` (
  `g` geometry NOT NULL,
  SPATIAL KEY `spatIdx` (`g`)
) ENGINE=MyISAM DEFAULT CHARSET=utf8mb4 COLLATE=utf8mb4_0900_ai_ci
```

可以看到,表t5的g字段上创建了名称为"spatIdx"的空间索引。注意在创建时指定空间类型字段值的非空约束,并且表的存储引擎为MyISAM。

7.2.2 在已经存在的表上创建索引

在已经存在的表上创建索引，可以使用ALTER TABLE语句或者CREATE INDEX语句。下面详细如何使用ALTER TABLE和CREATE INDEX语句在已知表的字段上创建索引。

1. 使用ALTER TABLE语句创建索引

使用ALTER TABLE语句创建索引的基本语法如下：

```
ALTER TABLE table_name  ADD [UNIQUE|FULLTEXT|SPATIAL]  [INDEX|KEY]
[index_name] (col_name[length],...) [ASC | DESC]
```

与创建表时创建索引的语法不同的是，在这里使用了ALTER TABLE和ADD关键字，ADD表示向表中添加索引。

【例7.7】在表book中的bookname字段上建立名为"BkNameIdx"的普通索引。

添加索引之前，使用SHOW INDEX语句查看指定表中创建的索引：

```
mysql> SHOW INDEX FROM book \G
*** 1. Row ***
        Table: book
   Non_unique: 1
     Key_name: year_publication
 Seq_in_index: 1
  Column_name: year_publication
    Collation: A
  Cardinality: 0
     Sub_part: NULL
       Packed: NULL
         Null:
   Index_type: BTREE
      Comment:
Index_comment:
```

其中各个主要参数的含义如下：

（1）Table表示创建索引的表。

（2）Non_unique表示索引非唯一，1代表是非唯一索引，0代表是唯一索引。

（3）Key_name表示索引的名称。

（4）Seq_in_index表示该字段在索引中的位置，如果是单列索引，则该值为1；组合索引，则为每个字段在索引定义中的顺序。

（5）Column_name表示定义索引的列字段。

（6）Sub_part表示索引的长度。

（7）Null表示该字段是否能为空值。

（8）Index_type表示索引类型。

可以看到，表book中已经存在了一个索引，即前面定义的名称为"year_publication"的索引，该索引为非唯一索引。

下面使用ALTER TABLE 在bookname字段上添加索引，SQL语句如下：

```
ALTER TABLE book ADD INDEX BkNameIdx( bookname(30) );
```

使用SHOW INDEX语句查看表中的索引：

```
mysql> SHOW INDEX FROM book \G
*** 1. Row ***
        Table: book
   Non_unique: 1
     Key_name: year_publication
 Seq_in_index: 1
  Column_name: year_publication
    Collation: A
  Cardinality: 0
     Sub_part: NULL
       Packed: NULL
         Null:
   Index_type: BTREE
      Comment:
Index_comment:
*** 2. Row ***
        Table: book
   Non_unique: 1
     Key_name: BkNameIdx
 Seq_in_index: 1
  Column_name: bookname
    Collation: A
  Cardinality: 0
     Sub_part: 30
       Packed: NULL
         Null:
   Index_type: BTREE
      Comment:
Index_comment:
```

可以看到，现在表中有了两个索引，一个为year_publication；另一个为通过ALTER TABLE语句添加的名称为"BkNameIdx"的索引，该索引为非唯一索引，长度为30。

【例7.8】 在表book的bookId字段上建立名称为"UniqidIdx"的唯一索引，SQL语句如下：

```
ALTER TABLE book ADD UNIQUE INDEX UniqidIdx ( bookId );
```

使用SHOW INDEX语句查看表中的索引：

```
mysql> SHOW INDEX FROM book \G
*** 1. Row ***
        Table: book
   Non_unique: 0
     Key_name: UniqidIdx
 Seq_in_index: 1
  Column_name: bookid
    Collation: A
```

```
    Cardinality: 0
      Sub_part: NULL
       Packed: NULL
         Null:
    Index_type: BTREE
       Comment:
Index_comment:
```

可以看到Non_unique属性值为0，表示名称为UniqidIdx的索引为唯一索引，说明创建唯一索引成功。

【例7.9】 在表book的comment字段上建立单列索引，SQL语句如下：

```
ALTER TABLE book ADD INDEX BkcmtIdx ( comment(50) );
```

使用SHOW INDEX语句查看表中的索引：

```
*** 3. Row ***
        Table: book
   Non_unique: 1
     Key_name: BkcmtIdx
 Seq_in_index: 1
  Column_name: comment
    Collation: A
  Cardinality: 0
     Sub_part: 50
       Packed: NULL
         Null: YES
   Index_type: BTREE
      Comment:
Index_comment:
```

可以看到，在表book的comment字段上建立了名称为"BkcmtIdx"的索引，长度为50，在查询时，只需要检索前50个字符。

【例7.10】 在表book的authors和info字段上建立组合索引，SQL语句如下：

```
ALTER TABLE book ADD INDEX BkAuAndInfoIdx ( authors(30),info(50) );
```

使用SHOW INDEX语句查看表中的索引：

```
mysql> SHOW INDEX FROM book \G
*** 4. Row ***
        Table: book
   Non_unique: 1
     Key_name: BkAuAndInfoIdx
 Seq_in_index: 1
  Column_name: authors
    Collation: A
  Cardinality: 0
     Sub_part: 30
       Packed: NULL
         Null:
```

```
        Index_type: BTREE
          Comment:
    Index_comment:
    *** 5. Row ***
            Table: book
       Non_unique: 1
         Key_name: BkAuAndInfoIdx
     Seq_in_index: 2
      Column_name: info
        Collation: A
      Cardinality: 0
         Sub_part: 50
           Packed: NULL
             Null: YES
       Index_type: BTREE
          Comment:
    Index_comment:
```

可以看到,名称为"BkAuAndInfoIdx"的索引由两个字段组成,authors字段长度为30,在组合索引中的序号为1,该字段不允许为NULL;info字段长度为50,在组合索引中的序号为2,该字段可以为NULL。

【例7.11】创建表t6,并在表t6上使用ALTER TABLE创建全文索引。

首先创建表t6,SQL语句如下:

```
CREATE TABLE t6
(
id    INT NOT NULL,
info  CHAR(255)
) ENGINE=MyISAM;
```

注意,需要修改ENGINE参数为MyISAM,MySQL的默认存储引擎InnoDB不支持全文索引。

然后使用ALTER TABLE语句在info字段上创建全文索引:

```
ALTER TABLE t6 ADD FULLTEXT INDEX infoFTIdx ( info );
```

最后使用SHOW INDEX语句查看索引:

```
mysql> SHOW index from t6 \G
*** 1. Row ***
        Table: t6
   Non_unique: 1
     Key_name: infoFTIdx
 Seq_in_index: 1
  Column_name: info
    Collation: NULL
  Cardinality: NULL
     Sub_part: NULL
       Packed: NULL
         Null: YES
```

```
    Index_type: FULLTEXT
      Comment:
ndex_comment:
```

可以看到，表t6中已经创建了名称为"infoFTIdx"的索引，该索引在info字段上创建，类型为FULLTEXT，允许为NULL。

【例7.12】 创建表t7，并在表t7的空间数据类型字段g上创建名称为"spatIdx"的空间索引。首先创建表t7，SQL语句如下：

```
CREATE TABLE t7 ( g GEOMETRY NOT NULL )ENGINE=MyISAM;
```

然后使用ALTER TABLE在表t7的g字段建立空间索引：

```
ALTER TABLE t7 ADD SPATIAL INDEX spatIdx(g);
```

最后使用SHOW INDEX语句查看索引：

```
mysql> SHOW index from t7 \G
*** 1. Row ***
        Table: t7
   Non_unique: 1
     Key_name: spatIdx
 Seq_in_index: 1
  Column_name: g
    Collation: A
  Cardinality: NULL
     Sub_part: 32
       Packed: NULL
         Null:
   Index_type: SPATIAL
      Comment:
Index_comment:
```

由结果可以看到，表t7的g字段上创建了名称为"spatIdx"的空间索引。

2. 使用CREATE INDEX语句创建索引

使用CREATE INDEX语句可以在已经存在的表上添加索引。在MySQL中，CREATE INDEX被映射到一个ALTER TABLE语句上，基本语法结构为：

```
CREATE [UNIQUE|FULLTEXT|SPATIAL] INDEX index_name
ON table_name (col_name[length],...) [ASC | DESC]
```

可以看到，CREATE INDEX语句和ALTER INDEX语句的语法基本一样，只是关键字不同。在这里，使用相同的表book，假设该表中没有任何索引值，创建表book的SQL语句如下：

```
CREATE TABLE book
(
bookid              INT NOT NULL,
bookname            VARCHAR(255) NOT NULL,
authors             VARCHAR(255) NOT NULL,
info                VARCHAR(255) NULL,
```

```
comment              VARCHAR(255) NULL,
year_publication     YEAR NOT NULL
);
```

> **提示** 读者可以将数据库中的表book删除，然后按上面的语句重新建立，再进行下面的操作。

【例7.13】在表book中的bookname字段上建立名称为"BkNameIdx"的普通索引，SQL语句如下：

```
CREATE INDEX BkNameIdx ON book(bookname);
```

语句执行完毕之后，将在表book中创建名称为"BkNameIdx"的普通索引。读者可以使用SHOW INDEX或者SHOW CREATE TABLE语句查看表book中的索引，其索引内容与前面介绍的相同。

【例7.14】在表book的bookId字段上建立名称为"UniqidIdx"的唯一索引，SQL语句如下：

```
CREATE UNIQUE INDEX UniqidIdx  ON book ( bookId );
```

语句执行完毕之后，将在表book中创建名称为"UniqidIdx"的唯一索引。

【例7.15】在表book的comment字段上建立单列索引，SQL语句如下：

```
CREATE INDEX BkcmtIdx ON book(comment(50) );
```

语句执行完毕之后，将在表book的comment字段上建立一个名称为"BkcmtIdx"的单列索引，长度为50。

【例7.16】在表book的authors和info字段上建立组合索引，SQL语句如下：

```
CREATE INDEX BkAuAndInfoIdx ON book ( authors(20),info(50) );
```

语句执行完毕之后，在表book的authors和info字段上建立了一个名称为"BkAuAndInfoIdx"的组合索引，authors的索引序号为1、长度为20，info的索引序号为2、长度为50。

【例7.17】先删除表t6，再重新建立表t6，在表t6中使用CREATE INDEX语句在CHAR类型的info字段上创建全文索引。

首先删除表t6，并重新建立该表，SQL语句如下：

```
mysql> drop table t6;
Query OK, 0 rows affected (0.00 sec)

mysql> CREATE TABLE t6
    (
    id    INT NOT NULL,
    info  CHAR(255)
    ) ENGINE=MyISAM;
Query OK, 0 rows affected (0.00 sec)
```

然后使用CREATE INDEX在表t6的info字段上创建名称为"infoFTIdx"的全文索引：

```
CREATE FULLTEXT INDEX infoFTIdx ON t6(info);
```

语句执行完毕之后，将在表t6中创建名称为"infoFTIdx"的索引，该索引在info字段上创建，类型为FULLTEXT，允许为NULL。

【例7.18】 删除表t7，重新创建表t7，在表t7中使用CREATE INDEX语句在空间数据类型字段g上创建名称为"spatIdx"的空间索引：

首先删除表t7，并重新建立该表，SQL语句如下：

```
mysql> drop table t7;
Query OK, 0 rows affected (0.00 sec)

mysql> CREATE TABLE t7 ( g GEOMETRY NOT NULL )ENGINE=MyISAM;
Query OK, 0 rows affected (0.00 sec)
```

然后使用CREATE INDEX语句在表t7的g字段建立空间索引：

```
CREATE SPATIAL INDEX spatIdx ON t7 (g);
```

语句执行完毕之后，将在表t7中创建名称为"spatIdx"的空间索引，该索引在g字段上创建。

7.3 删除索引

MySQL中删除索引使用ALTER TABLE或者DROP INDEX语句，两者可实现相同的功能，DROP INDEX语句在内部被映射到一个ALTER TABLE语句中。

1. 使用ALTER TABLE删除索引

使用ALTER TABLE删除索引的基本语法格式如下：

```
ALTER TABLE table_name DROP INDEX index_name;
```

【例7.19】 删除表book中名称为"UniqidIdx"的唯一索引。

首先查看表book中是否有名称为"UniqidIdx"的索引，SQL语句如下：

```
mysql> SHOW CREATE table book \G
*** 1. row ***
       Table: book
CREATE Table: CREATE TABLE `book` (
  `bookid` int NOT NULL,
  `bookname` varchar(255) NOT NULL,
  `authors` varchar(255) NOT NULL,
  `info` varchar(255) DEFAULT NULL,
  `year_publication` year(4) NOT NULL,
  UNIQUE KEY `UniqidIdx` (`bookid`),
  KEY `BkNameIdx` (`bookname`),
  KEY `BkAuAndInfoIdx` (`authors`(20),`info`(50))
```

```
) ENGINE=InnoDB DEFAULT CHARSET=utf8mb4 COLLATE=utf8mb4_0900_ai_ci
```

从查询结果中可以看到，表book中有名称为"UniqidIdx"的唯一索引，该索引在bookid字段上创建。

下面删除该索引，SQL语句如下：

```
mysql> ALTER TABLE book DROP INDEX UniqidIdx;
```

语句执行完毕，使用SHOW语句查看索引是否被删除：

```
mysql> SHOW CREATE table book \G
*** 1. row ***
       Table: book
CREATE Table: CREATE TABLE `book` (
  `bookid` int NOT NULL,
  `bookname` varchar(255) NOT NULL,
  `authors` varchar(255) NOT NULL,
  `info` varchar(255) DEFAULT NULL,
  `year_publication` year(4) NOT NULL,
  KEY `BkNameIdx` (`bookname`),
  KEY `BkAuAndInfoIdx` (`authors`(20),`info`(50))
) ENGINE=InnoDB DEFAULT CHARSET=utf8mb4 COLLATE=utf8mb4_0900_ai_ci
```

由结果可以看到，表book中已经没有名称为"uniqidIdx"的唯一索引，说明删除索引成功。

> **提示** 添加AUTO_INCREMENT约束字段的唯一索引不能被删除。

2. 使用DROP INDEX语句删除索引

使用DROP INDEX删除索引的基本语法格式如下：

```
DROP INDEX index_name ON table_name;
```

【例7.20】 删除表book中名称为"BkAuAndInfoIdx"的组合索引，SQL语句如下：

```
mysql> DROP INDEX BkAuAndInfoIdx ON book;
```

语句执行完毕，使用SHOW语句查看索引是否被删除：

```
mysql> SHOW CREATE table book \G
*** 1. row ***
       Table: book
CREATE Table: CREATE TABLE `book` (
  `bookid` int NOT NULL,
  `bookname` varchar(255) NOT NULL,
  `authors` varchar(255) NOT NULL,
  `info` varchar(255) DEFAULT NULL,
  `year_publication` year(4) NOT NULL,
  KEY `BkNameIdx` (`bookname`)
) ENGINE=InnoDB DEFAULT CHARSET=utf8
1 row in set (0.00 sec)
```

可以看到，表book中已经没有名称为"BkAuAndInfoIdx"的组合索引，说明删除索引成功。

> **提示**：删除表中的列时，如果要删除的列为索引的组成部分，则该列也会从索引中删除。如果组成索引的所有列都被删除，则整个索引将被删除。

7.4 使用降序索引

在MySQL 8.0之前，MySQL虽然在语法上已经支持降序索引，但实际上创建的仍然是升序索引。而从MySQL 8.0开始，MySQL支持创建降序索引。下面通过案例来对比不同的版本中对降序索引的支持情况。

分别在MySQL 5.7版本和MySQL 9.0版本中创建数据表ts1：

```
mysql> CREATE TABLE ts1(a int,b int,index idx_a_b(a,b desc));
Query OK, 0 rows affected (0.13 sec)
```

在MySQL 5.7版本中查看数据表ts1的结构，结果如下：

```
mysql> show create table ts1\G
*************************** 1. row ***************************
       Table: ts1
Create Table: CREATE TABLE `ts1` (
  `a` int(11) DEFAULT NULL,
  `b` int(11) DEFAULT NULL,
  KEY `idx_a_b` (`a`,`b`)
) ENGINE=MyISAM DEFAULT CHARSET=latin1
```

从结果中可以看出，索引仍然是默认的升序。

在MySQL 9.0版本中查看数据表ts1的结构，结果如下：

```
mysql>show create table ts1\G
*************************** 1. row ***************************
       Table: ts1
Create Table: CREATE TABLE `ts1` (
  `a` int(11) DEFAULT NULL,
  `b` int(11) DEFAULT NULL,
  KEY `idx_a_b` (`a`,`b` DESC)
) ENGINE=InnoDB DEFAULT CHARSET=utf8mb4 COLLATE=utf8mb4_0900_ai_ci
```

从结果中可以看出，索引已经是降序了。

下面继续测试降序索引在执行计划中的表现。

分别在MySQL 5.7版本和MySQL 9.0版本的数据表ts1中插入8万条随机数据，SQL语句如下：

```
mysql> DELIMITER ;;
mysql> CREATE PROCEDURE ts_insert ()
mysql> BEGIN
mysql> DECLARE i INT DEFAULT 1;
mysql> WHILE i<80000
mysql> DO
```

```
mysql> insert into ts1 select rand()*80000, rand()*80000;
mysql> SET i=i+1;
mysql> END WHILE ;
mysql> commit;
mysql> END;;
mysql> DELIMITER ;
mysql> CALL ts_insert();
```

在MySQL 5.7版本中查看数据表ts1的执行计划,结果如下:

```
mysql> explain select * from ts1 order by a , b desc limit 5;
+----+-------------+-------+------------+-------+---------------+---------+---------+------+-------+----------+-----------------------------+
| id | select_type | table | partitions | type  | possible_keys | key     | key_len | ref  | rows  | filtered | Extra                       |
+----+-------------+-------+------------+-------+---------------+---------+---------+------+-------+----------+-----------------------------+
|  1 | SIMPLE      | ts1   | NULL       | index | NULL          | idx_a_b | 10      | NULL | 79999 |   100.00 | Using index; Using filesort |
+----+-------------+-------+------------+-------+---------------+---------+---------+------+-------+----------+-----------------------------+
```

从结果中可以看出,执行计划中扫描数为79999,而且使用了Using filesort。

> **提示** Using filesort是MySQL里一种速度比较慢的外部排序,如果能避免使用它,当然是最好的。多数情况下,管理员可以通过优化索引来尽量避免出现Using filesort,从而提高数据库执行速度。

在MySQL 9.0版本中查看数据表ts1的执行计划,结果如下:

```
mysql> explain select * from ts1 order by a , b desc limit 5;
+----+-------------+-------+------------+-------+---------------+---------+---------+------+------+----------+-------------+
| id | select_type | table | partitions | type  | possible_keys | key     | key_len | ref  | rows | filtered | Extra       |
+----+-------------+-------+------------+-------+---------------+---------+---------+------+------+----------+-------------+
|  1 | SIMPLE      | ts1   | NULL       | index | NULL          | idx_a_b | 10      | NULL |    5 |   100.00 | Using index |
+----+-------------+-------+------------+-------+---------------+---------+---------+------+------+----------+-------------+
```

从结果中可以看出,执行计划中扫描数为5,而且没有使用Using filesort。

> **注意** 降序索引只对查询中特定的排列顺序有效,如果使用不当,反而使查询效率更低。例如上述查询排序条件如果改为"order by a desc, b desc",则MySQL 5.7的执行计划要明显好于MySQL 9.0。

将排序条件修改为"order by a desc, b desc"后,下面来对比不同版本中的执行计划的效果。

在MySQL 5.7版本中查看数据表ts1的执行计划，结果如下：

```
mysql> explain select * from ts1 order by a desc , b desc limit 5;
+----+-------------+-------+------------+-------+---------------+---------+---------+------+------+----------+-------------+
| id | select_type | table | partitions | type  | possible_keys | key     | key_len | ref  | rows | filtered | Extra       |
+----+-------------+-------+------------+-------+---------------+---------+---------+------+------+----------+-------------+
|  1 | SIMPLE      | ts1   | NULL       | index | NULL          | idx_a_b | 10      | NULL |    5 |   100.00 | Using index |
+----+-------------+-------+------------+-------+---------------+---------+---------+------+------+----------+-------------+
```

在MySQL 9.0版本中查看数据表ts1的执行计划，结果如下：

```
mysql> explain select * from ts1 order by a desc , b desc limit 5;
+----+-------------+-------+------------+-------+---------------+---------+---------+------+-------+----------+-----------------------------+
| id | select_type | table | partitions | type  | possible_keys | key     | key_len | ref  | rows  | filtered | Extra                       |
+----+-------------+-------+------------+-------+---------------+---------+---------+------+-------+----------+-----------------------------+
|  1 | SIMPLE      | ts1   | NULL       | index | NULL          | idx_a_b | 10      | NULL | 80133 |   100.00 | Using index; Using filesort |
+----+-------------+-------+------------+-------+---------------+---------+---------+------+-------+----------+-----------------------------+
```

对比结果可以看出，修改排序条件后，MySQL 5.7的执行计划要明显好于MySQL 9.0。

第 8 章 存储过程和存储函数

简单地说,存储过程就是一条或者多条SQL语句的集合,可视为批文件,但是其作用不仅限于批处理。存储函数允许将业务逻辑或数据处理逻辑封装在数据库中,从而提高应用程序的性能和可维护性。本章主要介绍如何创建、调用、查看、修改、删除存储过程和存储函数等。

8.1 创建存储过程和存储函数

存储程序可以分为存储过程和存储函数。在MySQL中,创建存储过程和存储函数使用的语句分别是CREATE PROCEDURE和CREATE FUNCTION。使用CALL语句来调用存储过程,只能用输出变量返回值。存储函数可以从语句外调用(引用函数名),也能返回标量值。一个存储过程可以调用其他存储过程。

8.1.1 创建存储过程

创建存储过程,需要使用CREATE PROCEDURE语句,基本语法格式如下:

```
CREATE PROCEDURE sp_name ( [proc_parameter] )
[characteristics ...] routine_body
```

各参数解释如下:

(1)CREATE PROCEDURE为用来创建存储过程的关键字。
(2)sp_name为存储过程的名称。
(3)proc_parameter为指定存储过程的参数列表,列表形式如下:

```
[ IN | OUT | INOUT ] param_name type
```

其中,IN表示输入参数,OUT表示输出参数,INOUT表示既可以输入也可以输出;param_name表示参数名称;type表示参数的类型,该类型可以是MySQL数据库中的任意类型。

（4）characteristics指定存储过程的特性，有以下取值：

- LANGUAGE SQL：说明routine_body部分是由SQL语句组成的，当前系统支持的语言为SQL。SQL是LANGUAGE特性的唯一值。
- [NOT] DETERMINISTIC：指明存储过程执行的结果是否确定。DETERMINISTIC表示结果是确定的，每次执行存储过程时，相同的输入会得到相同的输出；NOT DETERMINISTIC表示结果是不确定的，相同的输入可能得到不同的输出。如果没有指定值，则默认为NOT DETERMINISTIC。
- { CONTAINS SQL | NO SQL | READS SQL DATA | MODIFIES SQL DATA }：指明子程序使用SQL语句的限制。CONTAINS SQL表明子程序包含SQL语句，但是不包含读写数据的语句；NO SQL表明子程序不包含SQL语句；READS SQL DATA说明子程序包含读数据的语句；MODIFIES SQL DATA表明子程序包含写数据的语句。默认情况下，系统会指定为CONTAINS SQL。
- SQL SECURITY { DEFINER | INVOKER }：指明谁有权限来执行。DEFINER表示只有定义者才能执行；INVOKER表示拥有权限的调用者可以执行。默认情况下，系统指定为DEFINER。
- COMMENT 'string'：注释信息，可以用来描述存储过程或存储函数。

（5）routine_body是SQL代码的内容，可以用BEGIN...END来表示SQL代码的开始和结束。

编写存储过程并不是一件简单的事情，可能需要复杂的SQL语句，并且要有创建存储过程的权限。但是，使用存储过程将简化操作，减少冗余的操作步骤；同时，还可以减少操作过程中的失误，提高效率。因此，存储过程是非常有用的，而且应该尽可能地学会使用。

下面的代码演示了一个存储过程的创建，其名称为"AvgFruitPrice"，返回所有水果的平均价格：

```
CREATE PROCEDURE AvgFruitPrice ()
BEGIN
SELECT AVG(f_price) AS avgprice
FROM fruits;
END;
```

上述代码中，名为"AvgFruitPrice"的存储过程使用CREATE PROCEDURE AvgFruitPrice()语句定义。此存储过程没有参数，但是后面的()仍然需要。BEGIN和END语句用来限定存储过程体，过程本身仅是一个简单的SELECT语句（AVG()为求字段平均值的函数）。

【例8.1】创建查看表fruits的存储过程，SQL语句如下：

```
CREATE PROCEDURE Proc()
    BEGIN
        SELECT * FROM fruits;
    END ;
```

上述代码创建了一个查看表fruits的存储过程，每次调用这个存储过程的时候都会执行SELECT语句查看表的内容，代码的执行过程如下：

```
MySQL> DELIMITER //
MySQL> CREATE PROCEDURE Proc()
    -> BEGIN
    -> SELECT * FROM fruits;
    -> END //
Query OK, 0 rows affected (0.00 sec)

MySQL> DELIMITER ;
```

这个存储过程和使用SELECT语句查看表的效果得到的结果是一样的。

当然，存储过程也可以是很多复杂语句的组合，其本身也可以调用其他的函数来组成更加复杂的操作。

> **提示** "DELIMITER //"语句的作用是将MySQL的结束符设置为"//"。因为MySQL默认的语句结束符号为分号（;），为了避免与存储过程中SQL语句结束符相冲突，需要使用DELIMITER改变存储过程的结束符，并以"END //"结束存储过程。存储过程定义完毕之后再使用"DELIMITER ;"恢复默认结束符。DELIMITER也可以指定其他符号作为结束符。

【例8.2】创建名称为"CountProc"的存储过程，SQL语句如下：

```
CREATE PROCEDURE CountProc (OUT param1 INT)
BEGIN
SELECT COUNT(*) INTO param1 FROM fruits;
END;
```

上述代码的作用是创建一个获取表fruits中的记录条数的存储过程，其名称是CountProc；COUNT(*)计算后把结果放入参数param1中。执行结果如下：

```
mysql> DELIMITER //
mysql> CREATE PROCEDURE CountProc(OUT param1 INT)
 -> BEGIN
 -> SELECT COUNT(*) INTO param1 FROM fruits;
 -> END;
 -> //
Query OK, 0 rows affected (0.00 sec)
mysql> DELIMITER ;
```

> **提示** 当使用DELIMITER命令时，应该避免使用反斜杠（\）字符，因为反斜杠是MySQL的转义字符。

8.1.2 创建存储函数

创建存储函数，需要使用CREATE FUNCTION语句，基本语法格式如下：

```
CREATE FUNCTION func_name ( [func_parameter] )
 RETURNS type
[characteristic ...] routine_body
```

各参数解释如下:

(1) CREATE FUNCTION为用来创建存储函数的关键字。
(2) func_name表示存储函数的名称。
(3) func_parameter为存储函数的参数列表,参数列表形式如下:

```
[ IN | OUT | INOUT ] param_name type
```

其中,IN表示输入参数,OUT表示输出参数,INOUT表示既可以输入也可以输出;param_name表示参数名称;type表示参数的类型,该类型可以是MySQL数据库中的任意类型。

(4) RETURNS type语句表示函数返回数据的类型。
(5) characteristic指定存储函数的特性,取值与创建存储过程时的相同,这里不再赘述。

【例8.3】创建存储函数,名称为NameByZip,该函数返回SELECT语句的查询结果,数值类型为字符串型,SQL语句如下:

```
CREATE FUNCTION NameByZip ()
RETURNS CHAR(50)
RETURN  (SELECT s_name FROM suppliers WHERE s_call= '48075');
```

上述语句创建了一个存储函数NameByZip(),参数定义为空,返回一个CHAR类型的结果。代码的执行结果如下:

```
mysql> set global log_bin_trust_function_creators=TRUE;
mysql> DELIMITER //
mysql> CREATE FUNCTION NameByZip()
-> RETURNS CHAR(50)
-> RETURN  (SELECT s_name FROM suppliers WHERE s_call= '48075');
-> //

mysql> DELIMITER ;
```

如果存储函数中的RETURN语句返回一个类型不同于函数的RETURNS子句中的指定类型的值,则返回值将被强制为恰当的类型。例如,如果一个函数返回一个ENUM或SET值,但是RETURN语句返回一个整数,则对于SET成员集相应的ENUM成员,MySQL会将从函数返回的整数值转换为字符串。

> 提示 指定参数为IN、OUT或INOUT只对PROCEDURE是合法的(FUNCTION中总是默认为IN参数)。RETURNS子句只能对FUNCTION进行指定,对函数而言这是强制的。它用来指定函数的返回类型,而且函数体必须包含一个RETURN value语句。

8.1.3 变量的使用

变量可以在子程序中声明并使用,这些变量的作用范围是在BEGIN...END程序中。本小节主要介绍如何定义变量和为变量赋值。

1. 定义变量

在存储过程中使用DECLARE语句定义变量,语法格式如下:

```
DECLARE var_name[,varname]... date_type [DEFAULT value];
```

var_name为局部变量的名称;DEFAULT value子句给变量提供一个默认值,该值除了可以被声明为一个常数之外,还可以被指定为一个表达式。如果没有DEFAULT子句,则初始值为NULL。

【例8.4】定义名称为"myparam"的变量,类型为INT类型,默认值为100,SQL语句如下:

```
DECLARE myparam INT DEFAULT 100;
```

2. 为变量赋值

定义变量之后,为变量赋值可以改变变量的默认值。在MySQL中,使用SET语句为变量赋值,语法格式如下:

```
SET var_name = expr [, var_name = expr] ...;
```

存储程序中的SET语句是一般SET语句的扩展版本。被参考变量可能是子程序内声明的变量或者是全局服务器变量,如系统变量或者用户变量。

存储程序中的SET语句作为预先存在的SET语法的一部分来实现,允许SET a=x, b=y, ...这样的扩展语法。其中,不同的变量类型(局部变量和全局变量)可以被混合起来。这也允许把局部变量和一些只对系统变量有意义的选项合并起来。

【例8.5】声明3个变量,分别为var1、var2和var3,数据类型为INT,使用SET为变量赋值,SQL语句如下:

```
DECLARE var1, var2, var3 INT;
SET var1 = 10, var2 = 20;
SET var3 = var1 + var2;
```

在MySQL中,还可以通过SELECT ... INTO为一个或多个变量赋值,语法如下:

```
SELECT col_name[,...] INTO var_name[,...] table_expr;
```

这个SELECT语法把选定的列直接存储为对应位置的变量。col_name表示字段名称;var_name表示定义的变量名称;table_expr表示查询条件表达式,包括表名称和WHERE子句。

【例8.6】声明变量fruitname和fruitprice,通过SELECT ... INTO语句查询指定记录并为变量赋值,SQL语句如下:

```
DECLARE fruitname CHAR(50);
DECLARE fruitprice DECIMAL(8,2);

SELECT f_name,f_price INTO fruitname, fruitprice
FROM fruits WHERE f_id ='a1';
```

8.1.4 定义条件和处理程序

特定条件需要特定处理。这些条件可以联系到错误以及子程序中的一般流程控制。定义条件是事先定义程序执行过程中遇到的问题；处理程序定义了在遇到这些问题时应当采取的处理方式，并且保证存储过程或存储函数在遇到警告或错误时能继续执行。这样可以增强存储程序处理问题的能力，避免程序异常停止运行。本小节将介绍如何使用DECLARE关键字来定义条件和处理程序。

1. 定义条件

定义条件使用DECLARE语句，语法格式如下：

```
DECLARE condition_name CONDITION FOR [condition_type]

[condition_type]:
SQLSTATE [VALUE] sqlstate_value | mysql_error_code
```

其中，condition_name参数表示条件的名称；condition_type参数表示条件的类型；sqlstate_value和mysql_error_code都可以表示MySQL的错误，sqlstate_value为长度为5的字符串类型错误代码，mysql_error_code为数值类型错误代码。例如，在ERROR 1142（42000）中，sqlstate_value的值是42000，mysql_error_code的值是1142。

这个语句指定需要特殊处理的条件，它将一个名字和指定的错误条件关联起来。这个名字可以随后被用在定义处理程序的DECLARE HANDLER语句中。

【例8.7】定义"ERROR 1148(42000)"错误，其名称为command_not_allowed。可以用两种不同的方法来定义，SQL语句代码如下：

```
//方法一：使用sqlstate_value
DECLARE command_not_allowed CONDITION FOR SQLSTATE '42000';
//方法二：使用mysql_error_code
DECLARE command_not_allowed CONDITION FOR 1148
```

2. 定义处理程序

定义处理程序时，使用DECLARE语句的语法格式如下：

```
DECLARE handler_type HANDLER FOR condition_value[,...] sp_statement
handler_type:
    CONTINUE | EXIT | UNDO

condition_value:
    SQLSTATE [VALUE] sqlstate_value
 | condition_name
 | SQLWARNING
 | NOT FOUND
 | SQLEXCEPTION
 | mysql_error_code
```

各参数解释如下：

（1）handler_type为错误处理方式，有以下3个取值。

- CONTINUE表示遇到错误不处理，继续执行。
- EXIT表示遇到错误马上退出。
- UNDO表示遇到错误后撤回之前的操作，MySQL中暂时不支持这样的操作。

（2）condition_value表示错误类型，可以有以下6个取值：

- SQLSTATE [VALUE] sqlstate_value：包含5个字符的字符串错误值。
- condition_name：表示DECLARE CONDITION定义的错误条件名称。
- SQLWARNING：匹配所有以"01"开头的SQLSTATE错误代码。
- NOT FOUND：匹配所有以"02"开头的SQLSTATE错误代码。
- SQLEXCEPTION：匹配所有没有被SQLWARNING或NOT FOUND捕获的SQLSTATE错误代码。
- MySQL_error_code：匹配数值类型错误代码。

（3）sp_statement参数为程序语句段，表示在遇到定义的错误时需要执行的存储过程或存储函数。

【例8.8】 定义处理程序的几种方式，SQL语句如下：

```
//方法一：捕获sqlstate_value
DECLARE CONTINUE HANDLER FOR SQLSTATE '42S02' SET @info='NO_SUCH_TABLE';

//方法二：捕获mysql_error_code
DECLARE CONTINUE HANDLER FOR 1146 SET @info=' NO_SUCH_TABLE ';

//方法三：先定义条件，然后调用
DECLARE  no_such_table  CONDITION  FOR  1146;
DECLARE CONTINUE HANDLER FOR NO_SUCH_TABLE SET @info=' NO_SUCH_TABLE ';

//方法四：使用SQLWARNING
DECLARE EXIT HANDLER FOR SQLWARNING SET @info='ERROR';

//方法五：使用NOT FOUND
DECLARE EXIT HANDLER FOR NOT FOUND SET @info=' NO_SUCH_TABLE ';

//方法六：使用SQLEXCEPTION
DECLARE EXIT HANDLER FOR SQLEXCEPTION SET @info='ERROR';
```

上述代码演示了6种定义处理程序的方法：

- 方法一是捕获sqlstate_value值，为遇到sqlstate_value值为"42S02"时，则执行CONTINUE操作，并且输出"NO_SUCH_TABLE"信息。
- 方法二是捕获mysql_error_code值，当遇到mysql_error_code值为1146时，则执行CONTINUE操作，并且输出"NO_SUCH_TABLE"信息。
- 方法三是先定义条件，再调用条件。这里先定义no_such_table条件，如果遇到1146错误就执行CONTINUE操作。

- 方法四是使用SQLWARNING。SQLWARNING捕获所有以"01"开头的sqlstate_value值，然后执行EXIT操作，并且输出"ERROR"信息。
- 方法五是使用NOT FOUND。NOT FOUND捕获所有以"02"开头的sqlstate_value值，然后执行EXIT操作，并且输出"NO_SUCH_TABLE"信息。
- 方法六是使用SQLEXCEPTION。SQLEXCEPTION捕获所有没有被SQLWARNING或NOT FOUND捕获的sqlstate_value值，然后执行EXIT操作，并且输出"ERROR"信息。

【例8.9】定义条件和处理程序，具体的执行过程如下：

```
mysql> CREATE TABLE test_db.t (s1 int,primary key (s1));
mysql> DELIMITER //
mysql> CREATE PROCEDURE handlerdemo ()
    -> BEGIN
    ->  DECLARE CONTINUE HANDLER FOR SQLSTATE '23000' SET @x2 = 1;
    ->  SET @x = 1;
    ->  INSERT INTO test_db.t VALUES (1);
    ->  SET @x = 2;
    ->  INSERT INTO test_db.t VALUES (1);
    ->  SET @x = 3;
    -> END;
    -> //
mysql> DELIMITER ;
/*调用存储过程*/
mysql> CALL handlerdemo();
/*查看调用过程结果*/
mysql> SELECT @x;
+------+
| @x   |
+------+
| 3    |
+------+
```

@x是1个用户变量，执行结果是@x等于3，这表明MySQL被执行到程序的末尾。如果"DECLARE CONTINUE HANDLER FOR SQLSTATE '23000' SET @x2 = 1;"这一行不在，第二个INSERT因PRIMARY KEY强制而失败之后，MySQL可能已经采取默认（EXIT）路径，并且SELECT @x可能已经返回2。

> 提示 "@var_name"表示用户变量，使用SET语句为其赋值时，用户变量与连接有关，一个客户端定义的变量不能被其他客户端看到或使用。当客户端退出时，该客户端连接的所有变量将自动释放。

8.1.5 光标的使用

查询语句可能返回多条记录，如果数据量非常大，则需要在存储过程和存储函数中使用光标来逐条读取查询结果集中的记录。应用程序可以根据需要滚动或浏览其中的数据。本小节将介绍如何声明、打开、使用和关闭光标。

光标必须在声明处理程序之前被声明，并且变量和条件还必须在声明光标或处理程序之前被声明。

1. 声明光标

在MySQL中，使用DECLARE关键字来声明光标，其语法的基本形式如下：

```
DECLARE cursor_name CURSOR FOR select_statement
```

其中，cursor_name参数表示光标的名称；select_statement参数表示SELECT语句的内容，返回一个用于创建光标的结果集。

【例8.10】 声明名称为"cursor_fruit"的光标，SQL语句如下：

```
DECLARE cursor_fruit CURSOR FOR SELECT f_name, f_price FROM fruits ;
```

在本例中，光标的名称为"cur_fruit"，SELECT语句部分从表fruits中查询出f_name和f_price字段的值。

2. 打开光标

打开光标的语法如下：

```
OPEN cursor_name{光标名称}
```

【例8.11】 打开名称为"cursor_fruit"的光标，SQL语句如下：

```
OPEN  cursor_fruit ;
```

3. 使用光标

使用光标的语法如下：

```
FETCH cursor_name INTO var_name [, var_name] ...{参数名称}
```

其中，cursor_name参数表示光标的名称；var_name参数表示将光标中的SELECT语句查询出来的信息存入该参数中，var_name必须在声明光标之前就已经定义好。

【例8.12】 使用名称为"cursor_fruit"的光标将查询出来的数据存入fruit_name和fruit_price这两个变量中，SQL语句如下：

```
FETCH  cursor_fruit INTO fruit_name, fruit_price ;
```

在本例中，将光标cursor_fruit中用SELECT语句查询出来的信息存入fruit_name和fruit_price中，并且fruit_name和fruit_price必须在前面已经定义好。

4. 关闭光标

关闭光标的语法如下：

```
CLOSE cursor_name{光标名称}
```

这个语句关闭先前打开的光标。

如果光标未被明确地关闭，则它在被声明的复合语句的末尾关闭。

【例8.13】关闭名称为"cursor_fruit"的光标，SQL语句如下：

```
CLOSE cursor_fruit;
```

> 提示 在MySQL中，光标只能在存储过程和存储函数中使用。

8.1.6 流程控制的使用

流程控制语句用来根据条件控制语句的执行。MySQL中用来构造控制流程的语句有IF语句、CASE语句、LOOP语句、LEAVE语句、ITERATE语句、REPEAT语句和WHILE语句。

每个流程中可能包含一个单独语句，也可能是使用BEGIN ... END构造的复合语句，构造可以被嵌套。本小节将介绍这些控制流程语句。

1. IF语句

IF语句包含多个条件判断，根据判断的结果为TRUE或FALSE执行相应的语句，语法格式如下：

```
IF expr_condition THEN statement_list
    [ELSEIF expr_condition THEN statement_list] ...
    [ELSE statement_list]
END IF
```

IF实现了一个基本的条件构造。如果expr_condition求值为真（TRUE），则相应的SQL语句列表被执行；如果没有expr_condition匹配，则ELSE子句里的语句列表被执行。statement_list可以包括一个或多个语句。

> 提示 MySQL中还有一个IF()函数，它不同于这里描述的IF语句。

【例8.14】IF语句的示例如下：

```
IF val IS NULL
  THEN SELECT 'val is NULL';
  ELSE SELECT 'val is not NULL';
END IF;
```

该示例判断val值是否为空，如果val值为空，则输出字符串"val is NULL"；否则输出字符串"val is not NULL"。IF语句都需要使用END IF来结束。

2. CASE语句

CASE是另一个进行条件判断的语句，有两种格式。

第以种格式如下：

```
CASE case_expr
```

```
    WHEN when_value THEN statement_list
    [WHEN when_value THEN statement_list] ...
    [ELSE statement_list]
END CASE
```

其中，case_expr参数表示条件判断的表达式，决定了哪一个WHEN子句会被执行；when_value参数表示表达式可能的值，如果某个when_value表达式与case_expr表达式的结果相同，则执行对应THEN关键字后的statement_list中的语句；statement_list参数表示不同when_value值的执行语句。

【例8.15】使用CASE流程控制语句的第以种格式，判断val值是否等于1、等于2，或者两者都不等，SQL语句如下：

```
CASE val
  WHEN 1 THEN SELECT 'val is 1';
  WHEN 2 THEN SELECT 'val is 2';
  ELSE SELECT 'val is not 1 or 2';
END CASE;
```

当val值为1时，输出字符串"val is 1"；当val值为2时，输出字符串"val is 2"；否则输出字符串"val is not 1 or 2"。

CASE语句的第二种格式如下：

```
CASE
    WHEN expr_condition THEN statement_list
    [WHEN expr_condition THEN statement_list] ...
    [ELSE statement_list]
END CASE
```

其中，expr_condition参数表示条件判断语句；statement_list参数表示不同条件的执行语句。该语句中，WHEN语句将被逐个执行，直到某个expr_condition表达式为真，则执行对应THEN关键字后面的statement_list语句。如果没有条件匹配，则ELSE子句里的语句被执行。

> **提示** 这里介绍的用在存储程序里的CASE语句与"控制流程函数"里描述的SQL CASE表达式的CASE语句有轻微不同：这里的CASE语句不能有ELSE NULL子句，并且用END CASE代替END来终止。

【例8.16】使用CASE流程控制语句的第二种格式，判断val是否为空、小于0、大于0或者等于0，SQL语句如下：

```
CASE
  WHEN val is NULL THEN SELECT 'val is NULL';
  WHEN val < 0 THEN SELECT 'val is less than 0';
  WHEN val > 0 THEN SELECT 'val is greater than 0';
  ELSE SELECT 'val is 0';
END CASE;
```

当val值为空时，输出字符串"val is NULL"；当val值小于0时，输出字符串"val is less than 0"；当val值大于0时，输出字符串"val is greater than 0"；否则输出字符串"val is 0"。

3. LOOP语句

LOOP循环语句用来重复执行某些语句,与IF和CASE语句相比,LOOP只是创建一个循环操作的过程,并不进行条件判断。LOOP内的语句一直重复执行,直到循环被退出(使用LEAVE子句跳出循环过程)。LOOP语句的基本格式如下:

```
[loop_label:] LOOP
    statement_list
END LOOP [loop_label]
```

其中,loop_label表示LOOP语句的标注名称,该参数可以省略;statement_list参数表示需要循环执行的语句。

【例8.17】使用LOOP语句进行循环操作——id值小于10时将重复执行循环过程,SQL语句如下:

```
DECLARE id INT DEFAULT 0;
add_loop: LOOP
SET id = id + 1;
  IF id >= 10 THEN LEAVE add_loop;
  END IF;
END LOOP add_ loop;
```

该示例循环执行id加1的操作:当id值小于10时,循环重复执行;当id值大于或者等于10时,使用LEAVE语句退出循环。LOOP循环都以END LOOP结束。

4. LEAVE语句

LEAVE语句用来退出任何被标注的流程控制构造,基本格式如下:

```
LEAVE label
```

其中,label参数表示循环的标志。LEAVE和BEGIN...END或循环一起使用。

【例8.18】使用LEAVE语句退出循环,SQL语句如下:

```
add_num: LOOP
SET @count=@count+1;
IF @count=50 THEN LEAVE add_num ;
END LOOP add_num ;
```

该示例循环执行count加1的操作,当count的值等于50时,使用LEAVE语句跳出循环。

5. ITERATE语句

ITERATE语句将执行顺序跳转到语句块开头处,基本格式如下:

```
ITERATE label
```

ITERATE关键字只能出现在LOOP、REPEAT和WHILE语句内,其意思是"再次循环";label参数表示循环的标志。ITERATE关键字必须在循环的标志前面。

【例8.19】 ITERATE语句示例如下：

```
CREATE PROCEDURE doiterate()
BEGIN
DECLARE p1 INT DEFAULT 0;
my_loop: LOOP
  SET p1= p1 + 1;
  IF p1 < 10 THEN ITERATE my_loop;
  ELSEIF p1 > 20 THEN LEAVE my_loop;
  END IF;
  SELECT 'p1 is between 10 and 20';
END LOOP my_loop;
END
```

初始化p1=0，当p1的值小于10时，则重复执行p1加1操作；当p1大于或等于10并且小于或等于20时，打印消息"p1 is between 10 and 20"；当p1大于20时，退出循环。

6. REPEAT语句

REPEAT语句创建一个带条件判断的循环过程，每次语句执行完毕之后会对条件表达式进行判断，如果表达式为真，则循环结束，否则重复执行循环中的语句。REPEAT语句的基本格式如下：

```
[repeat_label:] REPEAT
    statement_list
UNTIL expr_condition
END REPEAT [repeat_label]
```

repeat_label为REPEAT语句的标注名称，该参数可以省略；REPEAT语句内的语句或语句块被重复执行，直至expr_condition为真。

【例8.20】 REPEAT语句示例——id值小于10时将重复执行循环过程，SQL语句如下：

```
DECLARE id INT DEFAULT 0;
REPEAT
SET id = id + 1;
UNTIL  id >= 10
END REPEAT;
```

该示例循环执行id加1的操作。当id值小于10时，循环重复执行；当id值大于或者等于10时，退出循环。REPEAT循环都以END REPEAT结束。

7. WHILE语句

WHILE语句创建一个带条件判断的循环过程。与REPEAT不同的是，WHILE在执行时，先对指定的表达式进行判断，如果为真，就执行循环内的语句，否则退出循环。WHILE语句的基本格式如下：

```
[while_label:] WHILE expr_condition DO
    statement_list
END WHILE [while_label]
```

while_label为WHILE语句的标注名称；expr_condition为进行判断的表达式，如果表达式结果为真，WHILE语句内的语句或语句群被执行，直至expr_condition为假，退出循环。

【例8.21】WHILE语句示例——i值小于10时，将重复执行循环过程，SQL语句如下：

```
DECLARE i INT DEFAULT 0;
WHILE i < 10 DO
SET i = i + 1;
END WHILE;
```

8.2 调用存储过程和存储函数

存储过程和存储函数已经定义好了，接下来需要知道如何调用这些过程和函数。存储过程和存储函数有多种调用方法：存储过程必须使用CALL语句调用，并且存储过程和数据库相关，如果要执行其他数据库中的存储过程，需要指定数据库名称，例如CALL dbname.procname；存储函数的调用与MySQL中预定义的函数的调用方式相同。本节将介绍存储过程和存储函数的调用，主要包括调用存储过程的语法、调用存储函数的语法，以及存储过程和存储函数的调用实例。

8.2.1 调用存储过程

存储过程是通过CALL语句进行调用的，语法如下：

```
CALL sp_name([parameter[,...]])
```

CALL语句调用一个先前用CREATE PROCEDURE创建的存储过程，其中sp_name为存储过程名称，parameter为存储过程的参数。

【例8.22】定义名为CountProc1的存储过程，然后调用这个存储过程。

定义存储过程的SQL语句如下：

```
mysql> use test_db;
mysql> DELIMITER //
mysql> CREATE PROCEDURE CountProc1 (IN sid INT, OUT num INT)
    -> BEGIN
    ->   SELECT COUNT(*) INTO num FROM fruits WHERE s_id = sid;
    -> END //
mysql> DELIMITER ;
```

调用存储过程的SQL语句如下：

```
mysql> CALL CountProc1 (101, @num);
```

查看返回结果：

```
mysql> SELECT @num;
+------+
```

```
| @num   |
+--------+
|   3    |
+--------+
```

该存储过程返回了s_id=101的水果供应商提供的水果种类，返回值存储在变量num中，使用SELECT查看，返回结果为3。

8.2.2 调用存储函数

在MySQL中，存储函数的使用方法与MySQL内部函数的使用方法是一样的。换言之，用户自定义的存储函数与MySQL内部函数是一个性质，区别在于，存储函数是用户自己定义的，而内部函数是MySQL的开发者定义的。

【例8.23】 定义存储函数CountProc2，然后调用这个函数，SQL语句如下：

```
mysql> DELIMITER //
mysql> CREATE FUNCTION CountProc2 (sid INT)
    -> RETURNS INT
    -> BEGIN
    -> RETURN (SELECT COUNT(*) FROM fruits WHERE s_id = sid);
    -> END;
    ->//
mysql> DELIMITER ;
```

注意 如果在创建存储函数中报错"you *might* want to use the less safe log_bin_trust_function_creators variable"，需要执行以下代码：

```
mysql> SET GLOBAL log_bin_trust_function_creators = 1;
```

调用存储函数：

```
mysql> SELECT CountProc2(101);
+--------------------+
| Countproc(101)     |
+--------------------+
|         3          |
+--------------------+
1 row in set (0.00 sec)
```

可以看到，本例与【例8.22】中返回的结果相同。存储函数和存储过程可以实现相同的功能，读者在实际应用中应该灵活选择。

8.3 查看存储过程和存储函数

MySQL中存储了存储过程和存储函数的状态信息，用户可以使用SHOW STATUS语句或

SHOW CREATE语句来查看，也可直接从系统的information_schema数据库中查询。本节将通过实例来介绍这3种方法。

8.3.1 使用 SHOW STATUS 语句查看存储过程和存储函数的状态

SHOW STATUS语句可以查看存储过程和存储函数的状态，其基本语法结构如下：

```
SHOW {PROCEDURE | FUNCTION} STATUS [LIKE 'pattern']
```

这个语句是一个MySQL的扩展，返回子程序的特征，如数据库、名字、类型、创建者及创建和修改日期。如果没有指定样式，那么根据使用的语句，所有存储程序或存储函数的信息都会被列出。其中，PROCEDURE和FUNCTION分别表示查看存储过程和存储函数；LIKE语句表示匹配存储过程或存储函数的名称。

【例8.24】SHOW STATUS语句示例如下：

```
mysql> SHOW PROCEDURE STATUS LIKE 'C%'\G
*** 1. row ***
                  Db: test_db
                Name: CountProc
Type: PROCEDURE
             Definer: root@localhost
            Modified: 2018-11-21 13:52:28
             Created: 2018-11-21 13:52:28
       Security_type: DEFINER
             Comment:
character_set_client: gbk
collation_connection: gbk_chinese_ci
  Database Collation: utf8mb4_0900_ai_ci
```

"SHOW PROCEDURE STATUS LIKE 'C%'\G"语句获取数据库中所有名称以字母"C"开头的存储过程的信息。通过上面的语句可以看到：这个存储过程所在的数据库为test_db、名称为CountProc等一些相关信息。

8.3.2 使用 SHOW CREATE 语句查看存储过程和存储函数的定义

除了SHOW STATUS之外，MySQL还可以使用SHOW CREATE语句查看存储过程和存储函数的状态。语法格式如下：

```
SHOW CREATE {PROCEDURE | FUNCTION} sp_name
```

这个语句是一个MySQL的扩展，类似于SHOW CREATE TABLE，它返回一个可用来重新创建已命名子程序的确切字符串。PROCEDURE和FUNCTION分别表示查看存储过程和存储函数；sp_name参数表示匹配存储过程或存储函数的名称。

【例8.25】SHOW CREATE语句示例如下：

```
mysql> SHOW CREATE FUNCTION test_db.CountProc2 \G
```

```
*************************** 1. row ***************************
            Function: CountProc2
            sql_mode: STRICT_TRANS_TABLES,NO_ENGINE_SUBSTITUTION
     Create Function: CREATE DEFINER=`root`@`localhost` FUNCTION `CountProc2`(sid
INT) RETURNS int(11)
BEGIN
RETURN (SELECT COUNT(*) FROM fruits WHERE s_id = sid);
END
 character_set_client: gbk
 collation_connection: gbk_chinese_ci
    Database Collation: utf8mb4_0900_ai_ci
```

通过上面的语句可以看到：存储函数的名称为CountProc2、sql_mode为sql，以及数据库设置的一些信息。

8.3.3 从 information_schema.Routines 表中查看存储过程和存储函数的信息

MySQL中存储过程和存储函数的信息存储在information_schema数据库下的Routines表中。可以通过查询该表的记录来查询存储过程和存储函数的信息。其基本语法形式如下：

```
SELECT * FROM information_schema.Routines
WHERE ROUTINE_NAME=' sp_name ' ;
```

其中，ROUTINE_NAME字段中存储的是存储过程和存储函数的名称；sp_name参数表示存储过程或存储函数的名称。

【例8.26】从Routines表中查询名称为CountProc2的存储函数的信息，SQL语句如下：

```
SELECT * FROM information_schema.Routines WHERE ROUTINE_NAME='CountProc2' AND
ROUTINE_TYPE = 'FUNCTION' \G
```

语句执行如下：

```
mysql> SELECT * FROM information_schema.Routines WHERE ROUTINE_NAME='CountProc2'
AND  ROUTINE_TYPE = 'FUNCTION' \G
*************************** 1. row ***************************
           SPECIFIC_NAME: CountProc2
          ROUTINE_CATALOG: def
           ROUTINE_SCHEMA: test_db
             ROUTINE_NAME: CountProc2
             ROUTINE_TYPE: FUNCTION
                DATA_TYPE: int
CHARACTER_MAXIMUM_LENGTH: NULL
  CHARACTER_OCTET_LENGTH: NULL
       NUMERIC_PRECISION: 10
           NUMERIC_SCALE: 0
      DATETIME_PRECISION: NULL
      CHARACTER_SET_NAME: NULL
          COLLATION_NAME: NULL
           DTD_IDENTIFIER: int(11)
             ROUTINE_BODY: SQL
```

```
        ROUTINE_DEFINITION: BEGIN
RETURN (SELECT COUNT(*) FROM fruits WHERE s_id = sid);
END
            EXTERNAL_NAME: NULL
        EXTERNAL_LANGUAGE: SQL
          PARAMETER_STYLE: SQL
         IS_DETERMINISTIC: NO
         SQL_DATA_ACCESS: CONTAINS SQL
                 SQL_PATH: NULL
            SECURITY_TYPE: DEFINER
                  CREATED: 2018-11-21 16:57:09
             LAST_ALTERED: 2018-11-21 16:57:09
                 SQL_MODE: STRICT_TRANS_TABLES,NO_ENGINE_SUBSTITUTION
          ROUTINE_COMMENT:
                  DEFINER: root@localhost
     CHARACTER_SET_CLIENT: gbk
     COLLATION_CONNECTION: gbk_chinese_ci
       DATABASE_COLLATION: utf8mb4_0900_ai_ci
```

在information_schema数据库下的Routines表中,存储了所有存储过程和存储函数的定义。使用SELECT语句查询Routines表中的存储过程和存储函数的定义时,一定要使用ROUTINE_NAME字段指定存储过程或存储函数的名称;否则,将查询出所有的存储过程或存储函数的定义。当存储过程和存储函数名称相同时,就需要同时指定ROUTINE_TYPE字段表明查询的是哪种类型的存储程序。

8.4 修改存储过程和存储函数

使用ALTER语句可以修改存储过程或存储函数的特性,本节将介绍如何使用ALTER语句修改存储过程和存储函数。语句格式如下:

```
ALTER {PROCEDURE | FUNCTION} sp_name [characteristic ...]
```

其中,sp_name参数表示存储过程或存储函数的名称;characteristic参数指定存储函数的特性,可能的取值有:

- CONTAINS SQL:表示子程序包含SQL语句,但不包含读或写数据的语句。
- NO SQL:表示子程序中不包含SQL语句。
- READS SQL DATA:表示子程序中包含读数据的语句。
- MODIFIES SQL DATA:表示子程序中包含写数据的语句。
- SQL SECURITY { DEFINER | INVOKER }:指明谁有权限来执行。
- DEFINER:表示只有定义者自己才能够执行。
- INVOKER:表示调用者可以执行。
- COMMENT 'string':表示注释信息。

> **提示** 修改存储过程使用ALTER PROCEDURE语句，修改存储函数使用ALTER FUNCTION语句。这两个语句的结构是一样的，语句中的所有参数也是一样的，而且它们与创建存储过程或存储函数的语句中的参数也基本一样。

【例8.27】 修改存储过程CountProc的定义，将读写权限改为MODIFIES SQL DATA，并指明调用者可以执行，SQL语句如下：

```
ALTER PROCEDURE CountProc MODIFIES SQL DATA SQL SECURITY INVOKER ;
```

执行语句，并查看修改后的信息，结果如下：

```
//执行ALTER PROCEDURE语句
mysql> ALTER PROCEDURE CountProc MODIFIES SQL DATA  SQL SECURITY INVOKER;
//查询修改后的CountProc表信息
mysql> SELECT SPECIFIC_NAME,SQL_DATA_ACCESS,SECURITY_TYPE
    -> FROM information_schema.Routines
    -> WHERE ROUTINE_NAME='CountProc' AND ROUTINE_TYPE='PROCEDURE';
+---------------+------------------+---------------+
| SPECIFIC_NAME | SQL_DATA_ACCESS  | SECURITY_TYPE |
+---------------+------------------+---------------+
| CountProc     |MODIFIES SQL DATA | INVOKER       |
+---------------+------------------+---------------+
```

结果显示，存储过程修改成功。从查询的结果可以看出，访问数据的权限（SQL_DATA_ACCESS）已经变成MODIFIES SQL DATA，安全类型（SECURITY_TYPE）已经变成INVOKER。

【例8.28】 修改存储函数CountProc2的定义，将读写权限改为READS SQL DATA，并加上注释信息"FIND NAME"，SQL语句如下：

```
ALTER FUNCTION CountProc2 READS SQL DATA COMMENT 'FIND NAME' ;
```

执行语句，并查看修改后的信息。结果如下：

```
//执行ALTER FUNCTION语句
mysql> ALTER FUNCTION CountProc2 READS SQL DATA  COMMENT 'FIND NAME' ;
Query OK, 0 rows affected (0.00 sec)
//查看修改后的信息
mysql> SELECT SPECIFIC_NAME,SQL_DATA_ACCESS,ROUTINE_COMMENT
FROM information_schema.Routines
WHERE ROUTINE_NAME='CountProc2' AND  ROUTINE_TYPE = 'FUNCTION';
+-------------------+-----------------+-----------------+
| SPECIFIC_NAME     | SQL_DATA_ACCESS | ROUTINE_COMMENT |
+-------------------+-----------------+-----------------+
| CountProc2        | READS SQL DATA  | FIND NAME       |
+-------------------+-----------------+-----------------+
```

存储函数修改成功。从查询的结果可以看出，访问数据的权限（SQL_DATA_ACCESS）已经变成READS SQL DATA，函数注释（ROUTINE_COMMENT）已经变成FIND NAME。

8.5 删除存储过程和存储函数

删除存储过程和存储函数，可以使用DROP语句，其语法结构如下：

```
DROP {PROCEDURE | FUNCTION} [IF EXISTS] sp_name
```

这个语句被用来移除一个存储过程或存储函数。sp_name为要移除的存储过程或存储函数的名称。IF EXISTS子句是一个MySQL的扩展。如果程序或函数不存储，它可以防止发生错误，产生一个用SHOW WARNINGS查看的警告。

【例8.29】删除存储过程和存储函数，SQL语句如下：

```
DROP PROCEDURE CountProc;
DROP FUNCTION CountProc2;
```

上面语句的作用就是删除存储过程CountProc和存储函数CountProc2。

8.6 全局变量的持久化

在MySQL数据库中，全局变量可以通过SET GLOBAL语句来设置。例如，设置服务器语句超时的限制，可以通过设置系统变量max_execution_time来实现：

```
SET GLOBAL max_execution_time=2000;
```

使用SET GLOBAL语句设置的变量值只会临时生效。数据库重启后，服务器又会从MySQL配置文件中读取变量的默认值。

MySQL 9.0版本中的SET PERSIST命令，用于设置或更新当前会话的持久化设置。持久化设置决定了一个会话在事务结束之前是否需要将所有更改保存到磁盘。例如，设置服务器的最大连接数为1000：

```
SET PERSIST max_connections = 1000;
```

MySQL会将该命令的配置保存到数据目录下的 mysqld-auto.cnf 文件中，下次启动时会读取该文件，用其中的配置来覆盖默认的配置文件。

下面通过一个案例来理解全部变量的持久化。

首先查看全局变量max_connections的值，结果如下：

```
mysql> SHOW VARIABLES LIKE '%max_connections%';
+------------------------+-------+
| Variable_name          | Value |
+------------------------+-------+
| max_connections        | 151   |
| mysqlx_max_connections | 100   |
+------------------------+-------+
```

然后设置全局变量max_connections的值：

```
mysql> SET PERSIST max_connections=1000;
```

最后重启MySQL服务器，再次查询max_connections的值：

```
mysql> SHOW VARIABLES LIKE '%max_connections%';
+------------------------+-------+
| Variable_name          | Value |
+------------------------+-------+
| max_connections        | 1000  |
| mysqlx_max_connections | 100   |
+------------------------+-------+
```

由结果可知，使用SET PERSIST命令设置的服务器被连接次数1000持久化了，覆盖了原来的151。

第 9 章 视 图

数据库中的视图是一张虚拟表。同真实的表一样，视图包含一系列带有名称的行和列数据。行和列数据来自由定义视图查询所引用的表，并且在引用视图时动态生成。本章将通过一些实例来介绍视图的含义、视图的作用、创建视图、查看视图、修改视图、更新视图和删除视图等MySQL的数据库知识。

9.1 视图概述

视图是一张虚拟表，是从数据库中一张或多张表中导出来的表。视图还可以在已经存在的视图的基础上进行定义。在视图中用户可以使用SELECT语句查询数据，以及使用INSERT、UPDATE和DELETE修改记录。从MySQL 5.0开始可以使用视图，视图可以方便用户操作，而且可以保障数据库系统的安全。

9.1.1 视图的含义

视图一经定义便存储在数据库中，与其相对应的数据并没有像表那样在数据库中再存储一份，通过视图看到的数据只是存放在基本表中的数据。对视图的操作与对表的操作一样，可以对其进行查询、修改和删除。当对通过视图看到的数据进行修改时，相应的基本表的数据也要发生变化；同时，若基本表的数据发生变化，则这种变化也可以自动反映到视图中。

下面有一张student表和一张stu_info表，在student表中包含了学生的id和姓名，在stu_info表中包含了学生的id、班级和家庭住址，代码如下：

```
CREATE TABLE student
(
s_id   INT,
name   VARCHAR(40)
);
```

```
CREATE TABLE stu_info
(
s_id    INT,
glass   VARCHAR(40),
addr    VARCHAR(90)
);
```

通过DESC命令可以查看表的设计，获得字段、字段的定义、是否为主键、是否为空、默认值和扩展信息等。

现在公布分班信息，只需要id号、姓名和班级，该如何解决呢？视图提供了一个很好的解决方法。使用表中的相关信息创建一个视图，其他的无关信息不采用，这样既能满足要求，也不破坏表原来的结构。

9.1.2 视图的作用

与直接从数据表中读取数据相比，视图有以下优点：

1．简单化

通过视图看到的就是需要的。视图不仅可以简化用户对数据的理解，还可以简化它们的操作。那些经常使用的查询可以定义为视图，从而使得用户不必为以后的操作每次都指定全部的条件。

2．安全性

通过视图用户只能查询和修改他们所能见到的数据，数据库中的其他数据则既看不见也取不到。数据库授权命令可以使每个用户对数据库的检索限制到特定的数据库对象上，但不能授权到数据库特定行和特定列上。通过视图，用户可以被限制在数据的不同子集上：

（1）使用权限可被限制在基表的行的子集上。
（2）使用权限可被限制在基表的列的子集上。
（3）使用权限可被限制在基表的行和列的子集上。
（4）使用权限可被限制在多张基表的连接所限定的行上。
（5）使用权限可被限制在基表中的数据的统计汇总上。
（6）使用权限可被限制在另一视图的一个子集上，或是一些视图和基表合并后的子集上。

3．逻辑数据独立性

视图可以帮助用户屏蔽真实表结构变化带来的影响。

9.2 创建视图

视图中包含了SELECT查询的结果，因此视图的创建基于SELECT语句和已存在的数据表。视图可以建立在一张表上，也可以建立在多张表上。本节主要介绍创建视图的方法。

9.2.1 创建视图的语法形式

创建视图使用CREATE VIEW语句，基本语法格式如下：

```
CREATE [OR REPLACE] [ALGORITHM = {UNDEFINED | MERGE | TEMPTABLE}]
VIEW view_name [(column_list)]
AS SELECT_statement
[WITH [CASCADED | LOCAL] CHECK OPTION]
```

各参数说明如下：

（1）CREATE表示创建新的视图。

（2）REPLACE表示替换已经创建的视图。

（3）ALGORITHM表示视图选择的算法，取值有以下3种：

- UNDEFINED表示MySQL将自动选择算法。
- MERGE表示将使用的视图语句与视图定义合并起来，使得视图定义的某一部分取代语句对应的部分。
- TEMPTABLE表示将视图的结果存入临时表，然后用临时表来执行语句。

（4）view_name为视图的名称，column_list为属性列。

（5）SELECT_statement表示SELECT语句。

（6）WITH [CASCADED | LOCAL] CHECK OPTION表示视图在更新时保证在视图的权限范围之内。CASCADED与LOCAL为可选参数，CASCADED为默认值，表示更新视图时要满足所有相关视图和表的条件；LOCAL表示更新视图时满足该视图本身定义的条件即可。

创建视图语句要求具有针对视图的CREATE VIEW权限，以及针对由SELECT语句选择的每一列上的某些权限。对于在SELECT语句中其他地方使用的列，必须具有SELECT权限。如果还有OR REPLACE子句，则必须在视图上具有DROP权限。

视图属于数据库。在默认情况下，将在当前数据库创建新视图。要想在给定数据库中明确创建视图，创建时应将名称指定为db_name.view_name。

9.2.2 在单表上创建视图

MySQL可以在单张数据表上创建视图。

【例9.1】在表t上创建一个名为"view_t"的视图。

首先创建基本表并插入数据，SQL语句如下：

```
CREATE TABLE tv (quantity INT, price INT);
INSERT INTO tv VALUES(3, 50);
```

然后创建视图，SQL语句如下：

```
CREATE VIEW view_t AS SELECT quantity, price, quantity *price FROM tv;
```

最后查询视图，SQL语句如下：

```
mysql> SELECT * FROM view_t;
+----------+-------+-----------------+
| quantity | price | quantity *price |
+----------+-------+-----------------+
|        3 |    50 |             150 |
+----------+-------+-----------------+
```

默认情况下创建的视图的字段和基本表的字段是一样的。

也可以通过指定视图字段的名称来创建视图。

【例9.2】 在表tv上创建一个名为"view_t2"的视图，SQL语句如下：

```
mysql> CREATE VIEW view_t2(qty, price, total ) AS SELECT quantity, price, quantity * price FROM tv;
```

语句执行成功后，查看view_t2视图中的数据：

```
mysql> SELECT * FROM view_t2;
+------+-------+-------+
| qty  | price | total |
+------+-------+-------+
|    3 |    50 |   150 |
+------+-------+-------+
```

可以看到，view_t2和view_t两个视图中的字段名称虽然不同，但数据却是相同的。因此，在使用视图的时候，用户根本就不需要了解基本表的结构，更接触不到实际表中的数据，从而保证了数据库的安全。

9.2.3 在多表上创建视图

在MySQL中也可以使用CREATE VIEW语句在两张或者两张以上的表上创建视图。

【例9.3】 在表student和表stu_info上创建视图stu_glass。

首先向两张表中插入数据，SQL语句如下：

```
mysql> INSERT INTO student VALUES(1,'wanglin1'),(2,'gaoli'),(3,'zhanghai');

mysql> INSERT INTO stu_info VALUES(1, 'wuban','henan'),(2,'liuban','hebei'),(3,'qiban','shandong');
```

然后创建视图stu_glass，SQL语句如下：

```
mysql> CREATE VIEW stu_glass (id,name, glass) AS SELECT student.s_id, student.name ,stu_info.glass FROM student ,stu_info WHERE student.s_id=stu_info.s_id;

mysql> SELECT * FROM stu_glass;
+------+----------+--------+
| id   | name     | glass  |
+------+----------+--------+
```

```
|   1 | wanglin1 | wuban  |
|   2 | gaoli    | liuban |
|   3 | zhanghai | qiban  |
+------+----------+--------+
```

这个例子就解决了9.1.1节提出的那个问题，通过这个视图可以很好地保护基本表中的数据。这个视图中的信息很简单，只包含了id、姓名和班级，id字段对应表student中的s_id字段，name字段对应表student中的name字段，glass字段对应表stu_info中的glass字段。

9.3 查看视图

查看视图是查看数据库中已存在的视图的定义。查看视图必须要有SHOW VIEW权限，MySQL数据库下的user表中保存着这个信息。查看视图的方法包括DESCRIBE、SHOW TABLE STATUS和SHOW CREATE VIEW，本节将介绍查看视图的各种方法。

9.3.1 使用 DESCRIBE 语句查看视图基本信息

DESCRIBE可以用来查看视图，具体的语法如下：

```
DESCRIBE 视图名;
```

【例9.4】通过DESCRIBE语句查看视图view_t的定义，SQL语句如下：

```
DESCRIBE view_t;
```

语句执行结果如下：

```
mysql> DESCRIBE view_t;
+----------------+-----------+------+-----+---------+-------+
| Field          | Type      | Null | Key | Default | Extra |
+----------------+-----------+------+-----+---------+-------+
| quantity       | int       | YES  |     | NULL    |       |
| price          | int       | YES  |     | NULL    |       |
| quantity*price | bigint(21)| YES  |     | NULL    |       |
+----------------+-----------+------+-----+---------+-------+
```

结果显示出了视图的字段定义、字段的数据类型、是否为空、是否为主/外键、默认值和额外信息。

DESCRIBE一般情况下都简写成DESC，DESC命令的执行结果和DESCRIBE命令的执行结果是一样的。

9.3.2 使用 SHOW TABLE STATUS 语句查看视图基本信息

查看视图的信息也可以通过SHOW TABLE STATUS语句完成，具体的语法如下：

```
SHOW TABLE STATUS LIKE '视图名';
```

【例9.5】 使用SHOW TABLE STATUS命令查看视图信息，SQL语句如下：

```
SHOW TABLE STATUS LIKE 'view_t' \G
```

语句执行结果如下：

```
mysql> SHOW TABLE STATUS LIKE 'view_t' \G
*************************** 1. row ***************************
           Name: view_t
         Engine: NULL
        Version: NULL
     Row_format: NULL
           Rows: NULL
 Avg_row_length: NULL
    Data_length: NULL
Max_data_length: NULL
   Index_length: NULL
      Data_free: NULL
 Auto_increment: NULL
    Create_time: 2027-07-25 09:06:24
    Update_time: NULL
     Check_time: NULL
      Collation: NULL
       Checksum: NULL
 Create_options: NULL
        Comment: VIEW
```

执行结果显示，表的Comment的值为VIEW，说明该表为视图；其他的信息为NULL，说明这是一张虚表。

用同样的语句来查看一下数据表tv的信息，执行结果如下：

```
mysql> SHOW TABLE STATUS LIKE 'tv' \G
*************************** 1. row ***************************
           Name: tv
         Engine: InnoDB
        Version: 10
     Row_format: Dynamic
           Rows: 1
 Avg_row_length: 16384
    Data_length: 16384
Max_data_length: 0
   Index_length: 0
      Data_free: 0
 Auto_increment: NULL
    Create_time: 2024-03-17 09:05:49
    Update_time: 2024-03-17 09:05:56
     Check_time: NULL
      Collation: utf8mb4_0900_ai_ci
       Checksum: NULL
 Create_options:
        Comment:
```

从查询的结果来看，这里的信息包含了存储引擎、创建时间等，并且Comment信息为空。这就是视图和表的区别。

9.3.3 使用 SHOW CREATE VIEW 语句查看视图详细信息

使用SHOW CREATE VIEW语句可以查看视图详细定义，语法如下：

```
SHOW CREATE VIEW 视图名;
```

【例9.6】使用SHOW CREATE VIEW查看视图的详细定义，SQL语句如下：

```
SHOW CREATE VIEW view_t \G
```

语句执行结果如下：

```
mysql> SHOW CREATE VIEW view_t \G
*************************** 1. row ***************************
                View: view_t
         Create View: CREATE ALGORITHM=UNDEFINED DEFINER=`root`@`localhost` SQL SECURITY DEFINER VIEW `view_t` AS select `tv`.`quantity` AS `quantity`,`tv`.`price` AS `price`,(`tv`.`quantity` * `tv`.`price`) AS `quantity*price` from `tv`
character_set_client: utf8mb4
collation_connection: utf8mb4_0900_ai_ci
```

执行结果中显示了视图的名称、创建视图的语句等信息。

9.3.4 在 views 表中查看视图详细信息

在MySQL中，information_schema数据库下的views表中存储了所有视图的定义。通过对views表进行查询，可以查看数据库中所有视图的详细信息，查询语句如下：

```
SELECT * FROM information_schema.views;
```

【例9.7】在views表中查看视图的详细定义，SQL语句如下：

```
mysql> SELECT * FROM information_schema.views\G
*** 1. row ***
       TABLE_CATALOG: def
        TABLE_SCHEMA: chapter11db
          TABLE_NAME: stu_glass
     VIEW_DEFINITION: select `chapter11db`.`student`.`s_id` AS `id`,`chapter11db`.`student`.`name` AS `name`,`chapter11db`.`stu_info`.`glass` AS `glass` from `chapter11db`.`student` join `chapter11db`.`stu_info` where (`chapter11db`.`student`.`s_id` = `chapter11db`.`stu_info`.`s_id`)
        CHECK_OPTION: NONE
        IS_UPDATABLE: YES
             DEFINER: root@localhost
       SECURITY_TYPE: DEFINER
CHARACTER_SET_CLIENT: gbk
COLLATION_CONNECTION: gbk_chinese_ci
```

```
    *** 2. row ***
       TABLE_CATALOG: def
        TABLE_SCHEMA: chapter11db
          TABLE_NAME: view_t
     VIEW_DEFINITION: select `chapter11db`.`t`.`quantity` AS
`quantity`,`chapter11db`.`t`.`price` AS `price`,(
   `chapter11db`.`t`.`quantity` * `chapter11db`.`t`.`price`) AS `quantity *price`
from `chapter11db`.`t`
        CHECK_OPTION: NONE
        IS_UPDATABLE: YES
            DEFINER: root@localhost
      SECURITY_TYPE: DEFINER
CHARACTER_SET_CLIENT: gbk
COLLATION_CONNECTION: gbk_chinese_ci
    *** 3. row ***
       TABLE_CATALOG: def
        TABLE_SCHEMA: chapter11db
          TABLE_NAME: view_t2
     VIEW_DEFINITION: select `chapter11db`.`t`.`quantity` AS
`qty`,`chapter11db`.`t`.`price` AS `price`,(`chap
ter11db`.`t`.`quantity` * `chapter11db`.`t`.`price`) AS `total` from
`chapter11db`.`t`
        CHECK_OPTION: NONE
        IS_UPDATABLE: YES
            DEFINER: root@localhost
      SECURITY_TYPE: DEFINER
CHARACTER_SET_CLIENT: gbk
COLLATION_CONNECTION: gbk_chinese_ci
3 rows in set (0.03 sec)
```

查询的结果中显示了当前已经定义的所有视图的详细信息，包括前面定义的3个名称为"stu_glass""view_t"和"view_t2"的视图的详细信息。

9.4 修改视图

修改视图是指修改数据库中存在的视图，当基本表的某些字段发生变化的时候，可以通过修改视图来保持其与基本表的一致性。在MySQL中通过CREATE OR REPLACE VIEW语句和ALTER语句来修改视图。

9.4.1 使用 CREATE OR REPLACE VIEW 语句修改视图

在MySQL中修改视图，可使用CREATE OR REPLACE VIEW语句，语法如下：

```
CREATE [OR REPLACE] [ALGORITHM = {UNDEFINED | MERGE | TEMPTABLE}]
    VIEW view_name [(column_list)]
    AS SELECT_statement
    [WITH [CASCADED | LOCAL] CHECK OPTION]
```

可以看到，修改视图的语句和创建视图的语句是完全一样的。当视图已经存在时，则修改语句对视图进行修改；当视图不存在时，则创建视图。下面通过一个实例来说明。

【例9.8】修改视图view_t。

首先通过DESC查看一下修改之前的视图，以便与修改之后的视图进行对比，SQL语句如下：

```
mysql> DESC view_t;
+----------------+-----------+------+-----+---------+-------+
| Field          | Type      | Null | Key | Default | Extra |
+----------------+-----------+------+-----+---------+-------+
| quantity       | int       | YES  |     | NULL    |       |
| price          | int       | YES  |     | NULL    |       |
| quantity*price | bigint(21)| YES  |     | NULL    |       |
+----------------+-----------+------+-----+---------+-------+
```

然后修改视图view_t，SQL语句如下：

```
mysql> CREATE OR REPLACE VIEW view_t AS SELECT * FROM tv;
```

最后查看修改后的视图view_t：

```
mysql> DESC view_t;
+----------+---------+------+-----+---------+-------+
| Field    | Type    | Null | Key | Default | Extra |
+----------+---------+------+-----+---------+-------+
| quantity | int     | YES  |     | NULL    |       |
| price    | int     | YES  |     | NULL    |       |
+----------+---------+------+-----+---------+-------+
```

从执行的结果来看，相比原来的视图view_t，新的视图view_t少了1个字段。

9.4.2 使用ALTER语句修改视图

ALTER语句是MySQL提供的另外一种修改视图的方法，语法如下：

```
ALTER [ALGORITHM = {UNDEFINED | MERGE | TEMPTABLE}]
    VIEW view_name [(column_list)]
    AS SELECT_statement
    [WITH [CASCADED | LOCAL] CHECK OPTION]
```

这个语法中的关键字和前面创建视图的关键字是一样的，这里就不再介绍了。

【例9.9】使用ALTER语句修改视图view_t。

首先使用DESC查看修改前的视图view_t，SQL语句如下：

```
mysql> DESC view_t;
+----------+---------+------+-----+---------+-------+
| Field    | Type    | Null | Key | Default | Extra |
+----------+---------+------+-----+---------+-------+
```

```
| quantity | int       | YES  |     | NULL    |       |
| price    | int       | YES  |     | NULL    |       |
+----------+-----------+------+-----+---------+-------+
```

然后，使用ALTER语句修改视图view_t：

```
mysql> ALTER VIEW view_t AS SELECT quantity FROM tv;
```

最后查看修改后的视图view_t：

```
mysql> DESC view_t;
+----------+-----------+------+-----+---------+-------+
| Field    | Type      | Null | Key | Default | Extra |
+----------+-----------+------+-----+---------+-------+
| quantity | int       | YES  |     | NULL    |       |
+----------+-----------+------+-----+---------+-------+
```

通过ALTER语句同样可以达到修改视图view_t的目的，从上面的执行结果来看，视图view_t中只剩下一个quantity字段，修改成功。

9.5 更新视图

更新视图是指通过视图来插入、更新、删除表中的数据。因为视图是一张虚拟表，其中没有数据，所以更新视图的操作都是转到基本表上进行的。如果对视图增加或者删除记录，实际上是对其基本表增加或者删除记录。本节将介绍视图更新的3种方法：INSERT、UPDATE和DELETE。

【例9.10】使用UPDATE语句更新视图view_t。

执行视图更新之前，先查看基本表和视图的信息，SQL语句如下：

```
mysql> SELECT * FROM view_t;
+----------+
| quantity |
+----------+
|    3     |
+----------+

mysql> SELECT * FROM tv;
+----------+-------+
| quantity | price |
+----------+-------+
|    3     |  50   |
+----------+-------+
```

然后使用UPDATE语句更新视图view_t，SQL语句如下：

```
mysql> UPDATE view_t SET quantity=5;
```

最后查看视图更新之后的基本表的数据：

```
mysql> SELECT * FROM tv;
+----------+-------+
| quantity | price |
+----------+-------+
|        5 |    50 |
+----------+-------+

mysql> SELECT * FROM view_t;
+----------+
| quantity |
+----------+
|        5 |
+----------+

mysql> SELECT * FROM view_t2;
+------+-------+-------+
| qty  | price | total |
+------+-------+-------+
|    5 |    50 |   250 |
+------+-------+-------+
```

对视图view_t进行更新后，基本表tv的内容也更新了；同样地，当对基本表tv进行更新后，另外一个视图view_t2中的内容也会更新。

【例9.11】使用INSERT语句在基本表tv中插入一条记录，SQL语句如下：

```
mysql> INSERT INTO tv VALUES(3,5);
Query OK, 1 row affected (0.04 sec)
```

查看基本表tv和视图view_t2中的数据：

```
mysql> SELECT * FROM tv;
+----------+-------+
| quantity | price |
+----------+-------+
|        5 |    50 |
|        3 |     5 |
+----------+-------+
2 rows in set (0.00 sec)

mysql> SELECT * FROM view_t2;
+------+-------+-------+
| qty  | price | total |
+------+-------+-------+
|    5 |    50 |   250 |
|    3 |     5 |    15 |
+------+-------+-------+
2 rows in set (0.00 sec)
```

由上面结果可知，向表tv中插入一条记录，通过SELECT查看表tv和视图view_t2，可以看到视图中的数据也跟着更新，并且视图更新的不仅仅是数量和单价，总价也会更新。

【例9.12】 使用DELETE语句删除视图view_t2中的一条记录，SQL语句如下：

```
mysql> DELETE FROM view_t2 WHERE price=5;
Query OK, 1 row affected (0.03 sec)
```

查看视图view_t2和表tv中的数据：

```
mysql> SELECT * FROM view_t2;
+------+-------+-------+
| qty  | price | total |
+------+-------+-------+
|   5  |   50  |  250  |
+------+-------+-------+
1 row in set (0.00 sec)

mysql> SELECT * FROM tv;
+----------+-------+
| quantity | price |
+----------+-------+
|     5    |   50  |
+----------+-------+
1 row in set (0.02 sec)
```

在视图view_t2中删除price=5的记录，该删除操作最终是通过删除基本表中相关的记录来实现的。查看执行删除操作之后的表tv和视图view_t2，可以看到通过视图删除了其所依赖的基本表tv中的数据。

当视图中包含有如下内容时，视图的更新操作将不能被执行：

（1）视图中不包含基表中被定义为非空的列。
（2）在定义视图的SELECT语句后的字段列表中使用了数学表达式。
（3）在定义视图的SELECT语句后的字段列表中使用了聚合函数。
（4）在定义视图的SELECT语句中使用了DISTINCT、UNION、TOP、GROUP BY或HAVING子句。

9.6 删除视图

当视图不再需要时，可以将其删除。删除一个或多个视图可以使用DROP VIEW语句，语法如下：

```
DROP VIEW [IF EXISTS]
    view_name [, view_name] ...
    [RESTRICT | CASCADE]
```

其中，view_name是要删除的视图名称，可以添加多个需要删除的视图名称，各个名称之间使用逗号分隔开。删除视图必须拥有DROP权限。

【例9.13】 删除stu_glass视图，SQL语句如下：

```
DROP VIEW IF EXISTS stu_glass;
```

如果名称为"stu_glass"的视图存在，那么该视图将被删除。

使用SHOW CREATE VIEW语句查看操作结果：

```
mysql> SHOW CREATE VIEW stu_glass;
ERROR 1146 (42S02): Table 'test_db.stu_glass' doesn't exist
```

可以看到，stu_glass视图已经不存在了，说明删除视图成功。

第 10 章 MySQL触发器

MySQL的触发器（Trigger）和存储过程一样，都是嵌入MySQL的一段程序。触发器是由事件来触发某个操作，这些事件包括INSERT、UPDATAE和DELETE。如果定义了触发程序，当数据库执行这些语句的时候，就会激发触发器执行相应的操作。触发程序是与表有关的命名数据库对象，当表上出现特定事件时，将激活该对象。本章通过实例来介绍触发器的含义、如何创建、查看、使用以及删除触发器。

10.1 创建触发器

触发器是一个特殊的存储过程，它与普通的存储过程不同的是：存储过程的执行要使用CALL语句来调用，而触发器的执行不需要使用CALL语句来调用，也不需要手工启动，只要当一个预定义的事件发生的时候，就会被MySQL自动调用。例如，当对表fruits进行操作（INSERT、DELETE或UPDATE）时，就会激活触发器。

触发器不仅可以查询其他表，而且可以包含复杂的SQL语句。它们主要用于满足复杂的业务规则或要求。例如，可以根据客户当前的账户状态，控制是否允许插入新订单。本节将介绍如何创建触发器。

10.1.1 创建只有一个执行语句的触发器

创建只有一个执行语句的触发器的语法如下：

```
CREATE TRIGGER trigger_name trigger_time trigger_event
ON tbl_name FOR EACH ROW trigger_stmt
```

其中，trigger_name表示触发器名称，用户自行指定；trigger_time表示触发时机，可以指定为before或after；trigger_event表示触发事件，包括INSERT、UPDATE和DELETE；tbl_name表示建立触发器的表名，即在哪张表上建立触发器；trigger_stmt是触发器的执行语句。

【例10.1】 创建一个单执行语句的触发器,SQL语句如下:

```
mysql> CREATE TABLE account (acct_num INT, amount DECIMAL(10,2));
mysql> CREATE TRIGGER ins_sum BEFORE INSERT ON account FOR EACH ROW SET @sum =
@sum + NEW.amount;
mysql>SET @sum =0;
mysql> INSERT INTO account VALUES(1,1.00), (2,2.00);
mysql> SELECT @sum;
+------+
| @sum |
+------+
| 3.00 |
+------+
```

对于上述语句,首先创建一张表account,表中有两个字段,分别为acct_num(定义为INT类型)和amount(定义成浮点类型);其次,创建一个名为ins_sum的触发器,触发的条件是向数据表account插入数据之前,对新插入的amount字段值进行求和计算。

10.1.2 创建有多个执行语句的触发器

创建有多个执行语句的触发器的语法如下:

```
CREATE TRIGGER trigger_name trigger_time trigger_event
ON tbl_name FOR EACH ROW
    BEGIN
      语句执行列表
    END
```

其中,trigger_name表示触发器的名称,用户自行指定;trigger_time表示触发时机,可以指定为before或after;trigger_event表示触发事件,包括INSERT、UPDATE和DELETE;tbl_name表示建立触发器的表名,即在哪张表上建立触发器;触发器程序可以使用BEGIN和END作为开始和结束,中间包含多条语句。

【例10.2】 创建一个包含多个执行语句的触发器,SQL语句如下:

```
CREATE TABLE test1(a1 INT);
CREATE TABLE test2(a2 INT);
CREATE TABLE test3(a3 INT NOT NULL AUTO_INCREMENT PRIMARY KEY);
CREATE TABLE test4(
  a4 INT NOT NULL AUTO_INCREMENT PRIMARY KEY,
  b4 INT DEFAULT 0
);
DELIMITER //
CREATE TRIGGER testref BEFORE INSERT ON test1
  FOR EACH ROW
BEGIN
    INSERT INTO test2 SET a2 = NEW.a1;
```

```
    DELETE FROM test3 WHERE a3 = NEW.a1;
    UPDATE test4 SET b4 = b4 + 1 WHERE a4 = NEW.a1;
  END
//
DELIMITER ;
INSERT INTO test3 (a3) VALUES
  (NULL), (NULL), (NULL), (NULL), (NULL),
  (NULL), (NULL), (NULL), (NULL), (NULL);
INSERT INTO test4 (a4) VALUES
  (0), (0), (0), (0), (0), (0), (0), (0), (0), (0);
```

上面的代码创建了一个名为"testref"的触发器。这个触发器的触发条件是在向表test1插入数据前执行触发器的语句，具体执行的代码如下：

```
mysql> INSERT INTO test1 VALUES (1), (3), (1), (7), (1), (8), (4), (4);
```

4张表中的数据如下：

```
mysql> SELECT * FROM test1;
+------+
| a1   |
+------+
|   1  |
|   3  |
|   1  |
|   7  |
|   1  |
|   8  |
|   4  |
|   4  |
+------+
8 rows in set (0.00 sec)

mysql> SELECT * FROM test2;
+------+
| a2   |
+------+
|   1  |
|   3  |
|   1  |
|   7  |
|   1  |
|   8  |
|   4  |
|   4  |
+------+
8 rows in set (0.00 sec)

mysql> SELECT * FROM test3;
+-----+
| a3  |
```

```
+----+
| 2  |
| 5  |
| 6  |
| 9  |
| 10 |
+----+
5 rows in set (0.00 sec)

mysql> SELECT * FROM test4;
+----+------+
| a4 | b4   |
+----+------+
| 1  |  3   |
| 2  |  0   |
| 3  |  1   |
| 4  |  2   |
| 5  |  0   |
| 6  |  0   |
| 7  |  1   |
| 8  |  1   |
| 9  |  0   |
| 10 |  0   |
+----+------+
10 rows in set (0.00 sec)
```

执行结果显示，在向表test1插入记录的时候，表test2、表test3、表test4都发生了变化。从这个例子中可以看到，表test1中的INSERT操作触发了触发器，向表test2中插入了表test1中的值，删除了表test3中相同的内容，同时更新了表test4中的b4，即与插入的值相同的个数。

10.2 查看触发器

查看触发器是指查看数据库中已存在的触发器的定义、状态和语法信息等。本节将介绍两种查看触发器的方法，分别是SHOW TRIGGERS和在triggers表中查看触发器信息。

10.2.1 利用 SHOW TRIGGERS 语句查看触发器信息

通过SHOW TRIGGERS命令查看触发器的语句如下：

```
SHOW TRIGGERS;
```

【例10.3】通过SHOW TRIGGERS命令查看一个触发器。

首先创建一个简单的触发器，名称为trig_update，每次向表account更新数据之后，都会向名称为"myevent"的数据表中插入一条记录。

创建触发器的SQL语句如下：

```
mysql> CREATE TRIGGER trig_update AFTER UPDATE ON account
    -> FOR EACH ROW INSERT INTO myevent VALUES (1,'after update');
Query OK, 0 rows affected (0.00 sec)
```

数据表myevent定义如下：

```
CREATE TABLE myevent
(
id int DEFAULT NULL,
evt_name char(20) DEFAULT NULL
) ;
```

使用SHOW TRIGGERS命令查看触发器：

```
 mysql> SHOW TRIGGERS;
+------+-----+------+---------+------+--------+---------+--------+-----------------
----+--------------------+------------------+
|Trigger|Event|Table|Statement|Timing|Created|sql_mode|Definer|character_set_c
lient|collation_connection|Database Collation|
+------+-----+------+---------+------+--------+---------+--------+-----------------
----+--------------------+------------------+
|ins_sum|INSERT|account| SET @sum = @sum + NEW.amount| BEFORE | NULL| |
root@localhost|latin1| latin1_wedish_ci    | latin1_swedish_ci |
| trig_update | UPDATE | account | INSERT INTO myevent VALUES (1,'after update')|
AFTER| NULL | |root@localhost | utf8mb4          | utf8mb4_0900_ai_ci    |
utf8mb4_0900_ai_ci|
+------+-----+------+---------+------+--------+---------+--------+-----------------
----+--------------------+------------------+
2 rows in set (0.00 sec)
```

可以看到，信息显示比较混乱。如果在SHOW TRIGGERS命令的后面添加上"\G"，显示信息会比较有条理，执行情况如下：

```
mysql> SHOW TRIGGERS \G
*************************** 1. row ***************************
             Trigger: ins_sum
               Event: INSERT
               Table: account
           Statement: SET @sum = @sum + NEW.amount
              Timing: BEFORE
             Created: 2024-07-17 09:27:50.92
            sql_mode: STRICT_TRANS_TABLES,NO_ENGINE_SUBSTITUTION
             Definer: root@localhost
character_set_client: utf8mb4
collation_connection: utf8mb4_0900_ai_ci
  Database Collation: utf8mb4_0900_ai_ci
*************************** 2. row ***************************
             Trigger: trig_update
               Event: UPDATE
               Table: account
           Statement: INSERT INTO myevent VALUES (1,'after update')
              Timing: AFTER
```

```
            Created: 2024-07-17 09:33:51.49
           sql_mode: STRICT_TRANS_TABLES,NO_ENGINE_SUBSTITUTION
            Definer: root@localhost
character_set_client: utf8mb4
collation_connection: utf8mb4_0900_ai_ci
  Database Collation: utf8mb4_0900_ai_ci
*************************** 3. row ***************************
            Trigger: testref
              Event: INSERT
              Table: test1
          Statement: BEGIN
  INSERT INTO test2 SET a2 = NEW.a1;
  DELETE FROM test3 WHERE a3 = NEW.a1;
  UPDATE test4 SET b4 = b4 + 1 WHERE a4 = NEW.a1;
  END
             Timing: BEFORE
            Created: 2024-03-17 09:31:33.40
           sql_mode: STRICT_TRANS_TABLES,NO_ENGINE_SUBSTITUTION
            Definer: root@localhost
character_set_client: utf8mb4
collation_connection: utf8mb4_0900_ai_ci
  Database Collation: utf8mb4_0900_ai_ci
```

其中，Trigger表示触发器的名称，在这里3个触发器的名称分别为ins_sum、trig_update和testref；Event表示激活触发器的事件，这里的触发事件包括插入操作INSERT和更新操作UPDATE；Table表示激活触发器的操作对象表；Statement表示触发器执行的操作；Timing表示触发器触发的时间，分别为插入操作之前（BEFORE）和更新操作之后（AFTER）。还有一些其他信息，比如sql的模式、触发器的定义账户和字符集等，这里不再一一介绍。

> **提示** 在触发器较少的情况下，使用SHOW TRIGGERS语句查看当前创建的所有触发器信息会很方便。如果要查看特定触发器的信息，可以直接从information_schema数据库中的triggers表中查找。在10.2.2节中，将介绍这种方法。

10.2.2 在triggers表中查看触发器信息

在MySQL中，所有触发器的定义都存在information_schema数据库的triggers表中，可以通过查询命令SELECT查看，具体的语法如下：

```
SELECT * FROM information_schema.triggers WHERE condition;
```

【例10.4】通过SELECT命令查看触发器，SQL语句如下：

```
SELECT * FROM information_schema.triggers WHERE TRIGGER_NAME= 'trig_update'\G
```

上述语句通过WHERE来指定查看特定名称的触发器。指定触发器名称的执行情况如下：

```
*************************** 1. row ***************************
         TRIGGER_CATALOG: def
          TRIGGER_SCHEMA: test_db
```

```
            TRIGGER_NAME: trig_update
      EVENT_MANIPULATION: UPDATE
    EVENT_OBJECT_CATALOG: def
     EVENT_OBJECT_SCHEMA: test_db
      EVENT_OBJECT_TABLE: account
            ACTION_ORDER: 1
        ACTION_CONDITION: NULL
        ACTION_STATEMENT: INSERT INTO myevent VALUES (1,'after update')
      ACTION_ORIENTATION: ROW
           ACTION_TIMING: AFTER
ACTION_REFERENCE_OLD_TABLE: NULL
ACTION_REFERENCE_NEW_TABLE: NULL
  ACTION_REFERENCE_OLD_ROW: OLD
  ACTION_REFERENCE_NEW_ROW: NEW
                 CREATED: 2022-03-17 09:33:51.49
                SQL_MODE: STRICT_TRANS_TABLES,NO_ENGINE_SUBSTITUTION
                 DEFINER: root@localhost
    CHARACTER_SET_CLIENT: utf8mb4
    COLLATION_CONNECTION: utf8mb4_0900_ai_ci
      DATABASE_COLLATION: utf8mb4_0900_ai_ci
```

执行结果中，TRIGGER_SCHEMA表示触发器所在的数据库；TRIGGER_NAME是触发器的名称；EVENT_OBJECT_TABLE表示在哪张数据表上触发；ACTION_STATEMENT表示触发器触发的时候执行的具体操作；ACTION_ORIENTATION是ROW，表示在每条记录上都触发；ACTION_TIMING表示触发的时刻是AFTER；剩下的是和系统相关的信息。

也可以不指定触发器名称，这样将查看所有的触发器，SQL语句如下：

```
SELECT * FROM information_schema.triggers \G
```

这条语句会显示triggers表中所有的触发器信息。

10.3 触发器的使用

触发程序是与表有关的命名数据库对象，当表上出现特定事件时，将激活该对象。某些触发程序可用于检查插入表中的值，或对更新涉及的值进行计算。

触发程序与表相关，当对表执行INSERT、DELETE或UPDATE语句时，将激活触发程序。可以将触发程序设置为在执行语句之前或之后激活。例如，可以在从表中删除每一行之前或在更新每一行之后激活触发程序。

【例10.5】创建一个在向表account插入记录之后更新数据表myevent的触发器，SQL语句如下：

```
CREATE TRIGGER trig_insert AFTER INSERT ON account FOR EACH ROW INSERT INTO myevent
VALUES (2,'after insert');
```

上面的代码创建了一个trig_insert触发器，在向表account插入数据之后会向表myevent插入一组数据，代码执行如下：

```
mysql> CREATE TRIGGER trig_insert AFTER INSERT ON account
    -> FOR EACH ROW INSERT INTO myevent VALUES (2, 'after insert ');
mysql>  INSERT INTO account VALUES (1,1.00), (2,2.00);
mysql> SELECT * FROM myevent;
+------+--------------+
| id   | evt_name     |
+------+--------------+
|    2 | after insert |
|    2 | after insert |
+------+--------------+
```

从执行的结果来看在向表account插入记录之后触发了trig_insert触发器，执行的操作是向表myevent插入一条记录。

10.4　删除触发器

使用DROP TRIGGER语句可以删除MySQL中已经定义的触发器，基本语法格式如下：

```
DROP TRIGGER [schema_name.]trigger_name
```

其中，schema_name表示数据库名称，是可选的。如果省略了schema，将从当前数据库中删除触发程序；trigger_name是要删除的触发器的名称。

【例10.6】删除一个触发器，SQL语句如下：

```
DROP TRIGGER test_db.ins_sum;
```

test_db是触发器所在的数据库，ins_sum是要删除的触发器的名称。代码执行如下：

```
mysql> DROP TRIGGER test_db.ins_sum;
Query OK, 0 rows affected (0.00 sec)
```

触发器ins_sum删除成功。

第 11 章 数据备份与恢复

尽管采取了一些管理措施来保证数据库的安全，但是意外情况总是有可能造成数据的损失，例如意外停电、管理员的操作失误等都可能会造成数据的丢失。保证数据安全最重要的一个措施是确保对数据进行定期备份。如果数据库中的数据丢失或者出现错误，可以使用备份的数据进行恢复，这样就最大限度地降低了意外原因导致的损失。MySQL提供了多种方法对数据进行备份和恢复。本章将介绍数据备份、数据恢复、数据迁移、数据的导出和导入的相关知识。

11.1 数据备份

数据备份是数据库管理员非常重要的工作之一。系统意外崩溃或者硬件的损坏都可能导致数据的丢失，因此MySQL管理员应该定期地备份数据，使得在意外情况发生时最大限度地减少损失。本节将介绍数据备份的3种方法。

11.1.1 使用 mysqldump 命令备份数据

mysqldump是MySQL提供的一个非常有用的数据库备份工具。mysqldump命令执行时，可以将数据库备份成一个文本文件，该文件中实际包含了多个CREATE和INSERT语句，使用这些语句可以重新创建表和插入数据。

mysqldump备份数据库的基本语法格式如下：

```
mysqldump -u user -h host -p password dbname[tbname, [tbname...]]> filename.sql
```

user表示用户名称；host表示登录用户的主机名称；password为登录密码；dbname为需要备份的数据库名称；tbname为dbname数据库中需要备份的数据表，可以指定多张需要备份的表，如果不指定，则表示备份所有数据表；右箭头符号（>）告诉mysqldump将备份数据表的定义和数据写入备份文件；filename.sql为备份文件的名称。

1. 使用mysqldump备份单个数据库中的所有表

【例11.1】使用mysqldump命令备份单个数据库中的所有表。

为了更好地理解mysqldump工具是如何工作的，这里给出一个完整的数据库例子。首先登录MySQL，按下面数据库结构创建booksDB数据库和各个表，并插入数据记录。数据库和表定义如下：

```
CREATE DATABASE booksDB;
use booksDB;

CREATE TABLE books
(
bk_id INT NOT NULL PRIMARY KEY,
bk_title VARCHAR(50) NOT NULL,
copyright YEAR NOT NULL
);
INSERT INTO books
VALUES (11078, 'Learning MySQL', 2010),
(11033, 'Study Html', 2011),
(11035, 'How to use php', 2003),
(11072, 'Teach yourself javascript', 2005),
(11028, 'Learning C++', 2005),
(11069, 'MySQL professional', 2009),
(11026, 'Guide to MySQL 9.0', 2024),
(11041, 'Inside VC++', 2011);

CREATE TABLE authors
(
auth_id INT NOT NULL PRIMARY KEY,
auth_name VARCHAR(20),
auth_gender CHAR(1)
);
INSERT INTO authors
VALUES (1001, 'WriterX' ,'f'),
(1002, 'WriterA' ,'f'),
(1003, 'WriterB' ,'m'),
(1004, 'WriterC' ,'f'),
(1011, 'WriterD' ,'f'),
(1012, 'WriterE' ,'m'),
(1013, 'WriterF' ,'m'),
(1014, 'WriterG' ,'f'),
(1015, 'WriterH' ,'f');

CREATE TABLE authorbook
(
auth_id INT NOT NULL,
bk_id INT NOT NULL,
PRIMARY KEY (auth_id, bk_id),
FOREIGN KEY (auth_id) REFERENCES authors (auth_id),
FOREIGN KEY (bk_id) REFERENCES books (bk_id)
```

```
);
INSERT INTO authorbook
VALUES (1001, 11033), (1002, 11035), (1003, 11072), (1004, 11028),
(1011, 11078), (1012, 11026), (1012, 11041), (1014, 11069);
```

完成数据插入后,打开操作系统命令行输入窗口,输入如下备份命令:

```
C:\ >mysqldump -u root -p booksdb > C:/backup/booksdb_20240301.sql
Enter password: **
```

> **提示** 这里要保证C盘下的backup文件夹存在,否则将提示错误信息:系统找不到指定的路径。

输入密码之后,MySQL便对数据库进行备份。使用文本查看器打开C:\backup文件夹下刚才备份过的文件,部分文件内容大致如下:

```
-- MySQL dump 10.13  Distrib 9.0.1, for Win64 (x86_64)
--
-- Host: localhost    Database: booksdb
-- ------------------------------------------------------
-- Server version    9.0.1
/*!40101 SET @OLD_CHARACTER_SET_CLIENT=@@CHARACTER_SET_CLIENT */;
/*!40101 SET @OLD_CHARACTER_SET_RESULTS=@@CHARACTER_SET_RESULTS */;
/*!40101 SET @OLD_COLLATION_CONNECTION=@@COLLATION_CONNECTION */;
/*!50503 SET NAMES utf8mb4 */;
/*!40103 SET @OLD_TIME_ZONE=@@TIME_ZONE */;
/*!40103 SET TIME_ZONE='+00:00' */;
/*!40014 SET @OLD_UNIQUE_CHECKS=@@UNIQUE_CHECKS, UNIQUE_CHECKS=0 */;
/*!40014 SET @OLD_FOREIGN_KEY_CHECKS=@@FOREIGN_KEY_CHECKS, FOREIGN_KEY_CHECKS=0 */;
/*!40101 SET @OLD_SQL_MODE=@@SQL_MODE, SQL_MODE='NO_AUTO_VALUE_ON_ZERO' */;
/*!40111 SET @OLD_SQL_NOTES=@@SQL_NOTES, SQL_NOTES=0 */;
--
-- Table structure for table `authorbook`
--
DROP TABLE IF EXISTS `authorbook`;
/*!40101 SET @saved_cs_client     = @@character_set_client */;
/*!50503 SET character_set_client = utf8mb4 */;
CREATE TABLE `authorbook` (
  `auth_id` int NOT NULL,
  `bk_id` int NOT NULL,
  PRIMARY KEY (`auth_id`,`bk_id`),
  KEY `bk_id` (`bk_id`),
  CONSTRAINT `authorbook_ibfk_1` FOREIGN KEY (`auth_id`) REFERENCES `authors` (`auth_id`),
  CONSTRAINT `authorbook_ibfk_2` FOREIGN KEY (`bk_id`) REFERENCES `books` (`bk_id`)
) ENGINE=InnoDB DEFAULT CHARSET=utf8mb4 COLLATE=utf8mb4_0900_ai_ci;
/*!40101 SET character_set_client = @saved_cs_client */;
--
```

```sql
-- Dumping data for table `authorbook`
--

LOCK TABLES `authorbook` WRITE;
/*!40000 ALTER TABLE `authorbook` DISABLE KEYS */;
INSERT INTO `authorbook` VALUES
(1012,11026),(1004,11028),(1001,11033),(1002,11035),(1012,11041),(1014,11069),(1003,11072),(1011,11078);
/*!40000 ALTER TABLE `authorbook` ENABLE KEYS */;
UNLOCK TABLES;

--
-- Table structure for table `authors`
--

DROP TABLE IF EXISTS `authors`;
/*!40101 SET @saved_cs_client     = @@character_set_client */;
/*!50503 SET character_set_client = utf8mb4 */;
CREATE TABLE `authors` (
  `auth_id` int NOT NULL,
  `auth_name` varchar(20) DEFAULT NULL,
  `auth_gender` char(1) DEFAULT NULL,
  PRIMARY KEY (`auth_id`)
) ENGINE=InnoDB DEFAULT CHARSET=utf8mb4 COLLATE=utf8mb4_0900_ai_ci;
/*!40101 SET character_set_client = @saved_cs_client */;

--
-- Dumping data for table `authors`
--

LOCK TABLES `authors` WRITE;
/*!40000 ALTER TABLE `authors` DISABLE KEYS */;
INSERT INTO `authors` VALUES
(1001,'WriterX','f'),(1002,'WriterA','f'),(1003,'WriterB','m'),(1004,'WriterC','f'),(1011,'WriterD','f'),(1012,'WriterE','m'),(1013,'WriterF','m'),(1014,'WriterG','f'),(1015,'WriterH','f');
/*!40000 ALTER TABLE `authors` ENABLE KEYS */;
UNLOCK TABLES;
--
-- Table structure for table `books`
--
DROP TABLE IF EXISTS `books`;
/*!40101 SET @saved_cs_client     = @@character_set_client */;
/*!50503 SET character_set_client = utf8mb4 */;
CREATE TABLE `books` (
  `bk_id` int NOT NULL,
  `bk_title` varchar(50) NOT NULL,
  `copyright` year NOT NULL,
  PRIMARY KEY (`bk_id`)
) ENGINE=InnoDB DEFAULT CHARSET=utf8mb4 COLLATE=utf8mb4_0900_ai_ci;
/*!40101 SET character_set_client = @saved_cs_client */;
--
-- Dumping data for table `books`
--
LOCK TABLES `books` WRITE;
```

```
    /*!40000 ALTER TABLE `books` DISABLE KEYS */;
    INSERT INTO `books` VALUES (11026,'Guide to MySQL 9.0',2024),(11028,'Learning
C++',2005),(11033,'Study Html',2011),(11035,'How to use php',2003),(11041,'Inside
VC++',2011),(11069,'MySQL professional',2009),(11072,'Teach yourself
javascript',2005),(11078,'Learning MySQL',2010);
    /*!40000 ALTER TABLE `books` ENABLE KEYS */;
    UNLOCK TABLES;
    /*!40103 SET TIME_ZONE=@OLD_TIME_ZONE */;
    /*!40101 SET SQL_MODE=@OLD_SQL_MODE */;
    /*!40014 SET FOREIGN_KEY_CHECKS=@OLD_FOREIGN_KEY_CHECKS */;
    /*!40014 SET UNIQUE_CHECKS=@OLD_UNIQUE_CHECKS */;
    /*!40101 SET CHARACTER_SET_CLIENT=@OLD_CHARACTER_SET_CLIENT */;
    /*!40101 SET CHARACTER_SET_RESULTS=@OLD_CHARACTER_SET_RESULTS */;
    /*!40101 SET COLLATION_CONNECTION=@OLD_COLLATION_CONNECTION */;
    /*!40111 SET SQL_NOTES=@OLD_SQL_NOTES */;
    -- Dump completed on 2024-07-25 18:39:42
```

可以看到，备份文件中包含了一些信息，文件开头首先表明了备份文件使用的mysqldump工具的版本号；然后是备份账户的名称和主机信息，以及备份的数据库的名称；最后是MySQL服务器的版本号，在这里为9.0.1。

接下来是一些SET语句，这些语句将一些系统变量值赋给用户自定义变量，以确保被恢复的数据库的系统变量和原来备份时的变量相同，例如：

```
    /*!40101 SET @OLD_CHARACTER_SET_CLIENT=@@CHARACTER_SET_CLIENT */;
```

该SET语句将当前系统变量character_set_client的值赋给用户自定义变量@old_character_set_client。其他变量与此类似。

备份文件的最后几行是使用SET语句恢复服务器系统变量原来的值，例如：

```
    /*!40101 SET CHARACTER_SET_CLIENT=@OLD_CHARACTER_SET_CLIENT */;
```

该语句将用户自定义的变量@old_character_set_client中保存的值赋给实际的系统变量character_set_client。

备份文件中以"--"字符开头的行为注释语句；以"/*!"开头、"*/"结尾的语句为可执行的MySQL注释，这些语句可以被MySQL执行，但在其他数据库管理系统中将被作为注释忽略，以提高数据库的可移植性。

另外，备份文件的一些可执行注释语句以数字开头，该数字代表的是MySQL版本号，表示这些语句只有在指定的MySQL版本或者比该版本高的情况下才能执行。例如"40101"，表明这些语句只有在MySQL版本号为4.01.01或者更高的条件下才可以被执行。

2. 使用mysqldump备份单个数据库中的某张表

mysqldump还可以备份数据库中的某张表。

备份某张表和备份数据库中所有表的语句的不同之处在于，要在数据库名称dbname之后指定需要备份的表名称。

【例11.2】备份booksDB数据库中的表books，SQL语句如下：

```
mysqldump -u root -p booksDB books > C:/backup/books_20240301.sql
```

该语句创建名称为"books_20240301.sql"的备份文件，文件中包含了前面介绍的SET语句等内容；不同的是，该文件只包含表books的CREATE和INSERT语句。

3. 使用mysqldump备份多个数据库

如果要使用mysqldump备份多个数据库，就需要使用--databases参数。备份多个数据库的语法格式如下：

```
mysqldump -u user -h host -p --databases [dbname, [dbname...]] > filename.sql
```

使用--databases参数之后，必须指定至少一个数据库的名称，多个数据库名称之间用空格隔开。

【例11.3】使用mysqldump备份数据库booksDB和test_db，SQL语句如下：

```
mysqldump -u root -p --databases booksDB test_db>C:\backup\books_testDB_20240301.sql
```

该语句创建名称为"books_testDB_20240301.sql"的备份文件，该文件中包含了创建两个数据库booksDB和test_db所必需的所有语句。

另外，使用--all-databases参数可以备份系统中所有的数据库，SQL语句如下：

```
mysqldump -u user -h host -p --all-databases > filename.sql
```

使用参数--all-databases时，不需要指定数据库名称。

【例11.4】使用mysqldump备份服务器中的所有数据库，SQL语句如下：

```
mysqldump -u root -p --all-databases > C:/backup/alldbinMySQL.sql
```

该语句创建名称为"alldbinMySQL.sql"的备份文件，文件中包含了系统中所有数据库的备份信息。

> 提示 如果在服务器上进行备份，并且表均为MyISAM表，就应该考虑使用MySQLhotcopy，因为可以更快地进行备份和恢复。

11.1.2 直接复制整个数据库目录

因为MySQL表保存为文件方式，所以可以直接复制MySQL数据库的存储目录和文件进行备份。MySQL的数据库目录位置不一定相同，在Windows平台下，MySQL 9.0存放数据库的目录通常为"C:\Documents and Settings\All Users\Application Data\MySQL\MySQL Server 9.0\data"或者其他用户自定义目录；在Linux平台下，数据库目录位置通常为"/var/lib/MySQL/"，不同Linux版本下目录会有所不同，读者应在自己使用的平台下查找该目录。

这是一种简单、快速、有效的备份方式。要想保持备份的一致性，备份前需要对相关表执行LOCK TABLES操作，然后对表执行FLUSH TABLES语句。FLUSH TABLES语句可以确保开始备份前将所有激活的索引页写入硬盘，这样当复制数据库目录中的文件时，将允许其他客户继续查询表。当然，也可以停止MySQL服务再进行备份操作。

这种方法虽然简单，但并不是最好的，因为这种方法对InnoDB存储引擎的表不适用。使用这种方法备份的数据最好恢复到相同版本的服务器中，不同的版本可能不兼容。

> **提示** 在MySQL版本号中，第一个数字表示主版本号，主版本号相同的MySQL数据库文件格式相同。

11.1.3 使用 MySQLhotcopy 工具快速备份

MySQLhotcopy是一个Perl脚本，最初由Tim Bunce编写并提供。它使用LOCK TABLES、FLUSH TABLES和cp或scp来快速备份数据库。它是备份数据库或单表最快的途径，但它只能运行在数据库目录所在的机器上，并且只能备份MyISAM类型的表。MySQLhotcopy在UNIX系统中运行。

MySQLhotcopy命令的语法格式如下：

```
mysqlhotcopy db_name_1, ... db_name_n  /path/to/new_directory
```

db_name_1,...,db_name_n分别为需要备份的数据库的名称；"/path/to/new_directory"指定备份文件目录。

【例11.5】使用MySQLhotcopy备份数据库test_db到"/usr/backup"目录下，SQL语句如下：

```
mysqlhotcopy -u root -p test_db /usr/backup
```

要想执行MySQLhotcopy，必须可以访问备份的表文件，具有那些表的SELECT权限、RELOAD权限（以便能够执行FLUSH TABLES）和LOCK TABLES权限。

> **提示** MySQLhotcopy只是将表所在的目录复制到另一个位置，只能用于备份MyISAM表和ARCHIVE表，备份InnoDB类型的数据表时会出现错误信息。由于它复制本地格式的文件，因此也不能移植到其他硬件或操作系统下。

11.2 数据恢复

管理人员操作的失误、计算机故障以及其他意外情况，都会导致数据的丢失和破坏。当数据丢失或意外破坏时，可以通过恢复已经备份的数据来最大限度地减少数据丢失和破坏造成的损失。本节将介绍数据恢复的方法。

11.2.1 使用 mysql 命令恢复数据

对于已经备份的包含CREATE、INSERT语句的sql文件，可以使用mysql命令导入数据库中。本小节将介绍使用mysql命令导入sql文件的方法。

备份的sql文件中包含CREATE、INSERT语句（有时也会有DROP语句），mysql命令可以直接执行文件中的这些语句，其语法如下：

```
mysql -u user -p [dbname] < filename.sql
```

user是执行filename.sql中的语句的用户名；-p表示输入用户密码；dbname是数据库名。如果filename.sql文件为mysqldump工具创建的包含创建数据库语句的文件，则执行的时候不需要指定数据库名。

【例11.6】使用mysql命令将C:\backup\booksdb_20240301.sql文件中的备份导入数据库中，SQL语句如下：

```
mysql -u root -p booksDB < C:/backup/booksdb_20240301.sql
```

执行该语句前，必须先在MySQL服务器中创建booksDB数据库，如果该数据库不存在，则恢复过程将会出错。命令执行成功之后，booksdb_20220301.sql文件中的语句就会在指定的数据库中恢复以前的表。

如果已经登录MySQL服务器，还可以使用source命令导入sql文件。Source命令的语法如下：

```
source filename
```

【例11.7】使用root用户登录到服务器，然后使用source导入本地的备份文件booksdb_20240301.sql，SQL语句如下：

```
--选择要恢复到的数据库
mysql> use booksDB;
Database changed

--使用source命令导入备份文件
mysql> source C:\backup\booksdb_20240301.sql
```

命令执行后，会列出备份文件booksdb_20240301.sql中每一条语句的执行结果。source命令执行成功后，booksdb_20240301.sql中的语句会全部导入现有数据库中。

> 提示 执行source命令前，必须使用use语句选择数据库；否则恢复过程中会出现"ERROR 1046 (3D000): No database selected"错误。

11.2.2 直接复制到数据库目录

在MySQL中，使用复制数据库文件的方法进行备份和恢复也是一种常见的方法。但是，通过这种方式恢复数据时，保存备份数据的数据库和待恢复的数据库的服务器的主版本号必须相同，而且这种方式只对MyISAM引擎的表有效，对InnoDB引擎的表不可用。

执行恢复前，先关闭MySQL服务，将备份的文件或目录覆盖MySQL的data目录，再启动MySQL服务。对于Linux/UNIX操作系统来说，复制完文件后，需要将文件的用户和组更改为MySQL运行的用户和组，通常用户是MySQL，组也是MySQL。

11.2.3　mysqlhotcopy 快速恢复

使用mysqlhotcopy备份的文件也可以用来恢复数据库。在MySQL服务器停止运行时，将备份的数据库文件复制到MySQL存放数据的位置（MySQL的data文件夹），再重新启动MySQL服务即可。如果以根用户执行该操作，则必须指定数据库文件的所有者，SQL语句如下：

```
chown -R mysql.mysql /var/lib/mysql/dbname
```

【例11.8】使用mysqlhotcopy复制的备份文件恢复数据库，SQL语句如下：

```
cp -R /usr/backup/test usr/local/mysql/data
```

执行完该语句，重启服务器，MySQL将恢复到备份状态。

> 提示　如果需要恢复的数据库已经存在，则在使用DROP语句删除已经存在的数据库之后，恢复才能成功。另外，MySQL不同版本之间必须兼容，恢复之后的数据才可以使用。

11.3　数据库迁移

数据库迁移就是把数据从一个系统移动到另一个系统上。数据迁移有以下原因：

（1）需要安装新的数据库服务器。
（2）MySQL版本更新。
（3）数据库管理系统的变更（如从Microsoft SQL Server迁移到MySQL）。

本节将讲解数据库迁移的方法。

11.3.1　相同版本的 MySQL 数据库之间的迁移

相同版本的MySQL数据库之间的迁移就是在主版本号相同的MySQL数据库之间进行数据库移动。迁移过程其实就是在源数据库备份和在目标数据库恢复的过程。

在讲解数据库备份和恢复时，已经知道最简单的方式是复制数据库文件目录，但是此种方法只适用于MyISAM引擎的表，而对于InnoDB表，不能用直接复制文件的方式备份数据库。因此，最常用和最安全的方式是使用mysqldump命令导出数据，然后在目标数据库服务器使用mysql命令导入。

【例11.9】将www.abc.com主机上的MySQL数据库全部迁移到www.bcd.com主机上。在www.abc.com主机上执行如下命令：

```
mysqldump -h www.bac.com -uroot -ppassword dbname |
mysql -h www.bcd.com -uroot -ppassword
```

mysqldump导入的数据直接通过管道符"｜"传给mysql命令导入主机www.bcd.com的数据库中，dbname为需要迁移的数据库名称。如果要迁移全部的数据库，可使用参数--all-databases。

11.3.2　不同版本的 MySQL 数据库之间的迁移

因为数据库升级等原因，需要将旧版本的MySQL数据库中的数据迁移到新版本的数据库中。不同版本的MySQL数据库之间迁移的方法是：

（1）备份旧版本的MySQL中的数据库。
（2）停止旧版本MySQL服务。
（3）卸载旧版本MySQL。
（4）安装新版本MySQL，并配置相应的设置，如数据目录、日志目录等。确保在安装过程中启用所需的存储引擎，如InnoDB。
（5）启动新版本MySQL服务。

旧版本与新版本的MySQL可能使用不同的默认字符集，例如MySQL 8.0版本之前，默认字符集为latin1，而从MySQL 8.0版本开始，默认字符集为utf8mb4。数据库中有中文数据的，迁移过程中需要对默认字符集进行修改，否则可能无法正常显示结果。

新版本会对旧版本有一定兼容性。从旧版本的MySQL向新版本的MySQL迁移时，对于MyISAM引擎的表，可以直接复制数据库文件，也可以使用mysqlhotcopy和mysqldump工具；对于InnoDB引擎的表，一般只能使用mysqldump将数据导出，然后使用mysql命令导入目标服务器上。从新版本的MySQL向旧版本的MySQL迁移数据时要特别小心，最好使用mysqldump命令导出，然后导入目标数据库中。

11.3.3　不同数据库之间的迁移

不同类型的数据库之间的迁移，是指把MySQL的数据库转移到其他类型的数据库，例如从MySQL迁移到Oracle，从Oracle迁移到MySQL，从MySQL迁移到SQL Server等。

迁移之前，需要了解不同数据库的架构，比较它们之间的差异。不同数据库中定义相同类型的数据的关键字可能会不同。例如，MySQL中日期字段分为DATE和TIME两种，而Oracle的日期字段只有DATE。另外，数据库厂商并没有完全按照SQL标准来设计数据库系统，导致不同的数据库系统的SQL语句有差别。例如，MySQL几乎完全支持标准SQL语言，而Microsoft SQL Server使用的是T-SQL语言，T-SQL中有一些非标准的SQL语句，因此在迁移时必须对这些语句进行语句映射处理。

数据库迁移可以使用一些工具。例如，在Windows系统下，可以使用MyODBC实现MySQL和SQL Server之间的迁移。MySQL官方提供的工具MySQL Migration Toolkit也可以在不同数据库之间进行数据迁移。

11.4 数据的导出和导入

有时需要将MySQL数据库中的数据导出到外部存储文件中，MySQL数据库中的数据可以导出成sql文本文件、xml文件或者html文件。同样地，这些导出文件也可以导入MySQL数据库中。本节将介绍数据导出和导入的常用方法。

11.4.1 使用 SELECT...INTO OUTFILE 导出文本文件

MySQL数据库导出数据时，允许使用包含导出定义的SELECT语句进行数据的导出操作。该文件被创建到服务器主机上，因此必须拥有文件写入权限（FILE权限）才能使用此语法。"SELECT...INTO OUTFILE 'filename'"形式的SELECT语句可以把被选择的行写入一个文件中，并且filename不能是一个已经存在的文件。SELECT...INTO OUTFILE语句的基本格式如下：

```
SELECT columnlist  FROM table WHERE condition  INTO OUTFILE 'filename'  [OPTIONS]
```

可以看到SELECT columnlist FROM table WHERE condition为一个查询语句，查询结果返回满足指定条件的一条或多条记录；INTO OUTFILE语句的作用就是把前面SELECT语句查询出来的结果，导出到名称为"filename"的外部文件中；[OPTIONS]为可选参数选项，OPTIONS部分的语法包括FIELDS和LINES子句，其可能的取值如下：

- FIELDS TERMINATED BY 'value'：设置字段之间的分隔字符，可以为单个或多个字符，默认情况下为制表符（\t）。
- FIELDS [OPTIONALLY] ENCLOSED BY 'value'：设置字段的包围字符，只能为单个字符，若使用了OPTIONALLY关键字，则只有CHAR和VERCHAR等字符数据字段被包围。
- FIELDS ESCAPED BY 'value'：设置如何写入或读取特殊字符，只能为单个字符，即设置转义字符，默认值为"\"。
- LINES STARTING BY 'value'：设置每行数据开头的字符，可以为单个或多个字符，默认情况下不使用任何字符。
- LINES TERMINATED BY 'value'：设置每行数据结尾的字符，可以为单个或多个字符，默认值为"\n"。

FIELDS和LINES两个子句都是自选的，但是如果两个都被指定了，则FIELDS必须位于LINES的前面。

使用SELECT...INTO OUTFILE语句，可以非常快速地把一张表转储到服务器上。如果想要在服务器主机之外的部分客户主机上创建结果文件，不能使用SELECT...INTO OUTFILE，而应该使用"MySQL –e "SELECT ..." > file_name"这类的命令来生成文件。

SELECT...INTO OUTFILE是LOAD DATA INFILE的补语，用于语句的OPTIONS部分的语法包括部分FIELDS和LINES子句，这些子句与LOAD DATA INFILE语句同时使用。

【例11.10】使用SELECT...INTO OUTFILE将test_db数据库中的person表中的记录导出到文本文件，SQL语句如下：

```
mysql> SELECT * FROM test_db.person INTO OUTFILE 'D:/person0.txt';
```

语句执行后报错信息如下：

```
ERROR 1290 (HY000): The MySQL server is running with the --secure-file-priv option so it cannot execute this statement
```

这是因为MySQL默认对导出的目录有权限限制，也就是说使用命令行进行导出的时候，需要指定目录。那么指定的目录是什么呢？

查询指定目录的命令如下：

```
show global variables like '%secure%';
```

执行结果如下：

```
+------------------------+------------------------------------------------+
| Variable_name          | Value                                          |
+------------------------+------------------------------------------------+
|require_secure_transport| OFF                                            |
|secure_file_priv        | D:\ProgramData\MySQL\MySQL Server 9.0\Uploads\ |
+------------------------+------------------------------------------------+
```

因为secure_file_priv配置的关系，所以必须导出到D:\ProgramData\MySQL\MySQL Server 9.0\Uploads\目录下，该目录就是指定目录。如果想自定义导出路径，需要修改my.ini配置文件。打开路径D:\ProgramData\MySQL\MySQL Server 9.0，用记事本打开my.ini文件，然后搜索以下代码：

```
secure-file-priv="D:/ProgramData/MySQL/MySQL Server 9.0/Uploads"
```

在上述代码前添加#注释掉，然后添加以下内容：

```
secure-file-priv="D:/"
```

结果如图11.1所示。

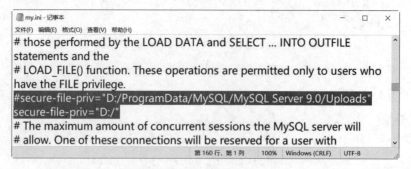

图11.1 设置数据表的导出路径

重启MySQL服务器后，再次使用SELECT…INTO OUTFILE将test_db数据库中的person表中的记录导出到文本文件，SQL语句如下：

```
mysql>SELECT * FROM test_db.person INTO OUTFILE 'D:/person0.txt';
Query OK, 1 row affected (0.01 sec)
```

由于指定了INTO OUTFILE子句，因此SELECT会将查询出来的3个字段值保存到C:\person0.txt文件中。打开该文件，内容如下：

```
1       Green           21      Lawyer
2       Suse            22      dancer
3       Mary            24      Musician
4       Willam          20      sports man
5       Laura           25      \N
6       Evans           27      secretary
7       Dale            22      cook
8       Edison          28      singer
9       Harry           21      magician
10      Harriet         19      pianist
```

默认情况下，MySQL使用制表符（\t）分隔不同的字段，字段没有被其他字符包围。另外，第5行中有一个字段值为"\N"，表示该字段的值为NULL。默认情况下，当遇到NULL时，会返回"\N"，代表空值，其中的反斜线（\）表示转义字符；如果使用ESCAPED BY选项，则N前面为指定的转义字符。

【例11.11】使用SELECT…INTO OUTFILE语句，将test_db数据库person表中的记录导出到文本文件，使用FIELDS选项和LINES选项，要求字段之间使用逗号分隔，所有字段值用双引号引起来，定义转义字符为单引号"\'"，SQL语句如下：

```
SELECT * FROM  test_db.person INTO OUTFILE "D:/person1.txt"
FIELDS
TERMINATED BY ','
ENCLOSED BY '\"'
ESCAPED BY '\''
LINES
TERMINATED BY '\r\n';
```

该语句将把person表中所有记录导入D盘目录下的person1.txt文本文件中。

"FIELDS TERMINATED BY ','"表示字段之间用逗号分隔；"ENCLOSED BY \""表示每个字段用双引号引起来；"ESCAPED BY '\'"表示将系统默认的转义字符替换为单引号；"LINES TERMINATED BY '\r\n'"表示每行以回车换行符结尾，保证每一条记录占一行。

语句执行成功后，在D盘下生成一个person1.txt文件。打开该文件，内容如下：

```
"1","Green","21","Lawyer"
"2","Suse","22","dancer"
"3","Mary","24","Musician"
"4","Willam","20","sports man"
"5","Laura","25",'N
"6","Evans","27","secretary"
```

```
"7","Dale","22","cook"
"8","Edison","28","singer"
"9","Harry","21","magician"
"10","Harriet","19","pianist"
```

可以看到，所有的字段值都被双引号引起来；第5条记录中空值的表示形式为"N"，即使用单引号替换了反斜线转义字符。

【例11.12】使用SELECT...INTO OUTFILE语句，将test_db数据库person表中的记录导出到文本文件，使用LINES选项，要求每行记录以字符串">"开始、以字符串"<end>"结尾，SQL语句如下：

```
SELECT * FROM test_db.person INTO OUTFILE "D:/person2.txt"
LINES
STARTING BY '> '
TERMINATED BY '<end>';
```

语句执行成功后，在D盘下生成一个person2.txt文件。打开该文件，内容如下：

```
> 1 Green    21   Lawyer <end>> 2 Suse22   dancer <end>> 3 Mary24   Musician
<end>> 4Willam  20  sports man <end>> 5 Laura   25   \N <end>> 6 Evans    27
secretary <end>> 7   Dale22  cook <end>> 8   Edison  28  singer <end>> 9
Harry   21  magician <end>> 10  Harriet 19  pianist <end>
```

可以看到，虽然将所有的字段值导出到文本文件中，但是所有的记录没有分行，出现这种情况是因为TERMINATED BY选项替换了系统默认的换行符。如果希望换行显示，则需要修改导出语句：

```
SELECT * FROM test_db.person INTO OUTFILE "D:/person3.txt"
LINES
STARTING BY '> '
TERMINATED BY '<end>\r\n';
```

执行完语句之后，换行显示每条记录，结果如下：

```
> 1 Green     21   Lawyer <end>
> 2 Suse      22   dancer <end>
> 3 Mary      24   Musician <end>
> 4 Willam    20   sports man <end>
> 5 Laura     25   \N <end>
> 6 Evans     27   secretary <end>
> 7 Dale      22   cook <end>
> 8 Edison    28   singer <end>
> 9 Harry     21   magician <end>
> 10Harriet   19   pianist <end>
```

11.4.2 使用 mysqldump 命令导出文本文件

除了使用SELECT... INTO OUTFILE语句导出文本文件之外，还可以使用mysqldump命令。11.1节开始介绍了使用mysqldump备份数据库，该工具不仅可以将数据导出为包含CREATE、INSERT的sql文件，也可以导出为纯文本文件。

mysqldump创建一个包含创建表的CREATE TABLE语句的tablename.sql文件和一个包含其数据的tablename.txt文件。mysqldump导出文本文件的基本语法格式如下：

```
mysqldump -T path-u root -p dbname [tables] [OPTIONS]
```

只有指定了-T参数才可以导出纯文本文件；path表示导出数据的目录；tables为指定要导出的表名称，如果不指定，将导出数据库dbname中所有的表；[OPTIONS]为可选参数选项，这些选项需要结合-T选项使用。OPTIONS常见的取值有：

- --fields-terminated-by=value：设置字段之间的分隔字符，可以为单个或多个字符，默认情况下为制表符（\t）。
- --fields-enclosed-by=value：设置字段的包围字符。
- --fields-optionally-enclosed-by=value：设置字段的包围字符，只能为单个字符，只能包括CHAR和VERCHAR等字符数据字段。
- --fields-escaped-by=value：控制如何写入或读取特殊字符，只能为单个字符，即设置转义符，默认值为\。
- --lines-terminated-by=value：设置每行数据结尾的字符，可以为单个或多个字符，默认值为"\n"。

> 提示 这里的OPTIONS的设置与SELECT...INTO OUTFILE语句中的OPTIONS不同，各个取值中等号后面的value值不要用引号引起来。

【例11.13】使用mysqldump将test_db数据库中的person表中的记录导出到文本文件，SQL语句如下：

```
mysqldump -T D:\ test_db person -u root -p
```

语句执行成功，系统D盘目录下面将会有两个文件，分别为person.sql和person.txt。person.sql包含创建person表的CREATE语句，其内容如下：

```
-- MySQL dump 10.13  Distrib 9.0.1, for Win64 (x86_64)
--
-- Host: localhost    Database: test_db
-- ------------------------------------------------------
-- Server version    9.0.1

/*!40101 SET @OLD_CHARACTER_SET_CLIENT=@@CHARACTER_SET_CLIENT */;
/*!40101 SET @OLD_CHARACTER_SET_RESULTS=@@CHARACTER_SET_RESULTS */;
/*!40101 SET @OLD_COLLATION_CONNECTION=@@COLLATION_CONNECTION */;
 SET NAMES utf8mb4 ;
/*!40103 SET @OLD_TIME_ZONE=@@TIME_ZONE */;
/*!40103 SET TIME_ZONE='+00:00' */;
/*!40101 SET @OLD_SQL_MODE=@@SQL_MODE, SQL_MODE='' */;
/*!40111 SET @OLD_SQL_NOTES=@@SQL_NOTES, SQL_NOTES=0 */;

--
-- Table structure for table `person`
--
```

```
DROP TABLE IF EXISTS `person`;
/*!40101 SET @saved_cs_client     = @@character_set_client */;
/*!40101 SET character_set_client = utf8 */;
CREATE TABLE `person` (
  `id` int(10) unsigned NOT NULL AUTO_INCREMENT,
  `name` char(40) NOT NULL DEFAULT '',
  `age` int(11) NOT NULL DEFAULT '0',
  `info` char(50) DEFAULT NULL,
  PRIMARY KEY (`id`)
) ENGINE=InnoDB AUTO_INCREMENT=11 DEFAULT CHARSET=utf8;
/*!40101 SET character_set_client = @saved_cs_client */;

/*!40103 SET TIME_ZONE=@OLD_TIME_ZONE */;

/*!40101 SET SQL_MODE=@OLD_SQL_MODE */;
/*!40101 SET CHARACTER_SET_CLIENT=@OLD_CHARACTER_SET_CLIENT */;
/*!40101 SET CHARACTER_SET_RESULTS=@OLD_CHARACTER_SET_RESULTS */;
/*!40101 SET COLLATION_CONNECTION=@OLD_COLLATION_CONNECTION */;
/*!40111 SET SQL_NOTES=@OLD_SQL_NOTES */;

-- Dump completed on 2024-07-25 16:40:55
```

备份文件中的信息已在11.1.1节中介绍了。

person.txt包含数据包中的数据，其内容如下：

```
1       Green        21      Lawyer
2       Suse         22      dancer
3       Mary         24      Musician
4       Willam       20      sports man
5       Laura        25      \N
6       Evans        27      secretary
7       Dale         22      cook
8       Edison       28      singer
9       Harry        21      magician
10      Harriet      19      pianist
```

【例11.14】使用mysqldump命令将test_db数据库中的person表中的记录导出到文本文件，使用FIELDS选项，要求字段之间使用","间隔，所有字符类型字段值用双引号引起来，定义转义字符为"?"，每行记录以"\r\n"结尾，SQL语句如下：

```
mysqldump -T D:\ test_db person -u root -p --fields-terminated-by=,
--fields-optionally-enclosed-by=\" --fields-escaped-by=? --lines-terminated-by=\r\n
Enter password: ******
```

上面语句要在一行中输入，语句执行成功后，系统D盘目录下面将会有两个文件，分别为person.sql和person.txt。person.sql包含创建person表的CREATE语句，其内容与【例11.13】中的相同；person.txt文件的内容与【例11.13】中的不同，显示如下：

```
1,"Green",21,"Lawyer"
2,"Suse",22,"dancer"
3,"Mary",24,"Musician"
4,"Willam",20,"sports man"
```

```
5,"Laura",25,?N
6,"Evans",27,"secretary"
7,"Dale",22,"cook"
8,"Edison",28,"singer"
9,"Harry",21,"magician"
10,"Harriet",19,"pianist"
```

可以看到，只有字符类型的值被双引号引起来了，而数值类型的值没有；第5行记录中的NULL表示为"?N"，使用"?"替代了系统默认的"\"。

11.4.3 使用 mysql 命令导出文本文件

mysql是一个功能丰富的工具命令，使用它还可以在命令行模式下执行SQL指令，将查询结果导入文本文件中。相比mysqldump，mysql工具导出的结果具有更强的可读性。

如果MySQL服务器是单独的机器，用户是在一个客户端上进行操作，要把数据结果导入客户端上。

使用mysql导出数据文本文件的基本语法格式如下：

```
mysql -u root -p --execute= "SELECT语句" dbname > filename.txt
```

该命令使用--execute选项，表示执行该选项后面的语句并退出，后面的语句必须用双引号引起来；dbname为要导出的数据库名称；导出的文件中不同列之间使用制表符分隔，第1行包含了各个字段的名称。

【例11.15】使用mysql命令，将test_db数据库中的person表中的记录导出到文本文件，SQL语句如下：

```
mysql -u root -p --execute="SELECT * FROM person;" test_db > D:\person3.txt
```

语句执行完毕之后，系统D盘目录下面将会有名称为"person3.txt"的文本文件，其内容如下：

```
id    name      age     info
1     Green     21      Lawyer
2     Suse      22      dancer
3     Mary      24      Musician
4     Willam    20      sports man
5     Laura     25      NULL
6     Evans     27      secretary
7     Dale      22      cook
8     Edison    28      singer
9     Harry     21      magician
10    Harriet   19      pianist
```

可以看到，person3.txt文件中包含了每个字段的名称和各条记录，该显示格式与MySQL命令行下SELECT查询结果的显示格式相同。

使用mysql命令还可以指定查询结果的显示格式，如果某条记录的字段很多，可能一行不能完全显示，可以使用--vartical参数将每条记录分为多行显示。

【例11.16】使用mysql命令将test_db数据库中的person表中的记录导出到文本文件，使用--vertical参数显示结果，SQL语句如下：

```
mysql -u root -p --vertical --execute="SELECT * FROM person;" test_db > D:\person4.txt
```

语句执行之后，D:\person4.txt文件中的内容如下：

```
*** 1. row ***
  id: 1
name: Green
 age: 21
info: Lawyer
*** 2. row ***
  id: 2
name: Suse
 age: 22
info: dancer
*** 3. row ***
  id: 3
name: Mary
 age: 24
info: Musician
*** 4. row ***
  id: 4
name: Willam
 age: 20
info: sports man
*** 5. row ***
  id: 5
name: Laura
 age: 25
info: NULL
*** 6. row ***
  id: 6
name: Evans
 age: 27
info: secretary
*** 7. row ***
  id: 7
name: Dale
 age: 22
info: cook
*** 8. row ***
  id: 8
name: Edison
 age: 28
info: singer
*** 9. row ***
  id: 9
name: Harry
```

```
    age: 21
    info: magician
*** 10. row ***
     id: 10
   name: Harriet
    age: 19
   info: pianist
```

可以看到，SELECT的查询结果导出到文本文件之后，显示格式发生了变化，如果person表中的记录内容很长，这样显示会让人更加容易阅读。

mysql还可以将查询结果导出到html文件中，使用--html选项即可。

【例11.17】使用MySQL命令将test_db数据库中的person表中的记录导出到html文件，SQL语句如下：

```
mysql -u root -p --html --execute="SELECT * FROM person;" test_db > D:\person5.html
```

语句执行成功，将在D盘创建文件person5.html，该文件在浏览器中的显示效果如图11.2所示。如果要将表数据导出到xml文件中，可使用--xml选项。

【例11.18】使用mysql命令将test_db数据库中的person表中的记录导出到xml文件，SQL语句如下：

```
mysql -u root -p --xml --execute="SELECT * FROM person;" test_db >D:\person6.xml
```

语句执行成功，将在D盘创建文件person6.xml，该文件在浏览器中的显示效果如图11.3所示。

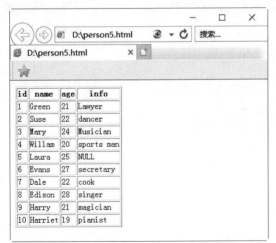
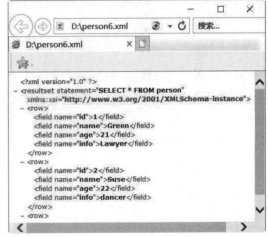

图 11.2　使用 mysql 命令导出数据到 html 文件　　　图 11.3　使用 mysql 命令导出数据到 xml 文件

11.4.4　使用 LOAD DATA INFILE 方式导入文本文件

MySQL允许将数据导出到外部文件，也可以从外部文件导入数据。MySQL提供了一些导入数据的工具，包括LOAD DATA语句、source命令和mysql命令。LOAD DATA INFILE语句用于高速地从一个文本文件中读取行，并输入一张表中。文件名称必须为文字字符串。

LOAD DATA语句的基本格式如下:

```
LOAD DATA INFILE 'filename.txt' INTO TABLE tablename [OPTIONS] [IGNORE number LINES]
```

在LOAD DATA语句中,关键字INFILE后面的filename文件为导入数据的来源;tablename表示待导入的数据表名称;[OPTIONS]为可选参数选项,OPTIONS部分的语法包括FIELDS和LINES子句,其可能的取值有:

- FIELDS TERMINATED BY 'value':设置字段之间的分隔字符,可以为单个或多个字符,默认情况下为"\t"。
- FIELDS [OPTIONALLY] ENCLOSED BY 'value':设置字段的包围字符,只能为单个字符。如果使用了OPTIONALLY,则只有CHAR和VERCHAR等字符数据字段被包围。
- FIELDS ESCAPED BY 'value':控制如何写入或读取特殊字符,只能为单个字符,即设置转义字符,默认值为"\"。
- LINES STARTING BY 'value':设置每行数据开头的字符,可以为单个或多个字符,默认情况下不使用任何字符。
- LINES TERMINATED BY 'value':设置每行数据结尾的字符,可以为单个或多个字符,默认值为"\n"。

IGNORE number LINES选项表示忽略文件开始处的行数,number表示忽略的行数。执行LOAD DATA语句需要FILE权限。

【例11.19】 使用LOAD DATA语句将D:\person0.txt文件中的数据导入test_db数据库的person表中,SQL语句如下:

```
LOAD DATA INFILE 'D:\person0.txt' INTO TABLE test_db.person;
```

导入数据之前,要将person表中的数据全部删除,即登录MySQL,使用DELETE语句删除person表中的数据:

```
mysql> USE test_db;
Database changed;
mysql> DELETE FROM person;
Query OK, 10 rows affected (0.00 sec)
```

从person0.txt文件中导入数据,SQL语句如下:

```
mysql> LOAD DATA INFILE 'D:\person0.txt' INTO TABLE test_db.person;
Query OK, 10 rows affected (0.00 sec)
Records: 10  Deleted: 0  Skipped: 0  Warnings: 0

mysql> SELECT * FROM person;
+----+---------+-----+------------+
| id | name    | age | info       |
+----+---------+-----+------------+
|  1 | Green   |  21 | Lawyer     |
|  2 | Suse    |  22 | dancer     |
|  3 | Mary    |  24 | Musician   |
```

```
|  4 | Willam  | 20 | sports man |
|  5 | Laura   | 25 | NULL       |
|  6 | Evans   | 27 | secretary  |
|  7 | Dale    | 22 | cook       |
|  8 | Edison  | 28 | singer     |
|  9 | Harry   | 21 | magician   |
| 10 | Harriet | 19 | pianist    |
+----+---------+----+------------+
```

可以看到，语句执行成功之后，person0.txt文件中的数据导入person表中了。

【例11.20】使用LOAD DATA语句将D:\person1.txt文件中的数据导入test_db数据库中的person表中，使用FIELDS选项和LINES选项，要求字段之间使用","间隔，所有字段值用双引号引起来，定义转义字符为"\"，每行记录以"\r\n"结尾，SQL语句如下：

```
LOAD DATA INFILE 'D:\person1.txt' INTO TABLE test_db.person
FIELDS
TERMINATED BY ','
ENCLOSED BY '\"'
ESCAPED BY '\''
LINES
TERMINATED BY '\r\n';
```

导入数据之前，使用DELETE语句将person表中的数据全部删除，执行过程如下：

```
mysql> DELETE FROM person;
Query OK, 10 rows affected (0.00 sec)
```

从person1.txt文件中导入数据，执行过程如下：

```
mysql> LOAD DATA  INFILE 'D:\person1.txt' INTO TABLE test_db  .person
    -> FIELDS
    -> TERMINATED BY ','
    -> ENCLOSED BY '\"'
    -> ESCAPED BY '\''
    -> LINES
    -> TERMINATED BY '\r\n';
Query OK, 10 rows affected (0.00 sec)
Records: 10  Deleted: 0  Skipped: 0  Warnings: 0
```

语句执行成功后，使用SELECT语句查看person表中的记录，结果与【例11.19】的相同。

11.4.5 使用 mysqlimport 命令导入文本文件

使用mysqlimport命令可以导入文本文件，并且不需要登录MySQL客户端。mysqlimport命令提供许多与LOAD DATA INFILE语句相同的功能，大多数选项直接对应LOAD DATA INFILE子句。使用mysqlimport命令需要指定所需的选项、导入的数据库名称以及导入的数据文件的路径和名称。mysqlimport命令的基本语法格式如下：

```
mysqlimport -u root-p dbname filename.txt [OPTIONS]
```

dbname为导入的表所在的数据库名称。注意，mysqlimport命令不指定导入数据库的表名称，数据表的名称由导入文件名称确定，即文件名作为表名，导入数据之前该表必须存在。[OPTIONS]为可选参数选项，其常见的取值有：

- --fields-terminated-by= 'value'：设置字段之间的分隔字符，可以为单个或多个字符，默认情况下为"\t"。
- --fields-enclosed-by= 'value'：设置字段的包围字符。
- --fields-optionally-enclosed-by= 'value'：设置字段的包围字符，只能为单个字符，包括CHAR和VERCHAR等字符数据字段。
- --fields-escaped-by= 'value'：控制如何写入或读取特殊字符，只能为单个字符，即设置转义字符，默认值为"\"。
- --lines-terminated-by= 'value'：设置每行数据结尾的字符，可以为单个或多个字符，默认值为"\n"。
- --ignore-lines=n：忽视数据文件的前n行。

【例11.21】使用mysqlimport命令将D盘目录下的person.txt文件内容导入test_db数据库中，字段之间使用","分隔，字符类型字段值用双引号引起来，将转义字符定义为"?"，每行记录以"\r\n"结尾，SQL语句如下：

```
C:\ >mysqlimport -u root -p test_db D:\person.txt --fields-terminated-by=,
--fields-optionally-enclosed-by=\"--fields-escaped-by=?--lines-terminated-by=\r\n
```

上面的语句要在一行中输入，语句执行成功后，将把person.txt中的数据导入数据库test_db中。除了前面介绍的几个选项之外，mysqlimport命令还支持许多别的选项，例如：

- --columns=column_list, -c column_list：采用逗号分隔的列名作为其值。列名的顺序指示如何匹配数据文件列和表列。
- --compress, -C：压缩在客户端和服务器之间发送的所有信息（如果二者均支持压缩）。
- -d, --delete：导入文本文件前清空表。
- --force, -f：忽视错误。例如，当某个文本文件的表不存在时，继续处理其他文件。不使用--force时，如果表不存在，则mysqlimport退出。
- --host=host_name, -h host_name：将数据导入给定主机上的MySQL服务器。默认主机是localhost。
- --ignore, -i：参见--replace选项的描述。
- --ignore-lines=n：忽视数据文件的前n行。
- --local, -L：从本地客户端读入输入文件。
- --lock-tables, -l：处理文本文件前锁定所有表，以便写入。这样可以确保所有表在服务器上保持同步。
- --password[=password], -p[password]：连接服务器时使用的密码。如果使用短选项形式(-p)，则选项和密码之间不能有空格。如果在命令行中--password或-p选项后面没有密码值，则提示输入一个密码。
- --port=port_num, -P port_num：用于连接的TCP/IP端口号。

- --protocol={TCP | SOCKET | PIPE | MEMORY}：使用的连接协议。
- --replace和--ignore选项控制复制唯一键值已有记录的输入记录的处理方式。如果指定--replace，新行替换有相同的唯一键值的已有行；如果指定--ignore，复制已有的唯一键值的输入行将被跳过，不发生替换；如果不指定这两个选项，当发现一个复制键值时会出现一个错误，并且忽视文本文件的剩余部分。
- --silent，-s：沉默模式。只有出现错误时才输出信息。
- --user=user_name，-u user_name：连接服务器时MySQL使用的用户名。
- --verbose，-v：冗长模式。打印出程序操作的详细信息。
- --version，-V：显示版本信息并退出。

第 12 章 MySQL日志

MySQL日志记录了MySQL数据库的日常操作和错误信息。MySQL有不同类型的日志文件（各自存储不同类型的日志），从日志当中可以查询到MySQL数据库的运行情况、用户操作、错误信息等，为MySQL管理和优化提供了必要的信息。对于MySQL的管理工作而言，这些日志文件是不可缺少的。本章将介绍MySQL各种日志的作用以及日志的管理。

12.1 日志简介

MySQL日志主要分为4类：二进制日志、错误日志、通用查询日志和慢查询日志，使用这些日志文件，可以查看MySQL内部发生的事情。

- 二进制日志：记录所有更改数据的语句，可以用于数据复制。
- 错误日志：记录MySQL服务的启动、运行或停止时出现的问题。
- 通用查询日志：记录建立的客户端连接和执行的语句。
- 慢查询日志：记录所有执行时间超过long_query_time的查询或不使用索引的查询。

默认情况下，所有日志创建于MySQL数据目录中。通过刷新日志，可以强制MySQL关闭和重新打开日志文件（或者在某些情况下切换到一个新的日志）。当执行一个FLUSH LOGS语句或执行mysqladmin flush-logs或mysqladmin refresh命令时，将刷新日志。

在MySQL中，当使用复制功能时，可以在复制服务器上维护更多的日志文件，这些日志文件被称为接替日志。接替日志记录了从主服务器接收到的所有二进制日志事件，这样复制服务器可以根据这些日志事件来执行相应的操作，以保持与主服务器的数据一致性。

启动日志功能会降低MySQL数据库的性能。例如，在查询非常频繁的MySQL数据库系统中，如果开启了通用查询日志和慢查询日志，那么MySQL数据库会花费很多时间记录日志。同时，日志会占用大量的磁盘空间。

12.2 二进制日志

二进制日志主要记录MySQL数据库的变化，以一种有效的格式并且是事务安全的方式包含更新日志中可用的所有信息。二进制日志包含了所有更新了数据或者已经潜在更新了数据（例如，没有匹配任何行的一个DELETE）的语句。语句以"事件"的形式保存，描述数据的更改。此外，二进制日志还包含每个更新数据库的语句的执行时间信息。

二进制日志不包含没有修改任何数据的语句。如果想要记录所有语句（例如，为了识别有问题的查询），需要使用通用查询日志。使用二进制日志的主要目的是最大可能地恢复数据库，因为二进制日志包含备份后进行的所有更新。本节将介绍二进制日志的相关内容。

12.2.1 启动和设置二进制日志

默认情况下，二进制日志是开启的，可以通过修改MySQL的配置文件来启动和设置二进制日志。

在my.ini配置文件中的[mysqld]组下面有关于二进制日志的设置：

```
log-bin [=path/ [filename] ]
```

- og-bin定义开启二进制日志；path表明日志文件所在的目录路径；filename指定日志文件的名称，如文件的全名为"filename.000001""filename.000002"等。除此之外，还有一个名称为"filename.index"的文件，文件内容为所有日志的清单，可以使用记事本打开该文件。添加以下几个参数与参数值：

```
[mysqld]
log-bin
expire_logs_days = 10
max_binlog_size = 100M
```

- expire_logs_days定义MySQL清除过期日志的时间，即二进制日志自动删除的天数。默认值为0，表示不自动删除。当MySQL启动或刷新二进制日志时，操作系统可能会删除过期的二进制日志文件。
- max_binlog_size定义单个文件的大小限制，如果二进制日志写入的内容大小超出了给定值，日志就会发生滚动（关闭当前文件，重新打开一个新的日志文件）。不能将该变量设置为大于1GB或小于4096B，默认值是1GB。

添加完毕之后，关闭并重新启动MySQL服务进程，即可打开二进制日志，然后可以通过SHOW VARIABLES语句来查询日志设置。

【例12.1】使用SHOW VARIABLES语句查询日志设置，执行的语句及结果如下：

```
mysql> SHOW VARIABLES LIKE 'log_%' ;
+--------------------------------+----------------------------------------+
|Variable_name                   | Value                                  |
```

```
+--------------------------------+------------------------------------+
|log_bin                         | ON                                 |
|log_bin_basename                | D:\ProgramData\MySQL\MySQL Server
9.0\Data\X0NHUNO7YDZVSSI-bin     |
|log_bin_index                   |D:\ProgramData\MySQL\MySQL Server
9.0\Data\X0NHUNO7YDZVSSI-bin.index |
|log_bin_trust_function_creators |OFF                                 |
|log_bin_use_v1_row_events       | OFF                                |
|log_error| .\X0NHUNO7YDZVSSI.err                                     |
|log_error_services              |log_filter_internal;log_sink_internal |
|log_error_suppression_list      |                                    |
|log_error_verbosity             |2                                   |
|log_output                      | FILE                               |
|log_queries_not_using_indexes   |OFF                                 |
|log_slave_updates               |ON                                  |
|log_slow_admin_statements       |OFF                                 |
|log_slow_slave_statements       |OFF                                 |
|log_statements_unsafe_for_binlog|ON                                  |
|log_throttle_queries_not_using_indexes|0                             |
| log_timestamps                 | UTC                                |
+--------------------------------+------------------------------------+
```

通过上面的查询结果可以看出，log_bin变量的值为ON，表明二进制日志已经打开。MySQL重新启动之后，读者可以在自己的计算机上的MySQL数据文件夹下面看到新生成的文件后缀为".000001"和".index"的两个文件，文件名称为默认主机名称。例如，在笔者的计算机上，文件名称为"X0NHUNO7YDZVSSI-bin.000001"和"X0NHUNO7YDZVSSI-bin.index"。

> **提示** 数据库文件最好不要与日志文件放在同一个磁盘上，这样当数据库文件所在的磁盘发生故障时，可以使用日志文件恢复数据。

12.2.2 查看二进制日志

MySQL二进制日志是经常用到的，它存储了所有的变更信息。当MySQL创建二进制日志文件时，先创建一个以"filename"为名称、以".index"为后缀的文件，再创建一个以"filename"为名称、以".000001"为后缀的文件。MySQL服务重新启动一次，以".000001"为后缀的文件会增加一个，并且后缀名加1递增；如果日志长度超过了max_binlog_size的上限（默认是1GB），就会创建一个新的日志文件。

SHOW BINARY LOGS语句可以查看当前的二进制日志文件的个数及其名称。MySQL二进制日志并不能直接查看，如果要查看日志内容，可以使用mysqlbinlog命令。

【例12.2】使用SHOW BINARY LOGS查看二进制日志文件的个数及其名称，SQL语句如下：

```
mysql> SHOW BINARY LOGS;
+------------------------------+-----------+-----------+
| Log_name                     | File_size | Encrypted +
+------------------------------+-----------+-----------+
```

```
| X0NHUN07YDZVSSI-bin.000001 |       178 | No        |
+----------------------------+-----------+-----------+
1 row in set (0.00 sec)
```

可以看到，当前只有一个二进制日志文件。日志文件的个数与MySQL服务启动的次数相同。每启动一次MySQL服务，就会产生一个新的日志文件。

【例12.3】使用mysqlbinlog查看二进制日志，SQL语句如下：

```
C:\> mysqlbinlog D:/mysql/log/binlog.000001
/*!40019 SET @@session.max_insert_delayed_threads=0*/;
/*!50003 SET @OLD_COMPLETION_TYPE=@@COMPLETION_TYPE,COMPLETION_TYPE=0*/;
DELIMITER /*!*/;
# at 4
#190130 15:27:48 server id 1  end_log_pos 107    Start: binlog v 4, server v 9.0.1-log
created 160330
# Warning: this binlog is either in use or was not closed properly.
ROLLBACK/*!*/;
BINLOG '
9JBcTg8BAAAAZwAAAGsAAAABAAQANS41LjEzLWxvZwAAAAAAAAAAAAAAAAAAAAAA
AAAAAAAAA
AAAAAAAAAAAAAAAAAAD0kFxOEzgNAAgAEgAEBAQEEgAAVAAEGggAAAAICAgCAA==
'/*!*/;
# at 107
#190330 15:34:17 server id 1  end_log_pos 175    Query   thread_id=2    exec_time=0
error_code=0
SET TIMESTAMP=1314689657/*!*/;
SET @@session.pseudo_thread_id=2/*!*/;
SET @@session.foreign_key_checks=1, @@session.sql_auto_is_null=0,
@@session.unique_checks=1,
@@session
SET @@session.sql_mode=0/*!*/;
SET @@session.auto_increment_increment=1, @@session.auto_increment_
offset=1/*!*/;
/*!\C gb2312 *///*!*/;
SET @@session.character_set_client=24,@@session.collation_connection=24,
@@session.collation_server=24/
SET @@session.lc_time_names=0/*!*/;
SET @@session.collation_database=DEFAULT/*!*/;
BEGIN
/*!*/;
# at 175
#190330 15:34:17 server id 1  end_log_pos 289    Query   thread_id=2    exec_time=0
error_code=0
use test/*!*/;
SET TIMESTAMP=1314689657/*!*/;
UPDATE fruits set f_price = 5.00  WHERE f_id = 'a1'
/*!*/;
# at 289
#190330 15:34:17 server id 1  end_log_pos 316    Xid = 14
COMMIT/*!*/;
```

```
DELIMITER ;
# End of log file
ROLLBACK /* added by mysqlbinlog */;
/*!50003 SET COMPLETION_TYPE=@OLD_COMPLETION_TYPE*/;
```

这是一个简单的日志文件，记录了一些用户的操作。从文件内容中可以看到，用户对表fruits进行了更新操作，语句为"UPDATE fruits set f_price = 5.00 WHERE f_id = 'a1';"。

12.2.3 删除二进制日志

MySQL的二进制日志文件可以配置为自动删除，但也提供了安全的手动删除二进制日志文件的方法：RESET MASTER删除所有的二进制日志文件；PURGE MASTER LOGS只删除部分二进制日志文件。本小节将介绍这两种删除二进制日志文件的方法。

1. 使用RESET MASTER语句删除所有二进制日志文件

RESTE MASTER语法如下：

```
RESET MASTER;
```

执行完该语句后，所有二进制日志文件将被删除，MySQL会重新创建二进制日志文件，新的日志文件扩展名将重新从000001开始编号。

2. 使用PURGE MASTER LOGS语句删除指定二进制日志文件

PURGE MASTER LOGS 语法如下：

```
PURGE {MASTER | BINARY} LOGS TO 'log_name'
PURGE {MASTER | BINARY} LOGS BEFORE 'date'
```

第一种方法指定文件名，执行该语句将删除文件编号比指定文件编号小的所有二进制日志文件。第二种方法指定日期，执行该语句将删除指定日期以前的所有二进制日志文件。

【例12.4】使用PURGE MASTER LOGS删除创建时间比X0NHUNO7YDZVSSI-bin.000003早的所有二进制日志文件。

为了演示语句的操作过程，准备多个日志文件，读者可以对MySQL服务进行多次重新启动。例如，这里有3个二进制日志文件：

```
mysql> SHOW BINARY LOGS;
+------------------------------+-----------+-----------+
| Log_name                     | File_size | Encrypted |
+------------------------------+-----------+-----------+
| X0NHUNO7YDZVSSI-bin.000001   |       178 | No        |
| X0NHUNO7YDZVSSI-bin.000002   |       641 | No        |
| X0NHUNO7YDZVSSI-bin.000003   |       345 | No        |
+------------------------------+-----------+-----------+
3 rows in set (0.00 sec)
```

执行删除命令：

第12章 MySQL 日志

```
mysql> PURGE MASTER LOGS TO " X0NHUNO7YDZVSSI-bin.000003";
Query OK, 0 rows affected (0.07 sec)
```

执行完成后，使用SHOW binary logs语句查看二进制日志：

```
mysql> SHOW binary logs;
+----------------------------+-----------+
| Log_name                   | File_size |
+----------------------------+-----------+
| X0NHUNO7YDZVSSI-bin.000003 |       345 |
+----------------------------+-----------+
1 rows in set (0.00 sec)
```

可以看到，X0NHUNO7YDZVSSI-bin.000001、X0NHUNO7YDZVSSI-bin 000002两个二进制日志文件被删除了。

【例12.5】 使用PURGE MASTER LOGS删除2024年7月30日前创建的所有二进制日志文件，SQL语句及结果如下：

```
mysql> PURGE MASTER LOGS BEFORE '20240730';
Query OK, 0 rows affected (0.05 sec)
```

语句执行之后，2024年7月30日之前创建的二进制日志文件都将被删除，但2024年7月30日的日志会被保留（读者可根据自己计算机中创建日志的时间修改命令参数）。使用mysqlbinlog可以查看指定日志的创建时间，如【例12.3】所示，部分日志内容如下：

```
/*!40019 SET @@session.max_insert_delayed_threads=0*/;
/*!50003 SET @OLD_COMPLETION_TYPE=@@COMPLETION_TYPE,COMPLETION_TYPE=0*/;
DELIMITER /*!*/;
# at 4
#240730 15:27:48 server id 1  end_log_pos 107   Start: binlog v 4, server v 9.0.1-log
created 160330
# Warning: this binlog is either in use or was not closed properly.
ROLLBACK/*!*/;
BINLOG '
```

其中，240730为日志创建的时间，即2024年7月30日。

12.2.4 使用二进制日志恢复数据库

如果MySQL服务器启用了二进制日志，那么当数据库意外丢失数据时，可以使用mysqlbinlog工具从指定的时间点开始（例如，最后一次备份），直到现在或另一个指定的时间点的日志中恢复数据。

要想使用二进制日志恢复数据，需要知道当前二进制日志文件的路径和文件名，一般可以从配置文件（my.cnf或者my.ini，文件名取决于MySQL服务器的操作系统）中找到路径。

mysqlbinlog恢复数据的语法如下：

```
mysqlbinlog [option] filename |mysql -uuser -ppass
```

option是一些可选的选项，filename是日志文件名。比较重要的两对option参数是--start-date

与--stop-date、--start-position与--stop-position。--start-date与--stop-date可以指定恢复数据库的起始时间点和结束时间点。--start-position与--stop-position可以指定恢复数据库的开始位置和结束位置。

【例12.6】使用mysqlbinlog恢复MySQL数据库到2024年7月30日15:27:48时的状态，执行命令及结果如下：

```
mysqlbinlog --stop-date="2024-07-30 15:27:48" D:\mysql\log\binlog\
X0NHUNO7YDZVSSI-bin.000003 | mysql -uuser -ppass
```

该命令执行成功后，会根据X0NHUNO7YDZVSSI-bin.000003日志文件恢复2024年7月30日15:27:48以前的所有操作。这种方法对于因意外操作而导致的数据丢失非常有效，比如因操作不当误删了数据表。

12.2.5　暂时停止二进制日志功能

如果在MySQL的配置文件中配置启动了二进制日志，那么MySQL会一直记录二进制日志。如果想停止记录二进制日志，可以修改配置文件，但是需要重启MySQL数据库。为此，MySQL提供了暂时停止记录二进制日志的功能：通过SET SQL_LOG_BIN语句可以使MySQL暂停或者启动二进制日志。

SET SQL_LOG_BIN的语法如下：

```
SET sql_log_bin = {0|1}
```

执行如下语句将暂停记录二进制日志：

```
mysql> SET sql_log_bin = 0;
Query OK, 0 rows affected (0.00 sec)
```

执行如下语句将恢复记录二进制日志：

```
mysql> SET sql_log_bin = 1;
Query OK, 0 rows affected (0.00 sec)
```

12.3　错 误 日 志

在MySQL系统运维中，错误日志也是非常有用的，MySQL会将启动和停止数据库时的信息以及一些错误信息记录到错误日志中。

12.3.1　启动和设置错误日志

在默认情况下，错误日志会记录到数据库的数据目录下。如果没有在配置文件中指定文件名，则文件名默认为hostname.err。例如，MySQL所在的服务器主机名为MySQL-db，记录错误信息的文件名为MySQL-db.err。如果执行了FLUSH LOGS，那么错误日志文件会被重新加载。

错误日志的启动、停止以及指定日志文件名，都可以通过修改my.ini（或者my.cnf）来配置。错误日志的配置项是log-error。在[mysqld]配置区下面配置log-error，则表示启动错误日志。如果需要指定文件名，则配置项如下：

```
[mysqld]
log-error=[path / [file_name] ]
```

path为错误日志文件所在的目录路径，file_name为错误日志文件名。修改配置项后，需要重启MySQL服务以使设置生效。

12.3.2 查看错误日志

通过错误日志可以监视系统的运行状态，便于及时发现故障、修复故障。MySQL错误日志是以文本文件形式存储的，可以使用文本编辑器直接查看。

如果不知道日志文件的存储路径，可以使用SHOW VARIABLES语句查询错误日志的存储路径。SHOW VARIABLES语句的语法格式如下：

```
SHOW VARIABLES LIKE 'log_error';
```

【例12.7】 使用记事本查看MySQL错误日志。

首先，通过SHOW VARIABLES语句查询错误日志的存储路径和文件名：

```
mysql> SHOW VARIABLES LIKE 'log_error';
+---------------+----------------------+
| Variable_name | Value                |
+---------------+----------------------+
| log_error     | .\X0NHUNO7YDZVSSI.err |
+---------------+----------------------+
1 row in set (0.00 sec)
```

可以看到错误日志文件是X0NHUNO7YDZVSSI.err，位于MySQL默认的数据目录下。

然后使用记事本打开该文件，可以看到MySQL的错误日志：

```
190130 16:45:14 [Note] Plugin 'FEDERATED' is disabled.
190130 16:45:14 InnoDB: The InnoDB memory heap is disabled
190130 16:45:14 InnoDB: Mutexes and rw_locks use Windows interlocked functions
190130 16:45:14 InnoDB: Compressed tables use zlib 1.2.3
190130 16:45:15 InnoDB: Initializing buffer pool, size = 46.0M
190130 16:45:15 InnoDB: Completed initialization of buffer pool
190130 16:45:15 InnoDB: highest supported file format is Barracuda.
190130 16:45:15 InnoDB: Waiting for the background threads to start
190130 16:45:16 InnoDB: 1.1.7 started; log sequence number 1679264
190130 16:45:16 [Note] Event Scheduler: Loaded 0 events
190130 16:45:16 [Note] C:\Program Files\MySQL\MySQL Server 9.0\bin\MySQLd: ready
for connections.
Version: '9.0.1-log'  socket: ''  port: 3306  MySQL Community Server (GPL)
```

以上是错误日志文件的一部分，里面记载了系统的一些错误。

12.3.3 删除错误日志

MySQL的错误日志是以文本文件的形式存储在文件系统中的,可以直接删除。

对于MySQL 5.5.7以前的版本,使用FLUSH LOGS语句可以将错误日志文件重命名为filename.err_old,并创建新的日志文件。但是从MySQL 5.5.7开始,FLUSH LOGS语句只是重新打开日志文件,并不做日志备份和创建的操作。如果日志文件不存在,则MySQL启动或者执行FLUSH LOGS时会创建新的日志文件。

在运行状态下删除错误日志文件后,MySQL并不会自动创建日志文件;使用FLUSH LOGS语句重新加载日志的时候,如果文件不存在,则会自动创建。因此,在删除错误日志之后,如果要重建日志文件,需要在服务器端执行以下命令:

```
mysqladmin -u root -p flush-logs
```

或者在客户端登录MySQL数据库,执行FLUSH LOGS语句:

```
mysql> FLUSH LOGS;
```

12.4 通用查询日志

通用查询日志记录MySQL的所有用户操作,包括启动和关闭服务、执行查询和更新语句等。本节将介绍通用查询日志的启动、查看、删除等内容。

12.4.1 启动通用查询日志

MySQL服务器默认情况下并没有开启通用查询日志。通过SHOW VARIABLES LIKE '%general%';语句,可以查询当前通用查询日志的状态。

```
mysql> SHOW VARIABLES LIKE '%general%';
+------------------+---------------------+
| Variable_name    | Value               |
+------------------+---------------------+
| general_log      | OFF                 |
| general_log_file | X0NHUNO7YDZVSSI.log |
+------------------+---------------------+
2 rows in set, 1 warning (0.11 sec)
```

从结果中可以看出,通用查询日志的状态为OFF,表示通用日志是关闭的。

开启通用日志的方法如下:

```
mysql> set @@global.general_log=1;
Query OK, 0 rows affected (0.04 sec)
```

再次查询通用日志的状态：

```
mysql> SHOW VARIABLES LIKE '%general%';
+------------------+----------------------+
| Variable_name    | Value                |
+------------------+----------------------+
| general_log      | ON                   |
| general_log_file | X0NHUNO7YDZVSSI.log  |
+------------------+----------------------+
2 rows in set, 1 warning (0.00 sec)
```

从结果中可以看出，通用查询日志的状态为ON，表示通用日志已经开启了。

如果想关闭通用查询日志，执行以下语句即可：

```
mysql> set @@global.general_log=0;
```

12.4.2 查看通用查询日志

通用查询日志中记录了用户的所有操作，通过查看通用查询日志，可以了解用户对MySQL进行的操作。通用查询日志是以文本文件的形式存储在文件系统中的，因此可以使用文本编辑器直接打开进行查看：Windows下可以使用记事本，Linux下可以使用vim、gedit等。

【例12.8】 使用记事本查看MySQL通用查询日志。

使用记事本打开 C:\ProgramData\MySQL\MySQL Server 9.0\Data\ 目录下的 X0NHUNO7YDZVSSI.log，可以看到如下内容：

```
C:\Program Files\MySQL\MySQL Server 9.0\bin\mysqld.exe, Version: 9.0.1 (MySQL
Community Server - GPL). started with:
TCP Port: 3306, Named Pipe: MySQL
Time                     Id Command    Argument
2024-07-18 17:24:32       1 Connect    root@localhost on
    1 Query        select @@version_comment limit 1
2024-07-18 17:24:36       1 Query      SELECT DATABASE()
    1 Init DB      test
2024-07-18 17:24:53       1 Query      SELECT * FROM fruits
2024-07-18 17:24:55       1 Quit
```

上面是笔者计算机上的通用查询日志的一部分，可以看到MySQL启动信息、用户root连接服务器与执行查询语句的记录。读者的文件内容可能与这里不同。

12.4.3 删除通用查询日志

通用查询日志会记录用户的所有操作，因此在用户查询、更新频繁的情况下，通用查询日志增长得很快。数据库管理员可以定期删除比较早的通用查询日志，以节省磁盘空间。

可以用直接删除日志文件的方式删除通用查询日志。要重新建立日志文件，可以使用 mysqladmin -flush logs命令。

【例12.9】直接删除MySQL通用查询日志。

首先在数据目录中找到日志文件所在目录C:\ProgramData\MySQL\MySQL Server 9.0\Data\，删除后缀为".log"的文件。

然后通过mysqladmin -flush logs命令建立新的日志文件：

```
C:\> mysqladmin -u root -p flush-logs
```

执行完该命令，可以看到C:\ProgramData\MySQL\MySQL Server 9.0\Data\目录中已经建立了新的日志文件。

12.5 慢查询日志

慢查询日志主要用来记录执行时间较长的查询语句。通过慢查询日志，可以找出执行时间较长、执行效率较低的语句，然后进行优化。本小节将讲解慢查询日志的相关内容。

12.5.1 启动和设置慢查询日志

MySQL中慢查询日志默认是关闭的，可以通过配置文件my.ini或者my.cnf中的log-slow-queries选项打开，也可以在MySQL服务启动的时候使用--log-slow-queries[=file_name]启动慢查询日志。启动慢查询日志时，需要在my.ini或者my.cnf文件中配置long_query_time选项指定记录阈值，如果某条查询语句的查询时间超过了这个值，这个查询过程就会被记录到慢查询日志文件中。

在my.ini或者my.cnf中开启慢查询日志的配置如下：

```
[mysqld]
log-slow-queries[=path / [filename] ]
long_query_time=n
```

path为日志文件所在目录路径，filename为日志文件名。如果不指定目录和文件名称，则默认存储在数据目录中，文件为hostname-slow.log，hostname是MySQL服务器的主机名。参数n是时间值，单位是秒。如果没有设置long_query_time选项，则默认时间为10秒。

12.5.2 查看慢查询日志

MySQL的慢查询日志是以文本形式存储的，可以直接使用文本编辑器查看。

【例12.10】查看慢查询日志。使用文本编辑器打开数据目录下的Kevin-slow.log文件，文件部分内容如下：

```
C:\Program Files\MySQL\MySQL Server 9.0\bin\mysqld.exe, Version: 9.0.1 (MySQL
Community Server - GPL). started with:
TCP Port: 3306, Named Pipe: MySQL
Time Id Command Argument
```

```
# Time: 181230 17:50:35
# User@Host: root[root] @ localhost [127.0.0.1]
# Query_time: 136.500000  Lock_time: 0.000000 Rows_sent: 1  Rows_examined: 0
SET timestamp=1314697835;
SELECT BENCHMARK(100000000, PASSWORD('newpwd'));
```

可以看到，这里记录了一条慢查询日志。执行该条查询语句的账户是root@localhost，查询时间是136.500000秒，查询语句是"SELECT BENCHMARK(100000000, PASSWORD('newpwd'));"，该语句的查询时间大大超过了默认值10秒，因此被记录在慢查询日志文件中。

> **提示** 借助慢查询日志分析工具，可以更加方便地分析慢查询语句。比较著名的慢查询工具有MySQL Dump Slow、MySQL SLA、MySQL Log Filter、MyProfi。关于这些慢查询分析工具的用法，可以参考相关软件的帮助文档。

12.5.3 删除慢查询日志

和错误日志一样，慢查询日志也可以直接删除。删除后在不重启服务器的情况下，需要执行mysqladmin -u root-p flush-logs重新生成日志文件，或者在客户端登录到MySQL服务器执行FLUSH LOGS语句重建日志文件。

第 13 章 MySQL权限与安全管理

MySQL是一个多用户数据库，具有功能强大的访问控制系统，可以为不同用户指定允许的权限。MySQL用户可以分为普通用户和root用户。root用户是超级管理员，拥有所有权限，包括创建用户、删除用户和修改用户的密码等；普通用户只拥有被授予的各种权限。用户管理包括管理用户账户、权限等。本章将介绍MySQL用户管理中的相关知识点，包括权限表、账户管理、权限管理、访问控制等。

13.1 权限表

MySQL服务器通过权限表来控制用户对数据库的访问，权限表存放在MySQL数据库中，由mysql_install_db脚本初始化。存储账户权限信息的表主要有user、db、host、tables_priv、columns_priv和procs_priv。本节将介绍这些表的内容和作用。

13.1.1 user 表

user表是MySQL中最重要的一张权限表，记录了允许连接到服务器的账号信息，里面的权限是全局级的。例如，一个用户在user表中被授予了DELETE权限，则该用户可以删除MySQL服务器上所有数据库中的任何记录。MySQL中user表的常用字段如表13.1所示。

表 13.1 user 表结构

字 段 名	数 据 类 型	默 认 值
Host	char(60)	
User	char(16)	
authentication_string	text	
Select_priv	enum('N','Y')	N

（续表）

字 段 名	数 据 类 型	默 认 值
Insert_priv	enum('N','Y')	N
Update_priv	enum('N','Y')	N
Delete_priv	enum('N','Y')	N
Create_priv	enum('N','Y')	N
Drop_priv	enum('N','Y')	N
Reload_priv	enum('N','Y')	N
Shutdown_priv	enum('N','Y')	N
Process_priv	enum('N','Y')	N
File_priv	enum('N','Y')	N
Grant_priv	enum('N','Y')	N
References_priv	enum('N','Y')	N
Index_priv	enum('N','Y')	N
Alter_priv	enum('N','Y')	N
Show_db_priv	enum('N','Y')	N
Super_priv	enum('N','Y')	N
Create_tmp_table_priv	enum('N','Y')	N
Lock_tables_priv	enum('N','Y')	N
Execute_priv	enum('N','Y')	N
Repl_slave_priv	enum('N','Y')	N
Repl_client_priv	enum('N','Y')	N
Create_view_priv	enum('N','Y')	N
Show_view_priv	enum('N','Y')	N
Create_routine_priv	enum('N','Y')	N
Alter_routine_priv	enum('N','Y')	N
Create_user_priv	enum('N','Y')	N
Event_priv	enum('N','Y')	N
Trigger_priv	enum('N','Y')	N
Create_tablespace_priv	enum('N','Y')	N
ssl_type	enum('','ANY','X509','SPECIFIED')	
ssl_cipher	blob	NULL
x509_issuer	blob	NULL
x509_subject	blob	NULL
plugin	char(64)	
password_expired	enum('N','Y')	
max_questions	int(11) unsigned	0

(续表)

字 段 名	数 据 类 型	默 认 值
max_updates	int(11) unsigned	0
max_connections	int(11) unsigned	0
max_user_connections	int(11) unsigned	0

这些字段可以分为4类，分别是用户列、权限列、安全列和资源控制列。

1. 用户列

user表的用户列包括Host、User、authentication_string，分别表示主机名、用户名和密码。其中User和Host为User表的联合主键。当用户与服务器之间建立连接时，输入的账户信息中的用户名称、主机名和密码必须匹配User表中对应的字段，只有3个值都匹配的时候，才允许连接的建立。这3个字段的值就是创建账户时保存的账户信息。修改用户密码时，实际就是修改user表的authentication_string字段的值。

2. 权限列

user表的权限列字段从Select_priv到Create_tablespace_priv。权限列中的字段决定了用户的权限，描述了在全局范围内允许对数据和数据库进行的操作。这些权限不仅包括查询权限、修改权限等普通权限，还包括关闭服务器、超级权限和加载用户等高级权限。普通权限用于操作数据库，高级权限用于数据库管理。

user表中对应的权限是针对所有用户数据库的。这些字段值的类型为ENUM，可以取的值只能为Y和N，Y表示该用户有对应的权限；N表示用户没有对应的权限。查看user表的结构可以看到，这些字段的值默认都是N。可以使用GRANT语句或UPDATE语句更改user表的这些字段来修改用户对应的权限。

3. 安全列

安全列只有6个字段，其中两个是ssl相关的，即ssl_type和ssl_cipher，用于加密；两个是x509相关的，即x509_issuer和x509_subject，x509标准可用于标识用户；plugin字段标识可以用于验证用户身份的插件，如果该字段为空，服务器使用内建授权验证机制验证用户身份；password_expired用于设置用户的过期时间。读者可以通过SHOW VARIABLES LIKE 'have_openssl'语句来查询服务器是否支持ssl功能。

4. 资源控制列

资源控制列的字段用来限制用户使用的资源，包含4个字段，分别为：

- max_questions：用户每小时允许执行的查询操作次数。
- max_updates：用户每小时允许执行的更新操作次数。
- max_connections：用户每小时允许执行的连接操作次数。
- max_user_connections：用户允许同时建立的连接次数。

若一个小时内用户查询或者连接数量超过资源控制限制,则该用户将被锁定,直到下一个小时,才可以再次执行对应的操作。可以使用GRANT语句更新user表中的字段值。

13.1.2　db 表

db表是MySQL数据中非常重要的权限表,表中存储了用户对某个数据库的操作权限,决定用户能从哪个主机存取哪个数据库。db表比较常用,其结构如表13.2所示。

表 13.2　db 表结构

字 段 名	数据类型	默 认 值
Host	char(60)	
Db	char(64)	
User	char(32)	
Select_priv	enum('N','Y')	N
Insert_priv	enum('N','Y')	N
Update_priv	enum('N','Y')	N
Delete_priv	enum('N','Y')	N
Create_priv	enum('N','Y')	N
Drop_priv	enum('N','Y')	N
Grant_priv	enum('N','Y')	N
References_priv	enum('N','Y')	N
Index_priv	enum('N','Y')	N
Alter_priv	enum('N','Y')	N
Create_tmp_table_priv	enum('N','Y')	N
Lock_tables_priv	enum('N','Y')	N
Create_view_priv	enum('N','Y')	N
Show_view_priv	enum('N','Y')	N
Create_routine_priv	enum('N','Y')	N
Alter_routine_priv	enum('N','Y')	N
Execute_priv	enum('N','Y')	N
Event_priv	enum('N','Y')	N
Trigger_priv	enum('N','Y')	N

1. 用户列

db表的用户列有3个字段,分别是Host、User、Db,标识从某个主机连接某个用户对某个数据库的操作权限,这3个字段的组合构成了db表的主键。host表不存储用户名称,用户列只有2个字段,分别是Host和Db,表示从某个主机连接的用户对某个数据库的操作权限,其主键包括Host和Db两个字段。host表很少用到,一般情况下使用db表就可以满足权限控制需求了。

2. 权限列

db表中Create_routine_priv和Alter_routine_priv这两个字段表明用户是否有创建和修改存储过程的权限。

user表中的权限是针对所有数据库的，如果希望用户只对某个数据库有操作权限，那么需要将user表中对应的权限设置为N，然后在db表中设置对应数据库的操作权限。例如，有一个名称为Zhangting的用户分别从名称为large.domain.com和small.domain.com的两个主机连接到数据库，并需要操作books数据库。这时，可以将用户名称Zhangting添加到db表中，而db表中的host字段值为空，然后将两个主机地址分别作为两条记录的host字段值添加到host表中，并将两张表的数据库字段设置为相同的值books。当有用户连接到MySQL服务器时，db表中没有用户登录的主机名称，则MySQL会从host表中查找相匹配的值，并根据查询的结果决定用户的操作是否被允许。

13.1.3 tables_priv 表和 columns_priv 表

tables_priv表用来对表设置操作权限，columns_priv表用来对表的某一列设置权限。tables_priv表和columns_priv表的结构分别如表13.3和表13.4所示。

表 13.3 tables_priv 表结构

字 段 名	数据类型	默 认 值
Host	char(60)	
Db	char(64)	
User	char(16)	
Table_name	char(64)	
Grantor	char(77)	
Timestamp	timestamp	CURRENT_TIMESTAMP
Table_priv	set('Select','Insert','Update','Delete','Create','Drop','Grant','References','Index','Alter','Create View','Show view','Trigger')	
Column_priv	set('Select','Insert','Update','References')	

表 13.4 columns_priv 表结构

字 段 名	数据类型	默 认 值
Host	char(60)	
Db	char(64)	
User	char(16)	
Table_name	char(64)	
Column_name	char(64)	
Timestamp	timestamp	CURRENT_TIMESTAMP
Column_priv	set('Select','Insert','Update','References')	

tables_priv表中有8个字段,分别是Host、Db、User、Table_name、Grantor、Timestamp、Table_priv和Column_priv,各个字段说明如下:

(1) Host、Db、User和Table_name字段分表示主机名、数据库名、用户名和表名。
(2) Grantor字段表示修改该记录的用户。
(3) Timestamp字段表示修改该记录的时间。
(4) Table_priv字段表示对表的操作权限,包括SELECT、INSERT、UPDATE、DELETE、CREATE、DROP、GRANT、REFERENCES、INDEX和ALTER。
(5) Column_priv字段表示对表中的列的操作权限,包括Select、Insert、Update和References。

columns_priv表中只有7个字段,分别是Host、Db、User、Table_name、Column_name、Timestamp、Column_priv。其中,Column_name用来指定对哪些数据列具有操作权限。

13.1.4 procs_priv 表

procs_priv表可以对存储过程和存储函数设置操作权限。procs_priv表的结构如表13.5所示。

表 13.5 procs_priv 表结构

字 段 名	数 据 类 型	默 认 值
Host	char(60)	
Db	char(64)	
User	char(16)	
Routine_name	char(64)	
Routine_type	enum('FUNCTION','PROCEDURE')	NULL
Grantor	char(77)	
Proc_priv	set('Execute','Alter Routine','Grant')	
Timestamp	timestamp	CURRENT_TIMESTAMP

procs_priv表中包含8个字段,分别是Host、Db、User、Routine_name、Routine_type、Grantor、Proc_priv和Timestamp,各个字段的说明如下:

(1) Host、Db和User字段分别表示主机名、数据库名和用户名。
(2) Routine_name字段表示存储过程或函数的名称。
(3) Routine_type字段表示存储过程或函数的类型。Routine_type字段有两个值,分别是FUNCTION和PROCEDURE:FUNCTION表示这是一个函数,PROCEDURE表示这是一个存储过程。
(4) Grantor字段表示插入或修改该记录的用户。
(5) Proc_priv字段表示拥有的权限,包括Execute、Alter Routine、Grant 3种。
(6) Timestamp字段表示记录更新时间。

13.2 账户管理

MySQL提供了许多语句来管理用户账号,包括登录和退出MySQL服务器、创建用户、删除用户、密码管理和权限管理等。MySQL数据库的安全性需要通过账户管理来保证。本节将介绍如何在MySQL中对账户进行管理。

13.2.1 登录和退出 MySQL 服务器

登录MySQL时,可以使用mysql命令并在后面指定登录主机、用户名和密码。本小节将详细介绍mysql命令的常用参数以及登录和退出MySQL服务器的方法。

通过mysql -help命令可以查看mysql命令帮助信息。mysql命令的常用参数如下:

(1) -h主机名,可以使用该参数指定主机名或IP地址,如果不指定,默认是localhost。

(2) -u用户名,可以使用该参数指定用户名。

(3) -p密码,可以使用该参数指定登录密码。如果该参数后面有一段字符串,则该段字符串将作为用户的密码直接登录;如果后面没有内容,则登录的时候会提示输入密码。注意:该参数后面的字符串和-p之间不能有空格。

(4) -P端口号,该参数后面接MySQL服务器的端口号,默认为3306。

(5) 数据库名,可以在命令的最后指定数据库名。

(6) -e SQL语句。如果指定了该参数,将在登录后执行-e后面的命令或SQL语句并退出。

【例13.1】使用root用户登录到本地MySQL服务器的mysql库中,命令如下:

```
mysql -h localhost -u root -p mysql
```

命令执行结果如下:

```
C:\Users\Administrator>mysql -h localhost -u root -p mysql
Enter password: ******
Welcome to the MySQL monitor.  Commands end with ; or \g.
Your MySQL connection id is 19
Server version: 9.0.1 MySQL Community Server - GPL

Copyright (c) 2000, 2024, Oracle and/or its affiliates. All rights reserved.

Oracle is a registered trademark of Oracle Corporation and/or its
affiliates. Other names may be trademarks of their respective
owners.

Type 'help;' or '\h' for help. Type '\c' to clear the current input statement.
```

执行命令时,会提示"Enter password:",如果没有设置密码,可以直接按Enter键。密码正确就可以直接登录到服务器下面的mysql数据库中了。

【例13.2】使用root用户登录到本地MySQL服务器的test_db数据库中，同时执行一条查询语句。命令如下：

```
MySQL -h localhost -u root -p test_db -e "DESC person;"
```

命令执行结果如下：

```
C:\ > MySQL -h localhost -u root -p test_db -e "DESC person;"
Enter password: **
+-------+------------------+------+-----+---------+----------------+
| Field | Type             | Null | Key | Default | Extra          |
+-------+------------------+------+-----+---------+----------------+
| id    | int    unsigned  | NO   | PRI | NULL    | auto_increment |
| name  | char(40)         | NO   |     |         |                |
| age   | int              | NO   |     | 0       |                |
| info  | char(50)         | YES  |     | NULL    |                |
+-------+------------------+------+-----+---------+----------------+
```

按照提示输入密码，命令执行完成后查询出person表的结构，查询返回之后会自动退出MySQL。

13.2.2 新建普通用户

要创建新用户，必须有相应的权限来执行创建操作。在MySQL数据库中，有两种方式创建新用户：一种是使用CREATE USER语句；另一种是直接操作MySQL授权表。下面分别介绍这两种创建新用户的方法。

1. 使用CREATE USER语句创建新用户

执行CREATE USER 或GRANT语句时，服务器会修改相应的用户授权表，添加或者修改用户及其权限。CREATE USER语句的基本语法格式如下：

```
CREATE USER user_specification
    [, user_specification] ...

user_specification:
    user@host
    [
        IDENTIFIED BY [PASSWORD] 'password'
      | IDENTIFIED WITH auth_plugin [AS 'auth_string']
    ]
```

user表示创建的用户的名称；host表示允许登录的主机名称；IDENTIFIED BY表示用来设置用户的密码；[PASSWORD]表示使用哈希值设置密码，该参数可选；'password'表示用户登录时使用的普通明文密码；IDENTIFIED WITH语句为用户指定一个身份验证插件；auth_plugin是插件的名称，插件的名称可以是一个带单引号或者双引号的字符串；auth_string是可选的字符串参数，该参数将传递给身份验证插件，由该插件解释该参数的意义。

使用CREATE USER语句的用户，必须有全局的CREATE USER权限或MySQL数据库的

INSERT权限。每添加一个用户，CREATE USER语句会在MySQL.user表中添加一条新记录，但是新创建的账户没有任何权限。如果添加的账户已经存在，则CREATE USER语句会返回一个错误。

【例13.3】使用CREATE USER语句创建一个用户，用户名是jeffrey，密码是mypass，主机名是localhost，SQL语句如下：

```
CREATE USER 'jeffrey'@'localhost' IDENTIFIED BY 'mypass';
```

如果只指定用户名部分"jeffrey"，主机名部分则默认为"%"（对所有的主机开放权限）。
user_specification告诉MySQL服务器当用户登录时怎么验证用户的登录授权。如果指定用户登录不需要密码，则可以省略IDENTIFIED BY部分：

```
CREATE USER 'jeffrey'@'localhost';
```

此种情况，MySQL服务器使用内建的身份验证机制，用户登录时不能指定密码。
如果要创建指定密码的用户，需要IDENTIFIED BY 指定明文密码值：

```
CREATE USER 'jeffrey'@'localhost' IDENTIFIED BY 'mypass';
```

此种情况，MySQL服务器使用内建的身份验证机制，用户登录时必须指定密码。
MySQL的某些版本中会引入授权表的结构变化，添加新的特权或功能。每当更新MySQL到一个新的版本时，应该更新授权表，以确保它们有最新的结构，可以使用任何新功能。

2. 直接操作MySQL用户表

通过前面的介绍可知，使用CREATE USER创建新用户时，实际上是在向user表中添加一条新的记录。因此，可以使用INSERT语句向user表中直接插入一条记录来创建一个新的用户。使用INSERT语句，必须拥有对MySQL.user表的INSERT权限。使用INSERT语句创建新用户的基本语法格式如下：

```
INSERT INTO MySQL.user(Host, User, authentication_string)
VALUES('host', 'username', MD5('password'));
```

Host、User、authentication_string分别为user表中的主机、用户名称和密码字段；MD5()函数为密码加密函数。

【例13.4】使用INSERT创建一个新用户，用户名称为customer1，主机名称为localhost，密码为aa123456，SQL语句如下：

```
INSERT INTO user (Host,User, authentication_string)
VALUES('localhost','customer1', MD5('aa123456'));
```

语句执行结果如下：

```
MySQL> use mysql;
MySQL> INSERT INTO user (Host,User, authentication_string)
    -> VALUES('localhost','customer1', MD5('aa123456'));
Query OK, 1 row affected, 3 warnings (0.00 sec)
```

语句执行成功，但是有3个警告信息，查看警告信息，结果如下：

```
MySQL> SHOW WARNINGS;
+-------+------+-----------------------------------------------------+
| Level | Code | Message                                             |
+-------+------+-----------------------------------------------------+
| Error | 1364 | Field 'ssl_cipher' doesn't have a default value     |
| Error | 1364 | Field 'x509_issuer' doesn't have a default value    |
| Error | 1364 | Field 'x509_subject' doesn't have a default value   |
+-------+------+-----------------------------------------------------+
```

因为ssl_cipher、x509_issuer和x509_subject这3个字段在user表中没有设置默认值，所以这里提示警告信息，但没有影响INSERT语句的执行。使用SELECT语句查看user表中的记录：

```
MySQL> SELECT host,user, authentication_string FROM user ;
+---------+-----------------+--------------------------------------------+
| host    | user            | authentication_string                      |
+---------+-----------------+--------------------------------------------+
|localhost| customer1       | 8a6f2805b4515ac12058e79e66539be9           |
|localhost|mysql.infoschema |$A$005$THISISACOMBINATIONOFIVALIDSALTANDP   |
|localhost|mysql.session    |$A$005$THISISACOMBINATIONOFINVALIDSALTANDPAS|
|localhost|mysql.sys        |$A$005$THISISACOMBINATIONOFINVAUSTNEVERREUSED|
|localhost|root             |*6BB4837EB74329105EE4568DDA7DC67ED2CA2AD9   |
+---------+-----------------+--------------------------------------------+
```

可以看到新用户customer1已经添加到user表中，表示添加新用户成功。

13.2.3　删除普通用户

在MySQL数据库中，即可以使用DROP USER语句删除用户，也可以使用DELETE语句从MySQL.user表中删除对应的记录来删除用户。

1. 使用DROP USER语句删除用户

DROP USER语句的语法如下：

```
DROP USER user [, user];
```

DROP USER语句用于删除一个或多个MySQL账户。要使用DROP USER，必须拥有MySQL数据库的全局CREATE USER权限或DELETE权限。使用与GRANT或REVOKE相同的格式为每个账户命名。例如，"'jeffrey'@'localhost'"账户名称的用户和主机部分与用户表记录的User和Host列值相对应。

使用DROP USER，可以删除一个账户及其权限，操作如下：

```
DROP USER 'user'@'localhost';
DROP USER;
```

第一条语句可以删除user在本地的登录权限；第二条语句可以删除所有来自授权表的账户权限记录。

【例13.5】使用DROP USER删除账户"'jeffrey'@'localhost'"，SQL语句如下：

```
DROP USER 'jeffrey'@'localhost';
```

执行过程如下：

```
MySQL> DROP USER 'jeffrey'@'localhost';
Query OK, 0 rows affected (0.00 sec)
```

可以看到语句执行成功。查看执行结果：

```
MySQL> SELECT host,user, authentication_string FROM user ;
+---------+----------------+------------------------------------------+
| host    | user           | authentication_string                    |
+---------+----------------+------------------------------------------+
|localhost|mysql.infoschema|$A$005$THISISACOMBINATIONOFIVALIDSALTANDP |
|localhost|mysql.session   |$A$005$THISISACOMBINATIONOFINVALIDSALTANDPAS|
|localhost|mysql.sys       |$A$005$THISISACOMBINATIONOFINVAUSTNEVERREUSED|
|localhost|root            |*6BB4837EB74329105EE4568DDA7DC67ED2CA2AD9 |
+---------+----------------+------------------------------------------+
```

user表中已经没有名称为jeffrey、主机名为localhost的账户，即"'jeffrey' @ 'localhost'"的账户已经被删除。

> **提示** DROP USER不能自动关闭任何已打开的用户对话。如果用户有打开的对话，则DROP USER不会生效，必须等到用户对话被关闭后才能生效。一旦对话被关闭，用户也被取消，则此用户再次试图登录时将会失败。

2. 使用DELETE语句删除用户

DELETE语句基本语法格式如下：

```
DELETE FROM MySQL.user WHERE Host='hostname' and User='username'
```

Host和User为user表中的两个字段，两个字段的组合确定所要删除的账户记录。

【例13.6】使用DELETE删除用户'customer1'@'localhost'。

首先创建用户customer1，SQL语句如下：

```
MySQL>CREATE USER 'customer1'@'localhost' IDENTIFIED BY 'my123';
Query OK, 0 rows affected (0.12 sec)
```

然后使用DELETE删除用户'customer1'@'localhost'，SQL语句如下：

```
mysql> DELETE FROM MySQL.user WHERE Host= 'localhost' and User='customer1';
Query OK, 1 row affected (0.01 sec)
```

可以看到语句执行成功，'customer1'@'localhost'的用户账号已经被删除。读者可以使用SELECT语句查询user表中的记录，确认删除操作是否成功。

13.2.4 root用户修改普通用户密码

root用户拥有很高的权限，可以修改其他用户的密码。root用户登录MySQL服务器后，可以通过SET语句修改MySQL.user表，通过UPDATE语句修改用户的密码。

创建用户user，SQL语句如下：

```
MySQL>CREATE USER 'user'@'localhost' IDENTIFIED BY 'my123';
Query OK, 0 rows affected (0.12 sec)
```

1. 使用SET语句修改普通用户的密码

使用SET语句修改普通用户的密码的语法格式如下：

```
SET PASSWORD FOR 'user'@'localhost' = 'sa123';
```

【例13.7】使用SET语句将用户user的密码修改为"sa123"。

使用root用户登录到MySQL服务器后，执行如下语句：

```
MySQL> SET PASSWORD FOR 'user'@'localhost' = 'sa123';
Query OK, 0 rows affected (0.00 sec)
```

SET语句执行成功，用户user的密码被设置为"sa123"。

2. 使用UPDATE语句修改普通用户的密码

使用root用户登录到MySQL服务器后，可以使用UPDATE语句修改MySQL数据库的user表的password字段，从而修改普通用户的密码。使用UPDATA语句修改用户密码的语法如下：

```
UPDATE MySQL.user SET authentication_string=MD5("123456")
WHERE User="username" AND Host="hostname";
```

MD5()函数用来加密用户密码。执行UPDATE语句后，需要执行FLUSH PRIVILEGES语句重新加载用户权限。

【例13.8】使用UPDATE语句将用户user的密码修改为"sns123"。

使用root用户登录到MySQL服务器后，执行如下语句：

```
MySQL> UPDATE MySQL.user SET authentication_string =MD5("sns123")
    -> WHERE User="user" AND Host="localhost";
Query OK, 1 row affected (0.00 sec)
Rows matched: 1  Changed: 1  Warnings: 0
MySQL> FLUSH PRIVILEGES;
Query OK, 0 rows affected (0.11 sec)
```

执行完UPDATE语句后，用户user的密码被修改成了"sns123"。使用FLUSH PRIVILEGES重新加载权限，用户user就可以使用新的密码登录了。

13.3 权限管理

权限管理主要是对登录到MySQL的用户进行权限验证。所有用户的权限都存储在MySQL的权限表中，不合理的权限规划会给MySQL服务器带来安全隐患。因此数据库管理员要对所有用户的权限进行合理地规划和管理。MySQL权限系统的主要功能是证实连接到一台给定主机的用户，并且赋予该用户在数据库上的SELECT、INSERT、UPDATE和DELETE权限。本节将介绍MySQL权限管理的相关内容。

13.3.1　MySQL 的各种权限

账户权限信息被存储在MySQL数据库的user、db、host、tables_priv、columns_priv和procs_priv表中。在启动MySQL时，服务器将这些数据库表中的权限信息读入内存。

GRANT和REVOKE语句所涉及的权限如表13.6所示。

表 13.6　GRANT 和 REVOKE 语句中可以使用的权限

权　　限	user 表中对应的列	权限的范围
CREATE	Create_priv	数据库、表或索引
DROP	Drop_priv	数据库、表或视图
GRANT OPTION	Grant_priv	数据库、表或存储过程
REFERENCES	References_priv	数据库或表
EVENT	Event_priv	数据库
ALTER	Alter_priv	数据库
DELETE	Delete_priv	表
INDEX	Index_priv	表
INSERT	Insert_priv	表
SELECT	Select_priv	表或列
UPDATE	Update_priv	表或列
CREATE TEMPORARY TABLES	Create_tmp_table_priv	表
LOCK TABLES	Lock_tables_priv	表
TRIGGER	Trigger_priv	表
CREATE VIEW	Create_view_priv	视图
SHOW VIEW	Show_view_priv	视图
ALTER ROUTINE	Alter_routine_priv	存储过程和存储函数
CREATE ROUTINE	Create_routine_priv	存储过程和存储函数
EXECUTE	Execute_priv	存储过程和存储函数
FILE	File_priv	访问服务器上的文件
CREATE TABLESPACE	Create_tablespace_priv	服务器管理
CREATE USER	Create_user_priv	服务器管理
PROCESS	Process_priv	存储过程和存储函数
RELOAD	Reload_priv	访问服务器上的文件
REPLICATION CLIENT	Repl_client_priv	服务器管理
REPLICATION SLAVE	Repl_slave_priv	服务器管理
SHOW DATABASES	Show_db_priv	服务器管理

权　　限	user 表中对应的列	权限的范围
SHUTDOWN	Shutdown_priv	服务器管理
SUPER	Super_priv	服务器管理

（1）CREATE和DROP权限，可以创建新数据库和表，或删除（移掉）已有数据库和表。如果将MySQL数据库中的DROP权限授予某用户，则该用户可以删掉MySQL访问权限保存的数据库。

（2）SELECT、INSERT、UPDATE和DELETE权限允许在一个数据库现有的表上实施操作。

（3）SELECT权限只有在它们真正从一张表中检索行时才被用到。

（4）INDEX权限允许创建或删除索引，INDEX适用已有表。如果具有某张表的CREATE权限，则可以在CREATE TABLE语句中包括索引定义。

（5）ALTER权限，可以使用ALTER TABLE来更改表的结构和重命名表。

（6）CREATE ROUTINE权限用来创建保存的程序（函数和过程），ALTER ROUTINE权限用来更改和删除保存的程序，EXECUTE权限用来执行保存的程序。

（7）GRANT权限允许授权给其他用户，可用于数据库、表和保存的程序。

（8）FILE权限给予用户使用LOAD DATA INFILE和SELECT ... INTO OUTFILE语句读或写服务器上的文件，任何被授予FILE权限的用户都能读或写MySQL服务器上的任何文件（说明用户可以读任何数据库目录下的文件，因为服务器可以访问这些文件）。FILE权限允许用户在MySQL服务器具有写权限的目录下创建新文件，但不能覆盖已有文件。

其余的权限用于管理性操作，它使用mysqladmin程序或SQL语句实施。表13.7显示每个权限允许执行的mysqladmin命令。

表 13.7　不同权限下可以使用的 mysqladmin 命令

权　　限	权限拥有者允许执行的命令
RELOAD	flush-hosts、flush-logs、flush-privileges、flush-status、flush-tables、flush-threads、refresh、reload
SHUTDOWN	shutdown
PROCESS	processlist
SUPER	kill

（1）reload命令告诉服务器将授权表重新读入内存；flush-privileges是reload的同义词；refresh命令清空所有表并关闭/打开记录文件；其他flush-xxx命令执行类似refresh的操作，但是范围更有限，不过在某些情况下可能更好用。例如，如果只是想清空记录文件，flush-logs是比refresh更好的选择。

（2）shutdown命令关掉服务器。只能从mysqladmin发出命令。

（3）processlist命令显示在服务器内执行的线程的信息（其他账户相关的客户端执行的语句）。

（4）kill命令杀死服务器线程。用户总是能显示或杀死自己的线程，但是需要PROCESS权限来显示或杀死其他用户和SUPER权限启动的线程。

总的来说，只授予权限给需要它们的那些用户。

13.3.2 授权

授权就是为某个用户授予权限。合理的授权可以保证数据库的安全。MySQL中可以使用GRANT语句为用户授予权限。

授予的权限可以分为以下5个层级：

1. 全局层级

全局权限适用于一张给定服务器中的所有数据库。这些权限存储在MySQL.user表中。GRANT ALL ON *.*和REVOKE ALL ON *.*只授予和撤销全局权限。

2. 数据库层级

数据库权限适用于一张给定数据库中的所有目标。这些权限存储在MySQL.db和MySQL.host表中。GRANT ALL ON db_name.和REVOKE ALL ON db_name.*只授予和撤销数据库权限。

3. 表层级

表权限适用于一张给定表中的所有列。这些权限存储在MySQL.talbes_priv表中。GRANT ALL ON db_name.tbl_name和REVOKE ALL ON db_name.tbl_name只授予和撤销表权限。

4. 列层级

列权限适用于一张给定表中的单一列。这些权限存储在MySQL.columns_priv表中。当使用REVOKE时，必须指定与被授权列相同的列。

5. 子程序层级

CREATE ROUTINE、ALTER ROUTINE、EXECUTE和GRANT权限适用于已存储的子程序。这些权限可以被授予为全局层级和数据库层级。除了CREATE ROUTINE之外，这些权限还可以被授予子程序层级，并存储在MySQL.procs_priv表中。CREATE ROUTINE、ALTER ROUTINE、EXECUTE和GRANT权限用于管理已存储的子程序。这些权限可以被授予为全局层级（GLOBAL）和数据库层级（DATABASE）。

除了CREATE ROUTINE 权限外，其他权限（ALTER ROUTINE、EXECUTE和GRANT）还可以被授予子程序层级（PROCEDURE），并存储在MySQL.procs_priv表中。

在MySQL中，要使用GRANT或REVOKE，必须拥有GRANT OPTION权限，并且必须用于正在授予或撤销的权限。GRANT的语法如下：

```
GRANT priv_type [(columns)] [, priv_type [(columns)]] ...
ON [object_type]  table1, table2,..., tablen
TO user  [WITH GRANT OPTION]

object_type = TABLE  |  FUNCTION  |  PROCEDURE
```

其中，priv_type参数表示权限类型；columns参数表示权限作用于哪些列上，不指定该参数，表示作用于整张表；table1,table2,...,tablen表示授予权限的列所在的表；object_type指定授权作用的对象类型包括TABLE（表）、FUNCTION（函数）和PROCEDURE（存储过程），当从旧版本的MySQL升级时，要使用object_tpye子句，必须升级授权表；user参数表示用户账户，由用户名和主机名构成，形式是"'username'@'hostname'"。

WITH关键字后可以跟一个或多个with_option参数。这个参数有5个选项，意义如下：

（1）GRANT OPTION：被授权的用户可以将这些权限赋予别的用户。
（2）MAX_QUERIES_PER_HOUR count：设置每小时可以执行count次查询。
（3）MAX_UPDATES_PER_HOUR count：设置每小时可以执行count次更新。
（4）MAX_CONNECTIONS_PER_HOUR count：设置每小时可以建立count个连接。
（5）MAX_USER_CONNECTIONS count：设置单个用户可以同时建立count个连接。

【例13.9】 创建一个新的用户grantUser。使用GRANT语句对用户grantUser赋予所有的表进行数据的查询、插入权限，并授予GRANT权限。GRANT语句及其执行结果如下：

```
MySQL> CREATE USER 'grantUser'@'localhost' IDENTIFIED BY 'mypass';
MySQL> GRANT SELECT,INSERT ON *.* TO 'grantUser'@'localhost' WITH GRANT OPTION;
Query OK, 0 rows affected (0.03 sec)
```

结果显示执行成功。使用SELECT语句查询用户grantUser的权限：

```
MySQL> SELECT Host,User,Select_priv,Insert_priv, Grant_priv FROM mysql.user where user='grantUser';
+-----------+-----------+-------------+-------------+------------+
| Host      | User      | Select_priv | Insert_priv | Grant_priv |
+-----------+-----------+-------------+-------------+------------+
| localhost | grantUser | Y           | Y           | Y          |
+-----------+-----------+-------------+-------------+------------+
1 row in set (0.00 sec)
```

查询结果显示用户grantUser被创建成功，并被赋予SELECT、INSERT和GRANT权限，其相应字段值均为"Y"。被授予GRANT权限的用户可以登录MySQL并创建其他用户账户。

13.3.3　收回权限

收回权限就是将赋予用户的某些权限取消。收回用户不必要的权限，可以在一定程度上保证系统的安全性。MySQL中使用REVOKE语句取消用户拥有的某些权限。使用REVOKE收回权限之后，用户账户的记录将从db、host、tables_priv和columns_priv表中删除，但是仍然在user表中保存（删除user表中的账户记录，要使用DROP USER语句，在13.2.3节已经介绍）。

在将用户账户从user表中删除之前，应该收回相应用户的所有权限。REVOKE语句有两种语法格式。

（1）第一种是收回所有用户的所有权限，用于取消已命名用户的所有全局层级、数据库层级、表层级和列层级的权限，具体如下：

```
REVOKE ALL PRIVILEGES, GRANT OPTION
FROM 'user'@'host' [, 'user'@'host' ...]
```

REVOKE语句必须和FROM语句一起使用。FROM语句指明需要收回权限的账户。

（2）第二种为长格式的REVOKE语句，基本语法如下：

```
REVOKE priv_type [(columns)] [, priv_type [(columns)]] ...
ON table1, table2,..., tablen
FROM 'user'@'host'[, 'user'@ 'host' ...]
```

该语法收回指定的权限。其中，priv_type参数表示权限类型；columns参数表示权限作用于哪些列上，如果不指定该参数，则表示作用于整张表；table1,table2,...,tablen表示从哪张表中收回权限；'user'@'host'参数表示用户账户，由用户名和主机名构成。

要使用REVOKE语句，必须拥有MySQL数据库的全局CREATE USER权限或UPDATE权限。

【例13.10】使用REVOKE语句取消用户grantUser的查询权限。REVOKE语句及其执行结果如下：

```
MySQL> REVOKE Select ON *.* FROM 'grantUser'@'localhost';
Query OK, 0 rows affected (0.00 sec)
```

执行结果显示执行成功。使用SELECT语句查询用户grantUser的权限：

```
mysql> SELECT Host,User,Select_priv,Insert_priv,Grant_priv FROM MySQL.user where user='grantUser';
+-----------+-----------+-------------+-------------+------------+
| Host      | User      | Select_priv | Insert_priv | Grant_priv |
+-----------+-----------+-------------+-------------+------------+
| localhost | grantUser | N           | Y           | Y          |
+-----------+-----------+-------------+-------------+------------+
```

查询结果显示用户grantUser的Select_priv字段值为"N"，说明SELECT权限已经被收回。

> 提示　当从旧版本的MySQL升级时，如果要使用EXECUTE、CREATE VIEW、SHOW VIEW、CREATE USER、CREATE ROUTINE和ALTER ROUTINE权限，必须首先升级授权表。

13.3.4　查看权限

使用SHOW GRANTS语句可以显示指定用户的权限信息，基本语法格式如下：

```
SHOW GRANTS FOR 'user'@ 'host' ;
```

其中，user表示登录用户的名称，host表示登录的主机名称或者IP地址。在使用该语句时，要确保指定的用户名和主机名都要用单引号引起来，并使用"@"符号将两个名字分隔开。

【例13.11】使用SHOW GRANTS语句查询用户grantUser的权限信息。SHOW GRANTS语句及其执行结果如下：

```
MySQL> SHOW GRANTS FOR 'grantUser'@'localhost';
+-------------------------------------------------------------------+
| Grants for grantUser@localhost                                    |
+-------------------------------------------------------------------+
| GRANT INSERT ON *.* TO `grantUser`@`localhost` WITH GRANT OPTION |
+-------------------------------------------------------------------+
```

返回的结果显示了grantUser表中的账户信息。接下来的行以"GRANT INSERT ON"关键字开头,表示用户被授予了INSERT权限;*.*表示INSERT权限作用于所有数据库的所有数据表。

在这里,只是定义了个别的用户权限,GRANT可以显示更加详细的权限信息,包括全局级的和非全局级的权限,如果表层级或者列层级的权限被授予用户,那么它们也能在结果中显示出来。

在前面创建用户时,查看新建的账户使用的是SELECT语句,也可以通过SELECT语句查看user表中的各个权限字段以确定用户的权限信息,其基本语法格式如下:

```
SELECT privileges_list FROM user WHERE user='username', host= 'hostname';
```

其中,privileges_list为想要查看的权限字段,可以为Select_priv、Insert_priv等。读者可以根据需要选择要查询的字段。

13.4 访问控制

正常情况下,并不希望每个用户都可以执行所有的数据库操作。当MySQL允许一个用户执行各种操作时,它首先核实该用户向MySQL服务器发送的连接请求,然后确认用户的操作请求是否被允许。本节将介绍MySQL中的访问控制过程。MySQL的访问控制分为两个阶段:连接核实阶段和请求核实阶段。

13.4.1 连接核实阶段

当连接MySQL服务器时,服务器基于用户的身份以及用户是否能提供正确的密码来接受或拒绝连接,即客户端用户连接请求中会提供用户名称、主机地址名和密码。MySQL使用user表中的3个字段(Host、User和authentication_string)执行身份检查,服务器只有在user表记录的Host和User字段匹配客户端主机名和用户名并且客户端提供正确的密码时才接受连接。如果连接核实没有通过,则服务器完全拒绝访问;否则,服务器接受连接,然后进入请求核定阶段等待用户请求。

13.4.2 请求核实阶段

建立了连接之后,服务器进入访问控制的请求核定阶段。对在此连接上的每个请求,服务器检查用户要执行的操作,然后检查是否有足够的权限来执行。这正是授权表中的权限列发挥作用的地方。这些权限可以来自user、db、host、tables_priv或columns_priv表。

确认权限时，MySQL首先检查user表，如果指定的权限没有在user表中被授权，MySQL将检查db表，db表是下一安全层级，其中的权限限定于数据库层级，在该层级的SELECT权限允许用户查看指定数据库所有表中的数据；如果在该层级没有找到限定的权限，则MySQL继续检查tables_priv表以及columns_priv表；如果所有权限表都检查完毕，但还是没有找到允许的权限操作，MySQL将返回错误信息，用户请求的操作不能执行，操作失败。

请求核实的过程如图13.1所示。

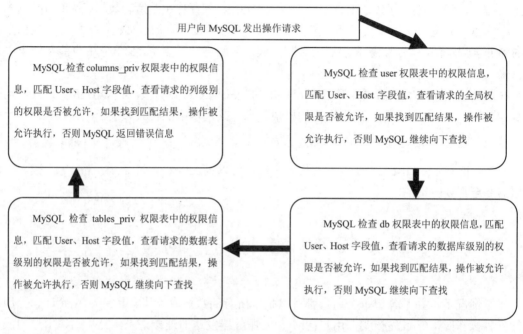

图 13.1　MySQL 请求核实过程

> **提示**　MySQL通过向下层级的顺序检查权限表（从user表到columns_priv表），但并不是所有的权限都要执行该过程。例如，一个用户登录到MySQL服务器之后只执行对MySQL的管理操作，此时只涉及管理权限，因此只检查user表。另外，如果请求的权限操作不被允许，MySQL也不会继续检查下一层级的表。

13.5　提升安全性

MySQL 进一步提升了安全性，主要表现在密码到期更换策略和安全模式。

13.5.1　密码到期更换策略

MySQL允许数据库管理员手动设置账户密码过期时间。任何密码超期的账户想要连接服务端时，都必须更改密码。通过设置default_password_lifetime参数可以设置账户过期时间。

【例13.12】 设置账户密码过期时间。

首先查看系统中的账户过期时间，SQL语句如下：

```
mysql> SELECT user,host, password_last_changed, password_lifetime,
password_expired FROM mysql.user \G
*************************** 1. row ***************************
            user: grantUser
            host: localhost
password_last_changed: 2024-07-27 12:02:09
    password_lifetime: NULL
    password_expired: N
*************************** 2. row ***************************
            user: jeffrey
            host: localhost
password_last_changed: 2024-07-27 12:00:40
    password_lifetime: NULL
    password_expired: N
*************************** 3. row ***************************
            user: mysql.infoschema
            host: localhost
password_last_changed: 2024-07-27 11:43:30
    password_lifetime: NULL
    password_expired: N
*************************** 4. row ***************************
            user: mysql.session
            host: localhost
password_last_changed: 2024-07-27 11:43:30
    password_lifetime: NULL
    password_expired: N
*************************** 5. row ***************************
            user: mysql.sys
            host: localhost
password_last_changed: 2024-07-27 11:43:30
    password_lifetime: NULL
    password_expired: N
*************************** 6. row ***************************
            user: root
            host: localhost
password_last_changed: 2024-07-27 11:43:50
    password_lifetime: NULL
    password_expired: N
```

结果中的"password_lifetime: NULL"表示密码永不过期。

下面设置root用户的密码过期时间为365天，SQL语句如下：

```
mysql> ALTER USER root@localhost PASSWORD EXPIRE INTERVAL 365 DAY;
Query OK, 0 rows affected (0.00 sec)
```

再次查看root用户的信息，SQL语句如下：

```
mysql>SELECT user,host, password_last_changed, password_lifetime,
password_expired FROM mysql.user WHERE user = 'root'\G
```

```
*************************** 1. row ***************************
                 user: root
                 host: localhost
password_last_changed: 2024-07-24 20:22:43
     password_lifetime: 365
     password_expired: N
1 row in set (0.00 sec)
```

从结果中可以看出，root用户的password_lifetime为365天。

将root用户的密码过期重新设置为永不过期，SQL语句如下：

```
mysql> ALTER USER root@localhost PASSWORD EXPIRE DEFAULT;
Query OK, 0 rows affected (0.00 sec)
```

再次查看root用户的信息，SQL语句如下：

```
mysql> SELECT user,host, password_last_changed, password_lifetime,
password_expired FROM mysql.user WHERE user = 'root'\G
*************************** 1. row ***************************
                 user: root
                 host: localhost
password_last_changed: 2024-07-24 20:22:43
     password_lifetime: NULL
     password_expired: N
1 row in set (0.00 sec)
```

从结果中可以看出，password_lifetime又被重新设置为NULL，此时该用户的密码将永不过期。

13.5.2 安全模式

在MySQL中，引入了名为"安全模式"的安装选项，旨在提高用户数据的安全性并减少潜在的泄露风险。安全模式安装在默认情况下是关闭的，但可以通过在安装过程中选择"安全模式"选项来启用。当使用安全模式安装MySQL时，以下安全功能将受到启用：

（1）默认情况下，安全模式会禁用远程root登录，以防止未经授权的用户通过远程连接访问数据库。可以通过在my.cnf配置文件中设置[mysqld]部分中的skip-networking参数为0来启用远程root登录，但这将使系统处于非安全模式。

（2）安全模式会启用审计功能，以记录数据库中的所有操作。这有助于跟踪用户活动并检测潜在的安全漏洞。可以使用--audit-log选项在安装过程中启用或禁用审计功能。

（3）安全模式还会启用二进制日志功能，以记录数据库中的所有更改。这有助于在发生数据泄露时进行故障恢复和审计。可以使用--log-bin选项在安装过程中启用或禁用二进制日志功能。

此外，用户还可以通过以下方式来提升MySQL安装的安全性：

（1）为root账户设置密码。
（2）移除能从本地主机以外的地址访问数据库的root账户。

（3）移除匿名账户。

（4）移除test数据库，该数据库默认可被任意用户甚至匿名账户访问。

使用mysqld -initialize命令来安装MySQL实例默认是安全的，主要原因如下：

（1）在安装过程只创建一个root账户'root'@'localhost'，自动为这个账户生成一个随机密码并标记密码过期。

（2）数据库管理员必须用root账户及该随机密码登录并设置一个新密码后才能对数据库进行正常操作。

（3）安装过程不创建任何匿名账户。

（4）安装过程不创建test数据库。

13.6 管理角色

在MySQL数据库中，角色可以看作一些权限的集合。为用户赋予统一的角色，权限的修改直接通过角色来进行，从而无须为每个用户单独授权。

【例13.13】管理角色。

首先创建角色，SQL语句如下：

```
mysql> CREATE ROLE role_tt; # 创建角色
Query OK, 0 rows affected (0.02 sec)
```

然后给角色授予权限，SQL语句如下：

```
mysql> GRANT SELECT ON db.* to 'role_tt'; # 给角色role_tt授予查询权限
Query OK, 0 rows affected (0.10 sec)
```

创建用户myuser1，SQL语句如下：

```
mysql> CREATE USER ' myuser1'@'%' identified by '123456';
Query OK, 0 rows affected (0.07 sec)
```

为用户myuser1赋予角色role_tt，SQL语句如下：

```
mysql> GRANT 'role_tt' TO 'myuser1'@'%';
Query OK, 0 rows affected (0.02 sec)
```

为角色role_tt增加INSERT权限，SQL语句如下：

```
mysql> GRANT INSERT ON db.* to 'role_tt';
Query OK, 0 rows affected (0.08 sec)
```

为角色role_tt删除INSERT权限，SQL语句如下：

```
mysql> REVOKE INSERT ON db.* FROM 'role_tt';
Query OK, 0 rows affected (0.10 sec)
```

查看角色与用户关系，SQL语句如下：

```
mysql> SELECT * FROM mysql.role_edges;
+-----------+-----------+---------+---------+-------------------+
| FROM_HOST | FROM_USER | TO_HOST | TO_USER | WITH_ADMIN_OPTION |
+-----------+-----------+---------+---------+-------------------+
| %         | role_tt   | %       | myuser1 | N                 |
+-----------+-----------+---------+---------+-------------------+
1 row in set (0.00 sec)
```

删除角色，SQL语句如下：

```
mysql> DROP ROLE role_tt;
Query OK, 0 rows affected (0.04 sec)
```

通过使用角色，可以更轻松地管理权限，而不需要为每个用户单独授权。这有助于简化权限的分配和管理，并提高数据库的安全性和可维护性。在实际操作中，要根据需求和安全要求来创建、分配和删除角色。

第 14 章 MySQL性能优化

MySQL性能优化就是通过合理安排资源、调整系统参数，使MySQL运行得更快、更节省资源。本章将讲解MySQL性能优化的相关内容，包括优化查询、优化数据库结构、优化临时表性能和创建全局通用表空间。

14.1 优化简介

优化MySQL数据库是数据库管理员和数据库开发人员的必备技能。MySQL优化，一方面是找出系统的瓶颈，提高MySQL数据库整体的性能；另一方面需要合理的结构设计和参数调整，以提高用户操作的响应速度，同时还要尽可能节省系统资源，以便系统可以提供更大负荷的服务。本节将介绍优化的基本知识。

MySQL数据库优化是多方面的，原则是减少系统的瓶颈，减少资源的占用，增加系统的反应速度。例如，通过优化文件系统，提高磁盘I\O的读写速度；通过优化操作系统调度策略，提高MySQL在高负荷情况下的负载能力；优化表结构、索引、查询语句等，使查询响应更快。

在MySQL中，可以使用SHOW STATUS语句查询数据库的性能参数。SHOW STATUS语句的语法如下：

```
SHOW STATUS LIKE 'value';
```

其中，value是要查询的参数值，一些常用的性能参数如下：

- Connections：连接MySQL服务器的次数。
- Uptime：MySQL服务器的上线时间。
- Slow_queries：慢查询的次数。
- Com_select：查询操作的次数。
- Com_insert：插入操作的次数。
- Com_update：更新操作的次数。

- Com_delete：删除操作的次数。

例如，查询MySQL服务器的连接次数，可以执行如下语句：

```
SHOW STATUS LIKE 'Connections';
```

查询MySQL服务器的慢查询次数，可以执行如下语句：

```
SHOW STATUS LIKE 'Slow_queries';
```

慢查询次数参数可以与慢查询日志结合，先找出慢查询语句，然后针对慢查询语句进行表结构优化或者查询语句优化。

14.2 优化查询

查询是数据库中最频繁的操作，提高查询速度可以有效地提高MySQL数据库的性能。本节将介绍优化查询的方法。

14.2.1 分析查询语句

通过对查询语句进行分析，可以了解查询语句的执行情况，找出查询语句执行的瓶颈，从而优化查询语句。MySQL中提供了EXPLAIN语句和DESCRIBE语句来分析查询语句。

1. EXPLAIN语句

EXPLAIN语句的基本语法如下：

```
EXPLAIN [EXTENDED] SELECT select_options
```

使用EXTENED关键字，EXPLAIN语句将产生附加信息；select_options是SELECT语句的查询选项，包括FROM WHERE子句等。

执行该语句，可以分析EXPLAIN后面SELECT语句的执行情况，并且能够分析出所查询表的一些特征。

【例14.1】使用EXPLAIN语句来分析一个查询语句，SQL语句如下：

```
mysql> EXPLAIN SELECT * FROM fruits;
+----+-------------+--------+------+---------------+------+---------+------+------+-------+
| id | select_type | table  | type | possible_keys | key  | key_len | ref  | rows | Extra |
+----+-------------+--------+------+---------------+------+---------+------+------+-------+
|  1 | SIMPLE      | fruits | ALL  | NULL          | NULL | NULL    | NULL |   16 |       |
+----+-------------+--------+------+---------------+------+---------+------+------+-------+
```

下面对查询结果进行解释。

（1）id：SELECT识别符。这是SELECT的查询序列号。
（2）select_type：表示SELECT语句的类型。它有以下几种取值：

- SIMPLE：表示简单查询，其中不包括连接查询和子查询。
- PRIMARY：表示主查询或者最外层的查询语句。
- UNION：表示连接查询的第二个或后面的查询语句。
- DEPENDENT UNION：连接查询中的第二个或后面的SELECT语句，取决于外面的查询。
- UNION RESULT：连接查询的结果。
- SUBQUERY，子查询中的第一个SELECT语句。
- DEPENDENT SUBQUERY，子查询中的第一个SELECT语句，取决于外面的查询。
- DERIVED，导出表的SELECT（FROM子句的子查询）。

（3）table：表示查询的表。
（4）type：表示表的连接类型。

下面按照从最佳类型到最差类型的顺序给出各种连接类型。

- system

该表是仅有一行的系统表。这是const连接类型的一个特例。

- const

数据表最多只有一个匹配行，将在查询开始时被读取，并在余下的查询优化中作为常量对待。const表查询速度很快，因为它们只读取一次。const用于使用常数值比较PRIMARY KEY或UNIQUE索引的所有部分的场合。

在下面的查询中，tbl_name可用于const表：

```
SELECT * from tbl_name WHERE primary_key=1;
SELECT * from tbl_name WHERE primary_key_part1=1AND primary_key_part2=2;
```

- eq_ref

eq-ref表示使用唯一性索引进行连接操作，使用索引查找来匹配连接条件，这种方式适用于连接条件中的列是主键或唯一性索引的情况。

eq_ref可以用于使用"="操作符比较带索引的列。比较值可以为常量或一个在该表前面所读取的表的列的表达式。

在下面的例子中，MySQL可以使用eq_ref连接来处理ref_tables：

```
SELECT * FROM ref_table,other_table
WHERE ref_table.key_column=other_table.column;
SELECT * FROM ref_table,other_table
WHERE ref_table.key_column_part1=other_table.column
AND ref_table.key_column_part2=1;
```

- ref

对于来自前面的表的任意行组合，将从该表中读取所有匹配的行。这种类型用于索引既不是UNIQUE也不是PRIMARY KEY的情况，或者查询中使用了索引列的左子集（即索引中左边的部分列组合）。ref可以用于使用"="或"<=>"操作符比较带索引的列。

在下面的例子中，MySQL可以使用ref连接来处理ref_tables：

```
SELECT * FROM ref_table WHERE key_column=expr;

SELECT * FROM ref_table,other_table
WHERE ref_table.key_column=other_table.column;

SELECT * FROM ref_table,other_table
WHERE ref_table.key_column_part1=other_table.column
AND ref_table.key_column_part2=1;
```

- ref_or_null

该连接类型如同ref，但是添加了MySQL可以专门搜索包含NULL值的行。在解决子查询中经常使用该连接类型的优化。

在下面的例子中，MySQL可以使用ref_or_null连接来处理ref_tables：

```
SELECT * FROM ref_table
WHERE key_column=expr OR key_column IS NULL;
```

- index_merge

该连接类型表示使用了索引合并优化方法。在这种情况下，key列包含了使用的索引的清单，key_len包含了所使用索引的最长关键元素。

- unique_subquery

该类型替换了下面形式的IN子查询的ref：

```
value IN (SELECT primary_key FROM single_table WHERE some_expr)
```

unique_subquery是一个索引查找函数，可以完全替换子查询，效率更高。

- index_subquery

该连接类型类似于unique_subquery，可以替换IN子查询，但只适合下列形式的子查询中的非唯一索引：

```
value IN (SELECT key_column FROM single_table WHERE some_expr)
```

- range

只检索给定范围的行，使用一个索引来选择行。key列显示使用了哪个索引，key_len包含所使用索引的最长关键元素。

当使用=、<>、>、>=、<、<=、IS NULL、<=>、BETWEEN或者IN操作符用常量比较关键字列时，类型为range。

下面介绍几种检索指定行的情况：

```
SELECT * FROM tbl_name WHERE key_column = 10;
SELECT * FROM tbl_name WHERE key_column BETWEEN 10 and 20;
SELECT * FROM tbl_name WHERE key_column IN (10,20,30);
SELECT * FROM tbl_name WHERE key_part1= 10 AND key_part2 IN (10,20,30);
```

- index

该连接类型与ALL相同,除了只扫描索引树。它通常比ALL快,因为索引文件通常比数据文件小。

- ALL

对于前面的表的任意行组合,进行完整的表扫描。如果表是第一个没标记const的表,使用ALL连接类型可能不是最佳选择,因为在其他情况下,它可能导致较差的查询性能。为了避免使用ALL连接类型,可以考虑增加更多的索引来优化查询性能。

(5)possible_keys:指出MySQL能使用哪个索引在该表中找到行。如果该列的值是NULL,则没有相关的索引。在这种情况下,可以检查WHERE子句,看是否可以创建适合的索引来提高查询性能。

(6)key:表示查询实际使用到的索引,如果没有选择索引,则该列的值是NULL。要想强制MySQL使用或忽视possible_keys列中的索引,可在查询中使用FORCE INDEX、USE INDEX或者IGNORE INDEX。参见SELECT语法。

(7)key_len:表示MySQL选择的索引字段按字节计算的长度,如果键是NULL,则长度为NULL。注意,通过key_len值可以确定MySQL将实际使用一个多列索引中的几个字段。

(8)ref:表示使用哪个列或常数与索引一起来查询记录。

(9)rows:显示MySQL在表中进行查询时必须检查的行数。

(10)Extra:表示MySQL在处理查询时的详细信息。

2. DESCRIBE语句

DESCRIBE语句的使用方法与EXPLAIN语句是一样的,并且分析结果也是一样的。DESCRIBE语句的语法形式如下:

```
DESCRIBE SELECT select_options
```

DESCRIBE可以简写成DESC。

14.2.2 索引对查询速度的影响

MySQL中提高性能的一个有效方式就是对数据表设计合理的索引。索引提供了高效访问数据的方法,并且可加快查询的速度,因此索引对查询的速度有着至关重要的影响。本小节将介绍索引对查询速度的影响。

如果查询时没有使用索引,查询语句将扫描表中的所有记录。在数据量大的情况下,这样的查询速度会很慢。如果使用索引进行查询,查询语句可以根据索引快速定位到待查询记录,从而减少查询的记录数,达到提高查询速度的目的。

【例14.2】下面是查询语句中不使用索引和使用索引的对比。

首先,分析未使用索引时的查询情况,EXPLAIN语句执行如下:

```
mysql> EXPLAIN SELECT * FROM fruits WHERE f_name='apple';
+----+-------------+--------+------+---------------+------+---------+------+------+-------------+
| id | select_type | table  | type | possible_keys | key  | key_len | ref  | rows | Extra       |
+----+-------------+--------+------+---------------+------+---------+------+------+-------------+
|  1 | SIMPLE      | fruits | ALL  | NULL          | NULL | NULL    | NULL |   15 | Using where |
+----+-------------+--------+------+---------------+------+---------+------+------+-------------+
1 row in set (0.00 sec)
```

可以看到，rows列的值是15，说明"SELECT * FROM fruits WHERE f_name='apple';"查询语句扫描了表中的15条记录。

然后，在表fruits的f_name字段上加上索引。执行添加索引的语句及结果如下：

```
mysql> CREATE INDEX index_name ON fruits(f_name);
Query OK, 0 rows affected (0.04 sec)
Records: 0  Duplicates: 0  Warnings: 0
```

现在，分析上面的查询语句，执行的EXPLAIN语句及结果如下：

```
mysql> EXPLAIN SELECT * FROM fruits WHERE f_name='apple';
+----+-------------+--------+------+---------------+------------+---------+-------+------+-------------+
| id | select_type | table  | type | possible_keys | key        | key_len | ref   | rows | Extra       |
+----+-------------+--------+------+---------------+------------+---------+-------+------+-------------+
|  1 | SIMPLE      | fruits | ref  | index_name    | index_name | 255     | const |    1 | Using where |
+----+-------------+--------+------+---------------+------------+---------+-------+------+-------------+
1 row in set (0.00 sec)
```

结果显示，rows列的值为1，表示这个查询语句只扫描了表中的一条记录，其查询速度自然比扫描15条记录的快；而且possible_keys和key的值都是index_name，说明查询时使用了index_name索引。

14.2.3 使用索引查询

虽然索引可以提高查询的速度，但并不是所有索引都会起作用。本小节将介绍索引的使用。

使用索引有几种特殊情况，在这些情况下使用带有索引的字段进行查询时索引可能并没有起作用。下面重点介绍这几种特殊情况。

1. 使用LIKE关键字的查询语句

在使用LIKE关键字进行查询的语句中，如果匹配字符串的第一个字符为"%"，则索引不会起作用，只有"%"不在第一个位置，索引才会起作用。下面举例说明。

【例14.3】 查询语句中使用LIKE关键字，并且匹配的字符串中含有"％"字符，EXPLAIN语句执行如下：

```
mysql> EXPLAIN SELECT * FROM fruits WHERE f_name like '%x';
+----+-------------+--------+------+---------------+------+---------+------+------+-------------+
| id | select_type | table  | type | possible_keys | key  | key_len | ref  | rows | Extra       |
+----+-------------+--------+------+---------------+------+---------+------+------+-------------+
|  1 | SIMPLE      | fruits | ALL  | NULL          | NULL | NULL    | NULL |   16 | Using where |
+----+-------------+--------+------+---------------+------+---------+------+------+-------------+
1 row in set (0.00 sec)

mysql> EXPLAIN SELECT * FROM fruits WHERE f_name like 'x%';
+----+-------------+--------+-------+---------------+------------+---------+------+------+-------------+
| id | select_type | table  | type  | possible_keys | key        | key_len | ref  | rows | Extra       |
+----+-------------+--------+-------+---------------+------------+---------+------+------+-------------+
|  1 | SIMPLE      | fruits | range | index_name    | index_name | 150     | NULL |    4 | Using where |
+----+-------------+--------+-------+---------------+------------+---------+------+------+-------------+
1 row in set (0.00 sec)
```

已知f_name字段上有索引index_name。第一条查询语句执行后，rows列的值为16，表示这次查询过程中扫描了表中所有的16条记录；第二条查询语句执行后，rows列的值为4，表示这次查询过程扫描了4条记录。第一条查询语句中的索引没有起作用，因为第一条查询语句中LIKE关键字后的字符串以"％"开头，而第二条查询语句使用了索引index_name。

2．使用多列索引的查询语句

MySQL可以为多个字段创建索引。一个索引可以包括16个字段。对于多列索引，只有查询条件中使用了这些字段中的第一个字段时索引才会被使用。

【例14.4】 在表fruits中的f_id、f_price字段上创建多列索引，验证多列索引的使用情况，SQL语句如下：

```
mysql> CREATE INDEX index_id_price ON fruits(f_id, f_price);
Query OK, 0 rows affected (0.39 sec)
Records: 0  Duplicates: 0  Warnings: 0
mysql> EXPLAIN SELECT * FROM fruits WHERE f_id='l2';
+----+-------------+--------+------+---------------+------+---------+------+------+-------+
| id | select_type | table  | type | possible_keys | key  | key_len | ref  | rows | Extra |
+----+-------------+--------+------+---------------+------+---------+------+------+-------+
```

```
    +----+-------------+--------+-------+----------------------+---------
---------+---------+-------+------+-------+
    |  1 | SIMPLE      | fruits | const | PRIMARY,index_id_price | PRIMARY | 20      | const
|    1 |       |
    +----+-------------+--------+-------+----------------------+---------
---------+---------+-------+------+-------+
    1 row in set (0.00 sec)

    mysql> EXPLAIN SELECT * FROM fruits WHERE f_price=5.2;
    +----+-------------+--------+------+---------------+------+---------+------
+------+-------------+
    | id | select_type | table  | type | possible_keys | key  | key_len | ref  | rows
| Extra       |
    +----+-------------+--------+------+---------------+------+---------+------
+------+-------------+
    |  1 | SIMPLE      | fruits | ALL  | NULL          | NULL | NULL    | NULL |   16 | Using where |
    +----+-------------+--------+------+---------------+------+---------+------
-+------+-------------+
    1 row in set (0.00 sec)
```

从第一条语句的查询结果中可以看出，"f_id='l2'"的记录有1条。第一条语句共扫描了1条记录，并且使用了索引index_id_price。从第二条语句的查询结果中可以看出，rows列的值是16，说明查询语句共扫描了16条记录，并且key列值为NULL，说明"SELECT * FROM fruits WHERE f_price=5.2;"语句并没有使用索引。因为f_price字段是多列索引的第二个字段，只有查询条件中使用了f_id字段才会使index_id_price索引起作用。

3. 使用OR关键字的查询语句

查询语句的查询条件中只有OR关键字，并且OR前后的两个条件中的列都有索引时，查询中才使用索引；否则，查询将不使用索引。

【例14.5】查询语句使用OR关键字的情况。

```
mysql> EXPLAIN SELECT * FROM fruits WHERE f_name='apple' or s_id=101 \G
*** 1. row ***
           id: 1
  select_type: SIMPLE
        table: fruits
         type: ALL
possible_keys: index_name
          key: NULL
      key_len: NULL
          ref: NULL
         rows: 16
        Extra: Using where
1 row in set (0.00 sec)

mysql> EXPLAIN SELECT * FROM fruits WHERE f_name='apple' or f_id='l2' \G
*** 1. row ***
           id: 1
```

```
       select_type: SIMPLE
            table: fruits
             type: index_merge
    possible_keys: PRIMARY,index_name,index_id_price
              key: index_name,PRIMARY
          key_len: 510,20
              ref: NULL
             rows: 2
            Extra: Using union(index_name,PRIMARY); Using where
1 row in set (0.00 sec)
```

因为s_id字段上没有索引，所以第一条查询语句没有使用索引，总共查询了16条记录；因为f_id字段和f_name字段上都有索引，所以第二条查询语句使用了索引，查询的记录数为2条。

14.2.4 优化子查询

MySQL从4.1版本开始支持子查询，使用子查询可以进行SELECT语句的嵌套查询，即一个SELECT查询的结果作为另一个SELECT语句的条件。子查询可以一次性完成很多逻辑上需要多个步骤才能完成的SQL操作。子查询虽然可以使查询语句更灵活，但执行效率不高。执行子查询时，MySQL需要为内层查询语句的查询结果建立一张临时表，然后外层查询语句从临时表中查询记录，查询完毕后，再撤销临时表。因此，子查询的速度会受到一定的影响。如果查询的数据量比较大，这种影响就会随之增大。

在MySQL中，可以使用连接（JOIN）查询来替代子查询，其速度比子查询快，如果查询中使用了索引，性能会更好。连接之所以更有效率，是因为MySQL不需要在内存中创建临时表来完成查询工作。

14.3 优化数据库结构

一个好的数据库设计方案对于数据库的性能常常会起到事半功倍的效果。合理的数据库结构不仅可以使数据库占用更小的磁盘空间，而且能够使查询速度更快。数据库结构的设计，需要考虑数据冗余、查询和更新的速度、字段的数据类型是否合理等多方面的内容。本节将介绍优化数据库结构的方法。

14.3.1 将字段很多的表分解成多张表

对于字段较多的表，如果有些字段的使用频率很低，可以将这些字段分离出来，形成新表。因为当一张表的数据量很大时，会由于使用频率低的字段的存在而变慢。下面用案例的形式来介绍这种优化表的方法。

【例14.6】假设members表存储会员登录认证信息，该表中有很多字段，如id、姓名、密码、地址、电话、个人描述等。其中，地址、电话、个人描述等字段并不常用，可以将这些不常用字段分解成另外一张表，并将这张表取名为"members_detail"。这样就把会员表分成两张表，分别为members表和members_detail表。

创建这两张表的SQL语句如下：

```
CREATE TABLE members (
  Id int NOT NULL AUTO_INCREMENT,
  username varchar(255) DEFAULT NULL ,
  password varchar(255) DEFAULT NULL ,
  last_login_time datetime DEFAULT NULL ,
  last_login_ip varchar(255) DEFAULT NULL ,
  PRIMARY KEY (Id)
) ;
CREATE TABLE members_detail (
  member_id int NOT NULL DEFAULT 0,
  address varchar(255) DEFAULT NULL ,
  telephone varchar(16) DEFAULT NULL ,
  description text
) ;
```

这两张表的结构如下：

```
mysql> desc members;
+-----------------+--------------+------+-----+---------+----------------+
| Field           | Type         | Null | Key | Default | Extra          |
+-----------------+--------------+------+-----+---------+----------------+
| Id              | int          | NO   | PRI | NULL    | auto_increment |
| username        | varchar(255) | YES  |     | NULL    |                |
| password        | varchar(255) | YES  |     | NULL    |                |
| last_login_time | datetime     | YES  |     | NULL    |                |
| last_login_ip   | varchar(255) | YES  |     | NULL    |                |
+-----------------+--------------+------+-----+---------+----------------+
5 rows in set (0.00 sec)

mysql> DESC members_detail;
+-------------+--------------+------+-----+---------+-------+
| Field       | Type         | Null | Key | Default | Extra |
+-------------+--------------+------+-----+---------+-------+
| member_id   | int          | NO   |     | 0       |       |
| address     | varchar(255) | YES  |     | NULL    |       |
| telephone   | varchar(16)  | YES  |     | NULL    |       |
| description | text         | YES  |     | NULL    |       |
+-------------+--------------+------+-----+---------+-------+
4 rows in set (0.00 sec)
```

如果需要查询会员的详细信息，可以用会员的id来查询。如果需要将会员的基本信息和详细信息同时显示，可以将members表和members_detail表进行联合查询，查询语句如下：

```
SELECT * FROM members LEFT JOIN members_detail ON members.id=members_detail.member_id;
```

通过这种分解，可以提高表的查询效率。对于字段很多且有些字段使用不频繁的表，可以通过这种分解的方式来优化数据库的性能。

14.3.2 增加中间表

对于经常需要联合查询的表，可以建立中间表，把经常需要联合查询的数据插入中间表中，然后将原来的联合查询改为对中间表的查询，以此来提高查询效率。下面通过案例来介绍增加中间表这种优化查询的方法。

【例14.7】 首先创建会员信息表和会员组信息表，SQL语句如下：

```
CREATE TABLE vip(
  Id int NOT NULL AUTO_INCREMENT,
  username varchar(255) DEFAULT NULL,
  password varchar(255) DEFAULT NULL,
  groupId INT DEFAULT 0,
  PRIMARY KEY (Id)
);
CREATE TABLE vip_group (
  Id int NOT NULL AUTO_INCREMENT,
  name varchar(255) DEFAULT NULL,
  remark varchar(255) DEFAULT NULL,
  PRIMARY KEY (Id)
);
```

然后查询会员信息表和会员组信息表：

```
mysql> DESC vip;
+----------+--------------+------+-----+---------+----------------+
| Field    | Type         | Null | Key | Default | Extra          |
+----------+--------------+------+-----+---------+----------------+
| Id       | int          | NO   | PRI | NULL    | auto_increment |
| username | varchar(255) | YES  |     | NULL    |                |
| password | varchar(255) | YES  |     | NULL    |                |
| groupId  | int          | YES  |     | 0       |                |
+----------+--------------+------+-----+---------+----------------+
4 rows in set (0.01 sec)

mysql> DESC vip_group;
+--------+--------------+------+-----+---------+----------------+
| Field  | Type         | Null | Key | Default | Extra          |
+--------+--------------+------+-----+---------+----------------+
| Id     | int          | NO   | PRI | NULL    | auto_increment |
| name   | varchar(255) | YES  |     | NULL    |                |
| remark | varchar(255) | YES  |     | NULL    |                |
+--------+--------------+------+-----+---------+----------------+
3 rows in set (0.01 sec)
```

假设现在有一个模块需要经常查询带有会员组名称、会员组备注（remark）、会员用户名信息的会员信息。根据这种情况可以创建一张temp_vip表，表中存储用户名（user_name）、会员组名称（group_name）和会员组备注（group_remark）信息。创建表的语句如下：

```
CREATE TABLE temp_vip (
  Id int NOT NULL AUTO_INCREMENT,
  user_name varchar(255) DEFAULT NULL,
  group_name varchar(255) DEFAULT NULL,
  group_remark varchar(255) DEFAULT NULL,
  PRIMARY KEY (Id)
);
```

接下来，从会员信息表和会员组表中查询相关信息存储到临时表中：

```
mysql> INSERT INTO temp_vip(user_name, group_name, group_remark)
    SELECT v.username,g.name,g.remark
    FROM vip as v ,vip_group as g
    WHERE v.groupId =g.Id;
Query OK, 0 rows affected (0.95 sec)
Records: 0  Duplicates: 0  Warnings: 0
```

以后，可以直接从temp_vip表中查询会员名称、会员组名称和会员组备注，而不用每次都进行联合查询。这样可以提高数据库的查询速度。

14.3.3　增加冗余字段

设计数据库表时应尽量遵循范式理论的规约，尽可能减少冗余字段，让数据库设计看起来精致、优雅。但是，合理地加入冗余字段可以提高查询速度。本小节将介绍通过增加冗余字段来优化查询速度的方法。

表的规范化程度越高，表与表之间的关系就越多，需要连接查询的情况也就越多。例如，员工的信息存储在staff表中，部门信息存储在department表中。通过staff表中的department_id字段与department表建立关联关系。如果要查询一个员工所在部门的名称，必须从staff表中查找员工所在部门的编号（department_id），然后根据这个编号去department表查找部门的名称。如果经常需要进行这个操作，连接查询会浪费很多时间。可以在staff表中增加一个冗余字段department_name，该字段用来存储员工所在部门的名称，这样就不用每次都进行连接操作了。

> **提示**　冗余字段会导致一些问题。比如，冗余字段的值在一张表中被修改了，就要想办法在其他表中更新该字段，否则就会使原本一致的数据变得不一致。虽然分解表、增加中间表和增加冗余字段都浪费了一定的磁盘空间，但从数据库性能来看，为了提高查询速度而增加少量的冗余大部分时候是可以接受的。是否通过增加冗余来提高数据库性能，这要根据实际需求综合分析。

14.3.4　优化插入记录的速度

插入记录时，影响插入速度的主要是索引、唯一性校验、一次插入记录条数等。根据这些情况，可以分别进行优化。本小节将介绍优化插入记录速度的几种方法。

1. 对于MyISAM引擎的表

对于MyISAM引擎的表，常见的优化方法如下：

1）禁用索引

对于非空表，插入记录时，MySQL会根据表的索引对插入的记录建立索引。如果要插入大量数据，建立索引会降低插入记录的速度。为了解决这种问题，可以在插入记录之前禁用索引，数据插入完毕后再开启索引。禁用索引的语句如下：

```
ALTER TABLE table_name DISABLE KEYS;
```

其中，table_name是禁用索引的表的名称。

重新开启索引的语句如下：

```
ALTER TABLE table_name ENABLE KEYS;
```

对于空表批量导入数据，则不需要进行此操作，因为MyISAM引擎的表是在导入数据之后才建立索引的。

2）禁用唯一性检查

插入数据时，MySQL会对插入的记录进行唯一性校验。这种唯一性校验也会降低插入记录的速度。为了降低这种情况对查询速度的影响，可以在插入记录之前禁用唯一性检查，等到记录插入完毕后再开启。禁用唯一性检查的语句如下：

```
SET UNIQUE_CHECKS=0;
```

开启唯一性检查的语句如下：

```
SET UNIQUE_CHECKS=1;
```

3）使用批量插入

插入多条记录时，可以使用一条INSERT语句插入一条记录；也可以使用一条INSERT语句插入多条记录。使用一条INSERT语句插入一条记录的情形如下：

```
INSERT INTO fruits VALUES('x1', '101', 'mongo2', '5.7');
INSERT INTO fruits VALUES('x2', '101', 'mongo3', '5.7');
INSERT INTO fruits VALUES('x3', '101', 'mongo4', '5.7')
```

使用一条INSERT语句插入多条记录的情形如下：

```
INSERT INTO fruits VALUES
('x1', '101', 'mongo2', '5.7'),
('x2', '101', 'mongo3', '5.7'),
('x3', '101', 'mongo4', '5.7');
```

第二种情形的插入速度要比第一种情形快。

4）使用 LOAD DATA INFILE 批量导入

当需要批量导入数据时，如果能用LOAD DATA INFILE语句，就尽量使用。因为LOAD DATA INFILE语句导入数据的速度比INSERT语句快。

2. 对于InnoDB引擎的表

对于InnoDB引擎的表，常见的优化方法如下：

1）禁用唯一性检查

插入数据之前执行set unique_checks=0来禁止对唯一索引的检查，数据导入完成之后，再运行set unique_checks=1恢复索引唯一性检查。这个和MyISAM引擎的使用方法一样。

2）禁用外键检查

插入数据之前执行禁止对外键的检查，数据插入完成之后再恢复对外键的检查。禁用外键检查的语句如下：

```
SET foreign_key_checks=0;
```

恢复外键检查的语句如下：

```
SET foreign_key_checks=1;
```

3）禁止自动提交

插入数据之前禁止事务的自动提交，数据导入完成之后，再执行恢复自动提交操作。禁止自动提交的语句如下：

```
set autocommit=0;
```

恢复自动提交的语句如下：

```
set autocommit=1;
```

14.3.5 分析表、检查表和优化表

MySQL提供了分析表、检查表和优化表的语句。分析表主要是分析关键字的分布；检查表主要是检查表中是否存在错误；优化表主要是消除删除或者更新造成的空间浪费。本小节将介绍分析表、检查表和优化表的方法。

1. 分析表

MySQL中提供了ANALYZE TABLE语句来分析表。ANALYZE TABLE语句的基本语法如下：

```
ANALYZE [LOCAL | NO_WRITE_TO_BINLOG] TABLE tbl_name[,tbl_name]...
```

LOCAL关键字是NO_WRITE_TO_BINLOG关键字的别名，二者都表示执行过程不写入二进制日志；tbl_name为分析表的表名，可以有一个或多个。

使用ANALYZE TABLE分析表的过程中，数据库系统会自动对表加一个只读锁。在分析期间，只能读取表中的记录，不能更新和插入记录。ANALYZE TABLE语句能够分析InnoDB、BDB和MyISAM类型的表。

【例14.8】 使用ANALYZE TABLE来分析message表，执行的语句及结果如下：

```
mysql> ANALYZE TABLE members;
+-----------------+---------+----------+----------+
| Table           | Op      | Msg_type | Msg_text |
+-----------------+---------+----------+----------+
| test_db.members | analyze | status   | OK       |
+-----------------+---------+----------+----------+
```

上面结果显示的信息说明如下：

- Table：表示分析的表的名称。
- Op：表示执行的操作。analyze表示进行分析操作。
- Msg_type：表示信息类型，其值通常是状态（status）、信息（info）、注意（note）、警告（warning）和错误（error）之一。
- Msg_text：显示信息。

2. 检查表

MySQL中可以使用CHECK TABLE语句来检查表。CHECK TABLE语句能够检查InnoDB和MyISAM类型的表是否存在错误。对于MyISAM类型的表，CHECK TABLE语句还会更新关键字统计数据。此外，CHECK TABLE也可以检查视图是否有错误，比如，在视图定义中被引用的表已不存在。该语句的基本语法如下：

```
CHECK TABLE tbl_name [, tbl_name] ... [option] ...
option = {QUICK | FAST | MEDIUM | EXTENDED | CHANGED}
```

其中，tbl_name 是表名。option参数有以下5个取值：

- QUICK：不扫描行，不检查错误的连接。
- FAST：只检查没有被正确关闭的表。
- CHANGED：只检查上次检查后被更改的表和没有被正确关闭的表。
- MEDIUM：扫描行，以验证被删除的连接是否有效。也可以计算各行的关键字校验和，并使用计算出的校验和验证这一点。
- EXTENDED：对每行的所有关键字进行一个全面查找。这可以确保表是百分百一致的，但是花的时间较长。

option只对MyISAM类型的表有效，对InnoDB类型的表无效。CHECK TABLE语句在执行过程中也会给表加上只读锁。

3. 优化表

MySQL中使用OPTIMIZE TABLE语句来优化表。该语句对InnoDB和MyISAM类型的表都有效。但是，OPTILMIZE TABLE语句只能优化表中VARCHAR、BLOB或TEXT类型的字段。OPTILMIZE TABLE语句的基本语法如下：

```
OPTIMIZE [LOCAL | NO_WRITE_TO_BINLOG] TABLE tbl_name [, tbl_name] ...
```

其中，LOCAL和NO_WRITE_TO_BINLOG关键字的意义与分析表相同，都是指定不写入二进制日志；tbl_name是表名。

通过OPTIMIZE TABLE语句可以消除删除和更新造成的文件碎片。OPTIMIZE TABLE语句在执行过程中也会给表加上只读锁。

> **提示** 假设一张表中使用了TEXT或者BLOB这样的数据类型，如果已经删除了表中的一大部分数据，或者已经对含有可变长度行的表（含有VARCHAR，BLOB或TEXT列的表）进行了大量更新，则应使用OPTIMIZE TABLE来重新利用未使用的空间，并整理数据文件的碎片。在多数的设置中，不需要运行OPTIMIZE TABLE，即使对可变长度的行进行了大量的更新，也不需要经常运行，每周一次或每月一次即可，并且只需对特定的表运行。

14.4 优化临时表性能

在MySQL中，用户可以把数据库和表归组到逻辑和物理表空间中，这样做可以提高资源的利用率。

在MySQL中，使用CREATE TABLESPACE语句可以创建一个通用表空间。通用表空间允许用户自由地选择表和表空间之间的映射，从而为数据表提供更大的灵活性。例如，可以创建一个表空间并设置该表空间应包含的表类型。这使得在同一个表空间中的用户可以对所有表进行分组，从而在文件系统中使用单独的文件来存储他们的数据。同时，通用表空间还实现了元数据锁，以支持更高效的数据管理和访问。

优化普通SQL 临时表性能是MySQL 的目标之一。首先，通过优化临时表在磁盘中的不必要的步骤，使得临时表的创建和移除成为一个轻量级的操作。将临时表移动到一个单独的表空间中，恢复临时表的过程就变得非常简单，在启动时重新创建临时表即可。

MySQL去掉了临时表中不必要的持久化。临时表仅仅在连接和会话内被创建，然后通过服务的生命周期绑定它们。通过移除不必要的UNDO（回滚）和REDO（重做）日志，改变缓冲和锁，从而为临时表做了优化操作。

MySQL引入了一种名为undo_tablespace的额外表空间类型，用于存储UNDO日志。undo_tablespace类型的日志存储在一个单独的临时表空间中，并且它在恢复期间不会被调用，而是在回滚操作中才会被调用。

MySQL为临时表设定了一个特别类型，称为"内在临时表"。内在临时表和普通临时表很像，只是内在临时表使用宽松的ACID和MVCC语义。

MYSQL为了提高临时表的相关性能，对临时表的相关部分进行了大幅修改，包括：引入新的临时表空间（ibtmp1）；对于临时表的DDL，不持久化相关表定义；对于临时表的DML，不写redo、关闭change buffer等。

InnoDB临时表元数据不再存储于InnoDB系统表，而是存储在innodb_temp_table_info中，包含所有用户和系统创建的临时表信息。该表在第一次运行SELECT时被创建，下面举例说明。

【例14.9】 MySQL的临时表。

```
mysql> SELECT * FROM information_schema.innodb_temp_table_info;
Empty set (0.00 sec)

mysql> create temporary table temp_1(id int,name varchar(100))default charset utf8;
Query OK, 0 rows affected (0.00 sec)

mysql> select * from information_schema.innodb_temp_table_info;
+----------+--------------+--------+------------+
| TABLE_ID | NAME         | N_COLS | SPACE      |
+----------+--------------+--------+------------+
|     1142 | #sql14a0_9_1 |      5 | 4243767290 |
+----------+--------------+--------+------------+
```

MySQL使用了独立的临时表空间来存储临时表数据，但不能是压缩表。临时表空间在实例启动的时候进行创建，在shutdown的时候进行删除，即为所有非压缩的InnoDB临时表提供一个独立的表空间。默认的临时表空间文件为ibtmp1，位于数据目录中。通过innodb_temp_data_file_path参数可指定临时表空间的路径和大小（默认为12MB）。只有重启实例才能回收临时表空间文件ibtmp1的大小。CREATE TEMPORARY TABLE和USING TEMPORARY TABLE将共用这个临时表空间。

```
mysql> SHOW VARIABLES LIKE 'innodb_temp_data_file_path';
+----------------------------+----------------------+
| Variable_name              | Value                |
+----------------------------+----------------------+
| innodb_temp_data_file_path | ibtmp1:12M:autoextend |
+----------------------------+----------------------+
```

在MySQL中，临时表在连接断开或者数据库实例关闭的时候被删除，从而提高了性能。只有临时表的元数据使用了redo保护，以保护元数据的完整性，以便异常启动后进行清理工作。

临时表的元数据在MySQL中使用了一张独立的表（innodb_temp_table_info）进行保存，不用使用redo保护，元数据也只保存在内存中。但这有一个前提，即必须使用共享的临时表空间，如果使用file-per-table，则仍然需要持久化元数据，以便异常恢复和清理。临时表需要UNDO日志，用于MySQL运行时的回滚。

在MySQL中有一个系统选项internal_tmp_disk_storage_engine，可定义磁盘临时表的引擎类型，默认为InnoDB，可选MyISAM。在这以前，只能使用MyISAM。在MySQL 5.6.3以后新增的参数default_tmp_storage_engine是控制CREATE TEMPORARY TABLE创建的临时表的存储引擎的，在以前默认是MEMORY。

查看结果如下：

```
mysql> SHOW VARIABLES LIKE '%engine%';
+----------------------------------+----------+
| Variable_name                    | Value    |
+----------------------------------+----------+
| default_storage_engine           | InnoDB   |
```

```
| default_tmp_storage_engine          | InnoDB        |
| disabled_storage_engines            |               |
| internal_tmp_disk_storage_engine    | InnoDB        |
| internal_tmp_mem_storage_engine     | TempTable     |
| secondary_engine_cost_threshold     | 100000.000000 |
| show_create_table_skip_secondary_engine | OFF       |
| use_secondary_engine                | ON            |
+-------------------------------------+---------------+
```

从结果中可以看出，当前MySQL实例中的默认存储引擎是InnoDB，默认临时存储引擎也是InnoDB。禁用的存储引擎为空，说明没有禁用的存储引擎。此外，内部临时磁盘存储引擎和内部临时内存存储引擎分别是InnoDB和TempTable。最后，使用secondary引擎的设置为ON，这意味着MySQL可以在InnoDB存储引擎不可用时使用其他存储引擎（如MyISAM）。

14.5 创建全局通用表空间

MySQL支持创建全局通用表空间，全局表空间可以被所有数据库的表共享，而且相比于独享表空间，手动创建共享表空间可以节约元数据方面的内存。可以在创建表的时候指定该表属于哪个表空间，也可以对已有表进行表空间修改。

下面创建名为dxy的共享表空间，SQL语句如下：

```
mysql> CREATE TABLESPACE dxy ADD datafile 'dxy.ibd' file_block_size=16k;
Query OK, 0 rows affected (0.10 sec)
```

指定表空间，SQL语句如下：

```
mysql> CREATE TABLE t1(id int,name varchar(10))engine = innodb default charset utf8mb4 tablespace dxy;
Query OK, 0 rows affected (0.01 sec)
```

也可以通过ALTER TABLE语句指定表空间，SQL语句如下：

```
mysql> alter table t1 tablespace dxy;
```

如何删除创建的共享表空间呢？因为是共享表空间，所以不能直接通过DROP TABLE tbname来删除，也不能回收空间。当确定共享表空间的数据都没用并且依赖该表空间的表均已被删除时，可以通过DROP TABLESPACE来删除共享表空间，以释放空间。如果依赖该共享表空间的表存在，则删除就会失败。

首先，删除依赖该表空间的数据表，SQL语句如下：

```
mysql> DROP TABLE t1;
Query OK, 0 rows affected (0.16 sec)
```

然后，删除表空间，SQL语句如下：

```
mysql> DROP TABLESPACE dxy;
Query OK, 0 rows affected (0.01 sec)
```

第 15 章
MySQL服务器性能优化

和大多数数据库一样，MySQL提供了很多的参数来进行服务器的优化设置。数据库服务器第一次启动的时候，很多参数都是默认设置的，这在实际生产环境中并不能完全满足需求，为此数据库管理员要进行必要的设置。本章主要讲解MySQL服务器优化方面的一些知识和技巧，其中包括MySQL服务器配置优化、I/O调优，以及其他的一些MySQL优化技巧。

15.1 优化MySQL服务器

MySQL服务器主要从两方面来优化：一方面是对硬件进行优化；另一方面是对MySQL服务的参数进行优化。这部分的内容需要较全面的知识，一般只有专业的数据库管理员才能进行这一类的优化。对于可以定制参数的操作系统，也可以针对MySQL进行操作系统优化。本章的操作系统为Red Hat Enterprise Linux 9。下面介绍优化MySQL服务器的方法。

15.1.1 优化服务器硬件

服务器的硬件性能直接决定了MySQL数据库的性能，硬件的性能瓶颈直接决定了MySQL数据库的运行速度和效率。针对性能瓶颈，提高硬件配置，可以提高MySQL数据库查询、更新的速度。优化服务器硬件的方法如下：

（1）配置较大的内存。足够大的内存是提高MySQL数据库性能的方法之一。内存的速度比磁盘I/O快得多，可以通过增加系统的缓冲区容量使数据在内存停留的时间更长，以减少磁盘I/O。

（2）配置高速磁盘系统，以减少读盘的等待时间，提高响应速度。

（3）合理分布磁盘I/O，把磁盘I/O分散在多个设备上，以减少资源竞争，提高并行操作能力。

（4）配置多处理器。MySQL是多线程的数据库，多处理器可以同时执行多个线程。

15.1.2 优化 MySQL 的参数

通过优化MySQL的参数可以提高资源利用率，从而达到提高MySQL服务器性能的目的。MySQL服务器的配置参数都在my.cnf或者my.ini文件的[mysqld]组中。下面针对几个对性能影响较大的参数进行详细介绍。

- key_buffer_size：表示索引缓冲区的大小。索引缓冲区所有的线程共享。增加索引缓冲区可以得到更好处理的索引（对所有读和多重写）。当然，这个值也不是越大越好，它的大小取决于内存的大小。如果这个值太大，导致操作系统频繁换页，也会降低系统性能。
- table_cache：表示同时打开的表的个数。这个值越大，能够同时打开的表的个数就越多。这个值不是越大越好，因为同时打开的表太多会影响操作系统的性能。
- query_cache_size：表示查询缓冲区的大小。该参数需要和query_cache_type配合使用。当query_cache_type=0时，所有的查询都不使用查询缓冲区，但是query_cache_type=0并不会让MySQL释放query_cache_size所配置的缓冲区内存。当query_cache_type=1时，所有的查询都将使用查询缓冲区，除非在查询语句中指定SQL_NO_CACHE，如SELECT SQL_NO_CACHE * FROM tbl_name。当query_cache_type=2时，只有在查询语句中使用SQL_CACHE关键字，查询才会使用查询缓冲区。使用查询缓冲区可以提高查询的速度，这种方式只适用于修改操作少且经常执行相同的查询操作的情况。
- sort_buffer_size：表示排序缓存区的大小。这个值越大，排序的速度越快。
- read_buffer_size：表示每个线程连续扫描时为扫描的每张表分配的缓冲区的大小（字节）。当线程从表中连续读取记录时需要用到这个缓冲区。SET SESSION read_buffer_size=n可以临时设置该参数的值。
- read_rnd_buffer_size：表示为每个线程保留的缓冲区的大小，与read_buffer_size相似，但主要用于存储按特定顺序读取出来的记录。也可以用SET SESSION read_rnd_buffer_size=n来临时设置该参数的值。如果频繁进行连续扫描，可以增加该值。
- innodb_buffer_pool_size：表示InnoDB类型的表和索引的最大缓存。这个值越大，查询的速度就越快，但是这个值太大会影响操作系统的性能。
- max_connections：表示数据库的最大连接数。这个连接数不是越大越好，因为这些连接会浪费内存的资源，过多的连接可能会导致MySQL服务器僵死。
- innodb_flush_log_at_trx_commit：表示何时将缓冲区的数据写入日志文件，并且将日志文件写入磁盘中。该参数对于InnoDB引擎非常重要。该参数有3个值，分别为0、1和2。值为0时表示每隔1秒将数据写入日志文件并将日志文件写入磁盘；值为1时表示每次提交事务时将数据写入日志文件并将日志文件写入磁盘；值为2时表示每次提交事务时将数据写入日志文件，每隔1秒将日志文件写入磁盘。该参数的默认值为1。默认值1的安全性最高，但是每次事务提交或事务外的指令都需要把日志写入硬盘，比较费时；值为0虽然更快，但在安全性方面较差；值为2在性能和安全性之间提供了一种平衡，日志每秒写入硬盘，即使出现故障，一般也不会丢失超过1~2秒的更新。
- back_log：表示在MySQL暂时停止回答新请求之前的短时间内，多少个请求可以被存在堆栈中。换句话说，该值表示对到来的TCP/IP连接的侦听队列的大小。只有期望在一个短时间内有很多连接时才需要增加该参数的值。操作系统在这个队列大小上也有限制，设定 back_log 高于操作系统的限制将是无效的。
- interactive_timeout：表示服务器在关闭连接前等待行动的秒数。
- sort_buffer_size：表示每个需要进行排序的线程分配的缓冲区的大小。增加这个参数的值可以提高 ORDER BY 或 GROUP BY 操作的速度，默认数值是 2097144 字节（约 2MB）。
- thread_cache_size：表示可以复用的线程的数量。如果有很多新线程，为了提高性能，可以增大该参数的值。
- wait_timeout：表示服务器在关闭一个连接时等待行动的秒数，默认数值是 28800。

合理地配置这些参数可以提高 MySQL 服务器的性能。配置完参数以后，需要重新启动，MySQL 服务器才会生效。

15.2 影响 MySQL 服务器性能的重要参数

在 MySQL 数据库系统中，有些参数直接影响到系统的整体性能。MySQL 默认的参数设置往往不能达到实际工作的要求，此时需要通过调整服务器的一些参数的来满足实际应用的需求。本节主要介绍一些 MySQL 服务器的参数调整方法，以优化 MySQL 服务器的性能。这里是一些通用的方法，不涉及具体的存储引擎。

15.2.1 查看性能参数的方法

MySQL 服务器启动后，可以使用 SHOW VARIABLES 语句来查询服务器一些静态参数，比如缓冲区大小、字符集、数据文件名称等信息。示例如下：

```
mysql> SHOW VARIABLES;
+--------------------------+-----------------+
| Variable_name            | Value           |
+--------------------------+-----------------+
| auto_increment_increment | 1               |
| auto_increment_offset    | 1               |
| autocommit               | ON              |
| automatic_sp_privileges  | ON              |
| back_log                 | 50              |
| basedir                  | /usr/           |
| big_tables               | OFF             |
| binlog_cache_size        | 32768           |
| binlog_format            | STATEMENT       |
| bulk_insert_buffer_size  | 8388608         |
| character_set_client     | utf8            |
| ...                      | ...             |
+--------------------------+-----------------+
662 rows in set (0.02 sec)
```

使用 SHOW RARIABLES 可以查看 MySQL 启动之前已经配置好的一些系统静态参数。

使用SHOW STATUS查询服务器运行中的状态信息，比如当前连接数、锁等待状态信息。示例如下：

```
mysql> SHOW STATUS;
+-----------------------------------+------------+
| Variable_name                     | Value      |
+-----------------------------------+------------+
| Aborted_clients                   | 0          |
| Aborted_connects                  | 0          |
| Binlog_cache_disk_use             | 0          |
| Binlog_cache_use                  | 0          |
| Bytes_received                    | 1567       |
| Bytes_sent                        | 24957      |
| Com_admin_commands                | 0          |
| Com_assign_to_keycache            | 0          |
| Com_alter_db                      | 0          |
| Com_alter_db_upgrade              | 0          |
| Com_alter_event                   | 0          |
| Com_alter_function                | 0          |
| Com_alter_procedure               | 0          |
| Com_alter_server                  | 0          |
| Com_alter_table                   | 0          |
| Com_alter_tablespace              | 0          |
| Com_analyze                       | 0          |
| Com_backup_table                  | 0          |
| Com_begin                         | 0          |
| Com_binlog                        | 0          |
| Com_call_procedure                | 0          |
| Com_change_db                     | 0          |
| Com_change_master                 | 0          |
| ...                               | ...        |
+-----------------------------------+------------+
507 rows in set (0.80 sec)
```

同时，可以在操作系统下直接查询数据库的状态，命令如下：

```
[root@localhost ~]$ mysqladmin -u root -p variables
+---------------------------+------------------------------------------+
| Variable_name             | Value                                    |
+---------------------------+------------------------------------------+
| auto_increment_increment  | 1                                        |
| auto_increment_offset     | 1                                        |
| autocommit                | ON                                       |
| automatic_sp_privileges   | ON                                       |
| back_log                  | 50                                       |
| basedir                   | /usr/                                    |
| big_tables                | OFF                                      |
| binlog_cache_size         | 32768                                    |
| binlog_format             | STATEMENT                                |
| bulk_insert_buffer_size   | 8388608                                  |
| character_set_client      | utf8                                     |
```

```
| ...                       | ...                                        |                 |
+---------------------------+--------------------------------------------+
267 rows in set (0.08 sec)
```

MySQL服务器参数比较多,如果需要了解某个参数的含义,可以通过如下命令查询:

```
[root@localhost ~]#mysqld -verbose -help|more

mysqld Ver 9.0.1-log for pc-linux-gnu on i686 (MySQL Community Server (GPL))
Copyright (C) 2024 MySQL AB, by Monty and others
This software comes with ABSOLUTELY NO WARRANTY. This is free software,
and you are welcome to modify and redistribute it under the GPL license

Starts the MySQL database server

Usage: mysqld [OPTIONS]

Default options are read from the following files in the given order:
/etc/my.cnf /usr/local/mysql/etc/my.cnf ~/.my.cnf
The following groups are read: mysql_cluster mysqld server mysqld-5.0
The following options may be given as the first argument:
--print-defaults  Print the program argument list and exit
--no-defaults     Don't read default options from any options file
--defaults-file=# Only read default options from the given file #
--defaults-extra-file=# Read this file after the global files are read
  -?, --help       Display this help and exit.
  --abort-slave-event-count=#
                   Option used by mysql-test for debugging and testing of
                   replication.
  --allow-suspicious-udfs
                   Allows use of UDFs consisting of only one symbol xxx()
                   without corresponding xxx_init() or xxx_deinit(). That
                   also means that one can load any function from any
                   library, for example exit() from libc.so
  -a, --ansi       Use ANSI SQL syntax instead of MySQL syntax. This mode
                   will also set transaction isolation level 'serializable'.
  --auto-increment-increment[=#]
                   Auto-increment columns are incremented by this
  --auto-increment-offset[=#]
                   Offset added to Auto-increment columns. Used when
                   auto-increment-increment != 1
  --automatic-sp-privileges
                   Creating and dropping stored procedures alters ACLs.
                   Disable with --skip-automatic-sp-privileges.
  -b, --basedir=name Path to installation directory. All paths are usually
                   resolved relative to this.
  --bdb            Enable Berkeley DB (if this version of MySQL supports
                   it). Disable with --skip-bdb (will save memory).
...
sync-frm                        TRUE
table_cache                     64
table_lock_wait_timeout         50
```

```
thread_cache_size              0
thread_concurrency             10
thread_stack                   196608
time_format                    (No default value)
tmp_table_size                 33554432
transaction_alloc_block_size   8192
transaction_prealloc_size      4096
updatable_views_with_limit     1
wait_timeout                   28800

To see what values a running MySQL server is using, type
'mysqladmin variables' instead of 'mysqld --verbose --help'.
```

查询的结果中包含了当前服务器参数的介绍。如果需要查询其中的某些参数的含义,可以通过下面的命令过滤查询:

```
[root@localhost ~]# mysqld --verbose --help|grep thread
  --log-slave-updates Tells the slave to log the updates from the slave thread
                      The number of seconds the slave thread will sleep before
                      set, the slave thread will not be started. Note that the
                      master and where the I/O replication thread is in the
                      The password the slave thread will authenticate with when
  --master-user=name  The username the slave thread will use for authentication
                      for other threads.
                      A dedicated thread is created to, at the given
                      the SQL replication thread is in the relay logs.
                      Tells the slave thread to restrict replication to the
                      Tells the slave thread to restrict replication to the
                      Tells the slave thread to not replicate to the specified
                      Tells the slave thread to not replicate to the specified
                      Tells the slave thread to restrict replication to the
                      Tells the slave thread to not replicate to the tables
  --skip-thread-priority
                      Don't give threads different priorities.
                      Tells the slave thread to continue replication when a
                      have. This comes into play when the main MySQL thread
                      that this is a limit per thread!
                      How long a INSERT DELAYED thread should wait for INSERT
  --flush_time=#      A dedicated thread is created to flush all tables at the
                      Number of times a thread is allowed to enter InnoDB
  --innodb_file_io_threads=#
                      Number of file I/O threads in InnoDB.
  --innodb_thread_concurrency=#
                      environments. Sets the maximum number of threads allowed
                      inside InnoDB. Value 0 will disable the thread
  --innodb_thread_sleep_delay=#
                      Time of innodb thread sleeping before joining InnoDB
  --max_delayed_threads=#
                      Don't start more than this number of threads to handle
  --myisam_repair_threads=#
                      Number of threads to use when repairing MyISAM tables.
```

```
                        Each thread that does a sequential scan allocates a
                        exception for replication (slave) threads and users with
                        Number of times the slave SQL thread will retry a
                        If creating the thread takes longer than this value (in
                        seconds), the Slow_launch_threads counter will be
                        Each thread that needs to do a sort allocates a buffer of
  --table_cache=#       The number of open tables for all threads.
  --thread_cache_size=#
                        How many threads we should keep in a cache for reuse.
  --thread_concurrency=#
                        Permits the application to give the threads system a hint
                        for the desired number of threads that should be run at
  --thread_stack=#      The stack size for each thread.
innodb_file_io_threads                  4
innodb_thread_concurrency               8
innodb_thread_sleep_delay               10000
max_delayed_threads                     20
myisam_repair_threads                   1
thread_cache_size                       0
thread_concurrency                      10
thread_stack                            196608
```

mysqld --verbose --help|grep thread命令会把包含thread的信息查询出来。

了解查询MySQL服务器参数的方法之后，接下来介绍对MySQL数据库性能有重要影响的参数。

15.2.2 key_buffer_size 的设置

在MySQL数据库中，key_buffer_size参数是对MyISAM表性能影响最大的一个参数。先来看看mysqld中是如何定义key_buffer_size参数的，如下所示：

```
[root@localhost ~]# mysqld --verbose --help|grep "\-\-key_buffer_size" -A 5
  --key_buffer_size=#  The size of the buffer used for index blocks for MyISAM
                        tables. Increase this to get better index handling (for
                        all reads and multiple writes) to as much as you can
                        afford; 64M on a 256M machine that mainly runs MySQL is
                        quite common.
```

该参数用来设置索引块缓存的大小，只适用于MyISAM存储引擎。MySQL 5.1版本以后提供了多个key_buffer，可以将制定的表索引缓存到指定的key_buffer，这样可以更好地降低线程之间的竞争。

【例15.1】设置key_buffer_size参数。

首先查询key_buffer_size的值，SQL语句如下：

```
mysql> SHOW VARIABLES LIKE 'key_buffer_size';
+-----------------+---------+
| Variable_name   | Value   |
+-----------------+---------+
```

```
| key_buffer_size | 8388608 |
+-----------------+---------+
```

然后修改该参数值为10240000，SQL语句如下：

```
mysql> SET GLOBAL key_buffer_size=10240000;
```

再次查看该值，发现修改成功，如下所示：

```
mysql> SHOW VARIABLES LIKE 'key_buffer_size';
+-----------------+----------+
| Variable_name   | Value    |
+-----------------+----------+
| key_buffer_size | 10240000 |
+-----------------+----------+
```

上面介绍的是默认的key_buffer，下面介绍如何设置多个key_buffer。

【例15.2】设置多个key_buffer。

首先建立一个索引缓存：

```
mysql> SET GLOBAL hot_cache2.key_buffer_size=128*1024;
```

然后将相关表的索引放到指定的索引缓存中，例如将表t1和表t2的索引存放到hot_cache2缓存中：

```
mysql> CREATE DATABASE mytest;
mysql> use mytest;
mysql> CREATE TABLE t1(id  INT);
mysql> CREATE TABLE t2(id  INT);
mysql> cache index t1,t2 in hot_cache2;
+---------+--------------------+----------+----------+
| Table   | Op                 | Msg_type | Msg_text |
+---------+--------------------+----------+----------+
| test.t  | assign_to_keycache | status   | OK       |
| test.t2 | assign_to_keycache | status   | OK       |
+---------+--------------------+----------+----------+
```

> **注意** MySQL的默认存储引擎为InnoDB，不支持索引缓存，这里需要将存储引擎修改为MyISAM。

如果想要把表t1的索引加载到默认的缓存（key_buffer）中，可以使用如下语句：

```
mysql> load index into cache t1;
+---------+--------------+----------+----------+
| Table   | Op           | Msg_type | Msg_text |
+---------+--------------+----------+----------+
| test.t  | preload_keys | status   | OK       |
+---------+--------------+----------+----------+
```

如果要删除索引缓存，则只需要设置该缓冲区大小为0即可：

```
mysql> SET GLOBAL hot_cache2.key_buffer_size=0;
```

值得注意的是，不能删除默认的索引缓冲区，下面看一下删除后的情况，如下所示。

```
mysql> SHOW VARIABLES LIKE 'key_buffer_size';
+-----------------+----------+
| Variable_name   | Value    |
+-----------------+----------+
| key_buffer_size | 10240000 |
+-----------------+----------+
mysql> set global key_buffer_size=0;
ERROR 1438 (HY000): Cannot drop default keycache
```

提示不能删除默认的索引缓存。接下来，查询key_buffer_size的大小，发现并没有删除成功，如下所示。

```
mysql> show variables like 'key_buffer_size';
+-----------------+----------+
| Variable_name   | Value    |
+-----------------+----------+
| key_buffer_size | 10240000 |
+-----------------+----------+
```

cache index命令可以将多张表的索引加载到指定的索引缓冲区中，但每次数据库重启后，索引缓冲区中的数据会被清空，此时可以考虑在配置文件/etc/my.cnf中添加init-file选项，以便每次服务器启动的时候自动将指定表的索引加载到缓冲区中，如下所示。

```
[root@localhost ~]# more /etc/my.cnf
[mysqld]
port    = 3309
...
key_buffer_size = 1G
hot_cache.key_buffer_size = 512M
init_file = /root/initSQL/mysqld_init.sql
...
```

数据库每次启动的时候执行/root/initSQL/mysqld_init.sql脚本文件，该文件可以将多个表索引加载到缓冲区中。

【例15.3】将表t1和表t2的索引加载到hot_cache缓冲区中，SQL语句如下：

```
cache index t1,t2 in hot_cache;
```

验证配置是否生效，SQL语句如下：

```
mysql> SHOW VARIABLES LIKE 'key_buffer_size';
+-----------------+------------+
| Variable_name   | Value      |
+-----------------+------------+
| key_buffer_size | 1073741824 |
+-----------------+------------+

mysql> select @@global.hot_cache.key_buffer_size;
```

```
+----------------------------------+
| @@global.hot_cache.key_buffer_size |
+----------------------------------+
|                        536870912 |
+----------------------------------+
```

从结果中可以看出，表t1和表t2的索引已加载到hot_cache缓冲区中。

15.2.3　内存参数的设置

对于数据库系统，内存设置会直接影响到系统的性能，因为内存的读写速度要远远高于磁盘的速度，所以应尽量让数据的读写直接在内存中进行。然而，由于系统内存资源的限制，过多地使用内存空间必将降低系统的整体性能，因此内存的设置需要兼顾服务器的配置。

1. innodb_buffer_pool_size的设置

在[mysqld]中，对innodb_buffer_pool_size参数的定义如下：

```
[root@localhost ~]# mysqld --verbose --help|grep "\-\-innodb_buffer_pool_size" -A 2
  --innodb_buffer_pool_size= #
          The size of the memory buffer InnoDB uses to cache data
          and indexes of its tables.
```

该参数的作用是缓存InnoDB表和索引数据。这个值设置得越高，访问表中数据需要的磁盘I/O就越少。该参数的默认值是128MB，建议不要将该参数设置过大，设置该参数应该遵循以下分配原则。

（1）如果数据库服务器是一个独立的服务器，可以考虑使用物理内存的70%～80%。

（2）由于该参数不能动态更改，因此要修改这个值，需要重启mysqld服务。

（3）不要把该参数设置太大，如果分配过大，会导致操作系统的虚拟空间被占用，导致操作系统变慢，从而减低SQL查询的效率。

2. innodb_additional_mem_pool的设置

在[mysqld]中，对innodb_additional_mem_pool参数的定义如下：

```
[root@localhost~]# mysqld --verbose --help|grep "\-\-innodb_additional_mem_pool" -A 3
  --innodb_additional_mem_pool_size=#
              Size of a memory pool InnoDB uses to store data
              dictionary information and other internal data
              structures.
```

这个参数用来设置InnoDB存储的数据目录信息和其他内部数据结构的内存池大小。应用程序里的表越多，就需要在这里分配越多的内存。对于一个相对稳定的应用，这个参数的大小也是相对稳定的，没有必要预留非常大的值。如果InnoDB用光了这个池内的内存，就开始从操作系统分配内存，并且往MySQL错误日志中写警告信息。其默认值是1MB，当发现错误日志中已经有相关的警告信息时，就应该适当增加该参数的大小。

管理员也可以使用show engine innodb status\G;查询运行中的数据库的状态,如下所示。

```
mysql> show engine innodb status\G;
...
----------------------
BUFFER POOL AND MEMORY
----------------------
Total memory allocated 17445316; in additional pool allocated 869760
Dictionary memory allocated 25752
Buffer pool size    512
Free buffers        488
Database pages      24
Modified db pages   0
Pending reads 0
Pending writes: LRU 0, flush list 0, single page 0
Pages read 24, created 0, written 0
0.00 reads/s, 0.00 creates/s, 0.00 writes/s
No buffer pool page gets since the last printout
...
```

这些信息有助于了解InnoDB缓冲池的使用情况和性能。根据这些数据,可以调整innodb_buffer_pool_size参数以优化数据库性能。

15.2.4 日志和事务参数的设置

关系数据库的日志与事务管理也是影响性能的重要因素之一,尤其是以事务为主的系统,日志读写比较频繁,进行合理的设置可以降低系统I/O和其他资源开销,提升系统性能。

1. innodb_log_file_size的设置

在[mysqld]中,对innodb_log_file_size参数的定义如下:

```
[root@localhost ~]# mysqld --verbose --help|grep "\-\-innodb_log_file_size" -A 1
  --innodb_log_file_size=#
                      Size of each log file in a log group.
```

该参数的作用是设置日志组中每个日志文件的大小。该参数在高写入负载尤其是大数据集的情况下很重要,这个值越大,则性能相对越高。该参数的默认值是5MB。该参数的设置应该遵循以下原则:

(1)如果日志文件大小过小,日志切换就会比较频繁,会影响服务器性能;如果日志文件过大,当系统发生灾难时,恢复时间会加大。

(2)在MySQL 5.5及以前的InnoDB中,logfile的最大设置为4GB,在MySQL 5.6以后的版本中,logfile最大可以设为512GB。一般控制在几个日志文件大小相加总和在2GB以内最佳,具体情况还需要以数据大小为依据。

2. innodb_log_files_in_group的设置

在[mysqld]中,对innodb_log_files_in_group参数的定义如下:

```
[root@localhost ~]# mysqld --verbose --help|grep "\-\-innodb_log_files_in_group"
-A 3
    --innodb_log_files_in_group=#
                    Number of log files in the log group. InnoDB writes to
                    the files in a circular fashion. Value 3 is recommended
                    here.
```

该参数用来指定数据库有几个日志组,默认为2个,因为有可能出现跨日志的大事务,所以一般建议使用3~4个日志组。

3. innodb_log_buffer_size的设置

在[mysqld]中,对innodb_log_buffer_size参数的定义如下:

```
[root@localhost ~]# mysqld --verbose --help|grep "\-\-innodb_log_buffer_size" -A2
    --innodb_log_buffer_size=#
                    The size of the buffer which InnoDB uses to write log to
                    the log files on disk.
```

该参数的作用是设置日志缓存的大小,一旦提交事务,就将该缓存池中的内容写到磁盘的日志文件上。该参数的设置在中等强度写入负载以及较短事务的情况下,一般都可以满足服务器的性能要求。如果服务器负载较大,可以考虑加大该参数的值。一般缓存池中的内存每秒写入磁盘一次,所以设置该参数较大则会浪费内存空间,一般设置为8~16MB就足够了。

另外,可以参考Innodb_os_log_written的值,如果这个值增加过快,可以适当地增加innodb_log_buffer_size参数的值。

4. innodb_flush_log_at_trx_commit的设置

在[mysqld]中,对innodb_flush_log_at_trx_commit参数的定义如下:

```
[root@localhost~]#mysqld --verbose --help|grep "\-\-innodb_flush_log_at_trx_commit" -A 3
    --innodb_flush_log_at_trx_commit[=#]
                    Set to 0 (write and flush once per second), 1 (write and
                    flush at each commit) or 2 (write at commit, flush once
                    per second).
```

该参数用来控制将缓冲区的数据写入日志文件和将日志文件数据刷新到磁盘的操作时间。对该参数的设置可以在数据库的性能与数据库的安全之间进行折中。

(1)当该参数设置为0时,日志缓存每秒一次地被写入日志文件中,并且将日志文件数据刷新到磁盘。

(2)当该参数设置为1时,InnoDB的事务日志在每次提交后写入日志文件,并对日志做刷新到磁盘的操作。这个可以做到不丢失任何一个事务。

(3)当该参数设置为2时,在每个事务提交时,日志缓存被写入日志文件,但不对日志文件做刷新到磁盘的操作,对日志文件的刷新也是每秒发生一次。需要注意的是,由于进程调度方面的原因,并不能保证日志文件的刷新操作每秒一定会发生。

innodb_flush_log_at_trx_commit参数只有3个值,默认值为1,也是最安全的设置。出于性能考虑，可以设置为0或者2，但会在数据库崩溃的时候丢失一秒钟的事务。为了保证事务的持久性和复制设置的一致性，建议将这个参数设置为1。

15.2.5 存储和 I/O 相关参数的设置

数据库数据存储和I/O的性能直接影响到数据库数据文件读写执行的效率，合理地配置存储和I/O方面的参数，将会提升数据库整体的性能。

1. innodb_open_files的设置

在[mysqld]中，对innodb_open_files参数的定义如下：

```
[root@localhost ~]# mysqld --verbose --help|grep "\-\-innodb_open_files" -A 2
  --innodb_open_files=#
                      How many files at the maximum InnoDB keeps open at the
                      same time.
```

该参数的作用是限制InnoDB存储引擎同时能打开的表的数量。该参数默认值为300，如果数据库里面的表特别多，可以考虑增加该参数的值。

2. innodb_flush_method的设置

在[mysqld]中，对innodb_flush_method参数的定义如下：

```
[root@localhost ~]# mysqld --verbose --help|grep "\-\-innodb_flush_method" -A 1
  --innodb_flush_method=name
                      With which method to flush data.
```

该参数的作用是设置InnoDB存储引擎与操作系统进行I/O交互的模式，即刷新数据和日志的方法。该参数有3个可选项，含义如下：

（1）Fdatasync（默认值）：InnoDB使用fsync()函数去更新日志和数据文件。

（2）O_DSYNC：InnoDB使用O_SYNC模式打开并更新日志文件，用fsync()函数去更新数据文件。

（3）O_DIRECT：InnoDB使用O_DIRECT模式打开数据文件，用fsync()函数去更新日志和数据文件。

3. innodb_max_dirty_pages_pct的设置

在[mysqld]中，对innodb_max_dirty_pages_pct参数的定义如下：

```
[root@localhost ~]# mysqld --verbose --help|grep "\-\-innodb_max_dirty_pages_pct" -A 1
  --innodb_max_dirty_pages_pct=#
                      Percentage of dirty pages allowed in bufferpool.
```

该参数的作用是控制InnoDB的脏页在缓冲中的百分比，将脏页的比例控制在所设定的百分比之下。如果该参数的值是50，则脏页在缓冲中最多占50%。建议该参数设置为15~90之间，如果设置太大，则缓存中每次更新需要置换的数据页就太多；如果设置过小，则可以存放脏数据页的缓冲区内存空间会很小，性能会受到一定的影响。

15.2.6　其他重要参数的设置

除了前面介绍的参数外，还有一些其他的参数也会影响服务器的性能。

1. max_connect_errors的设置

在[mysqld]中，对max_connect_errors参数的定义如下：

```
[root@localhost ~]# mysqld --verbose --help|grep '\-\-wait_timeout' -A 2
  --wait_timeout=#    The number of seconds the server waits for activity on a
                      connection before closing it.
```

max_connect_errors的默认值为10，表示mysqld线程没重新启动过，一台物理服务器只要连接异常中断累计超过10次，就再也无法连接上mysqld服务。建议设置此值至少大小或等于10。若异常中断累计超过该参数设置的值，有两种解决方法：重启mysqld服务或者执行FLUSH HOSTS命令。

2. interactive_timeout的设置

在[mysqld]中，对interactive_timeout参数的定义如下：

```
[root@localhost ~]# mysqld --verbose --help|grep '\-\-interactive_timeout' -A 2
  --interactive_timeout=#
                      The number of seconds the server waits for activity on an
                      interactive connection before closing it.
```

该参数用于设置交互式会话在没有任何活动的情况下能够保持连接的最长时间。

3. wait_timeout

在[mysqld]中，对wait_timeout参数的定义如下：

```
[root@localhost ~]# mysqld --verbose --help|grep '\-\-wait_timeout' -A 2
  --wait_timeout=#    The number of seconds the server waits for activity on a
                      connection before closing it.
```

该参数用于设置非交互式会话在没有任何活动的情况下能够保持连接的最长时间。此参数只对基于TCP/IP或基于Socket通信协议建立的连接有效。

4. query_cache_type

在[mysqld]中，对query_cache_type参数的定义如下：

```
[root@localhost ~]# mysqld --verbose --help|grep '\-\-query_cache_type' -A 3
  --query_cache_type=#
                      0 = OFF = Don't cache or retrieve results. 1 = ON = Cache
                      all results except SELECT SQL_NO_CACHE ... queries. 2 =
                      DEMAND = Cache only SELECT SQL_CACHE ... queries.
```

该参数用于控制是否将查询结果存放到查询缓存中。查询缓存类型的可选值只能是0、1、2，具体选项的含义如下：

（1）该参数选项设置为0：表示禁止查询缓存的功能。
（2）该参数选项设置为1：表示启用查询缓存的功能，缓存所有符合要求的查询结果集，除SELECT SQL_NO_CACHE...以及不符合查询缓存设置的结果集之外。
（3）该参数选项设置为2：表示仅缓存SELECT SQL_CACHE ...子句的查询结果集，除不符合查询缓存设置的结果集之外。

5. query_cache_size

在[mysqld]中，对query_cache_size参数的定义如下：

```
[root@localhost ~]# mysqld --verbose --help|grep '\-\-query_cache_size' -A 1
 --query_cache_size=#
                    The memory allocated to store results from old queries.
```

该参数用来设置查询缓存区的大小。需要从以下几个方面考虑如何设置该参数的大小：

（1）查询缓存区对DDL和DML语句的性能影响。
（2）查询缓存区的内部维护成本。
（3）查询缓存区的命中率以及内存使用率等因素。

15.3　MySQL日志设置优化

MySQL安装之后，可以通过修改各个配置参数来优化数据库的性能。数据库日志对MySQL的I/O性能有很大的影响，本节主要介绍通过设置日志参数来提升数据库性能的方法。下面先来看二进制日志文件的相关参数，如下所示。

```
mysql> SHOW VARIABLES LIKE '%binlog%';
+-----------------------------------------------+----------------+
| binlog_cache_size                             | 32768          |
| binlog_checksum                               | CRC32          |
| binlog_direct_non_transactional_updates       | OFF            |
| binlog_encryption                             | OFF            |
| binlog_error_action                           | ABORT_SERVER   |
| binlog_expire_logs_seconds                    | 2592000        |
| binlog_format                                 | ROW            |
| binlog_group_commit_sync_delay                | 0              |
| binlog_group_commit_sync_no_delay_count       | 0              |
| binlog_gtid_simple_recovery                   | ON             |
| binlog_max_flush_queue_time                   | 0              |
| binlog_order_commits                          | ON             |
| binlog_rotate_encryption_master_key_at_startup| OFF            |
| binlog_row_event_max_size                     | 8192           |
| binlog_row_image                              | FULL           |
| binlog_row_metadata                           | MINIMAL        |
| binlog_row_value_options                      |                |
| binlog_rows_query_log_events                  | OFF            |
```

```
| binlog_stmt_cache_size                           | 32768                |
| binlog_transaction_compression                   | OFF                  |
| binlog_transaction_compression_level_zstd        | 3                    |
| binlog_transaction_dependency_history_size       | 25000                |
| binlog_transaction_dependency_tracking           | COMMIT_ORDER         |
| innodb_api_enable_binlog                         | OFF                  |
| log_statements_unsafe_for_binlog                 | ON                   |
| max_binlog_cache_size                            | 18446744073709547520 |
| max_binlog_size                                  | 1073741824           |
| max_binlog_stmt_cache_size                       | 18446744073709547520 |
| sync_binlog                                      | 1                    |
+--------------------------------------------------+----------------------+
```

（1）binlog_cache_size参数的默认大小是32768，即32K。该参数表示在事务中允许二进制日志缓存的大小。如果要设置该参数，首先需要了解binlog_cache_size的使用情况，如下所示。

```
mysql> SHOW STATUS LIKE '%binlog%';
+----------------------------+-------+
| Variable_name              | Value |
+----------------------------+-------+
| Binlog_cache_disk_use      | 16    |
| Binlog_cache_use           | 339   |
| Binlog_stmt_cache_disk_use | 0     |
| Binlog_stmt_cache_use      | 54    |
| Com_binlog                 | 0     |
| Com_show_binlog_events     | 0     |
| Com_show_binlogs           | 0     |
+----------------------------+-------+
```

其中影响binlog_cashe_size性能的参数主要如下：

- Binlog_cache_use：表示使用二进制日志缓存事务的数量。
- Binlog_cache_disk_use：表示使用二进制日志缓存，并且当值达到binlog_cache_size设置的值时，用临时文件存储事务的数量。

（2）max_binlog_cache_size参数表示二进制日志所使用缓存的最大值，该参数的默认值是18446744073709547520，该默认值已经足够大了。

（3）max_binlog_size参数代表二进制日志使用的最大值，如果系统中事务过多，而此参数的数值设置过小，则会报错，该参数默认值是1073741824（1GB），该参数一般设置为512MB或者1GB。

（4）sync_binlog参数比较重要，它不仅对数据完整性有影响，而且对数据库的性能也有影响，该参数的选项如下：

- sync_binlog=0：表示当事务提交之后，不做文件系统之类的磁盘同步指令刷新binlog_cache中的信息到磁盘，而会让文件系统自行决定同步，或者缓存满了之后才同步到磁盘。
- sync_binlog=n：表示当事务提交n次之后，将进行一次同步，将binlog_cache中的数据写入磁盘。

值得注意的是，sync_binlog=0是数据库默认的设置，此时性能是最好的，不过也是最危险的，一旦系统崩溃，binlog_cache中的数据就会丢失。sync_binlog=1的时候，数据库消耗最大，不过安全性比较高，当事务提交后，数据库会将缓存中的数据同步到磁盘中。

MySQL的复制过程中实际上就是将客户端的日志通过I/O拷贝到服务器端，然后在服务器端将日志中的数据解析出来应用到数据库中。

15.4　MySQL I/O设置优化

目前，大多数应用程序都以I/O密集型为主，存储技术的发展远没有计算机中其他系统那么迅速。尽管市面上也有不少高端存储设备，但其价格昂贵，一般用户难以承受。当前，大多数用户采用SAS硬盘并结合不同的RAID（Redundant Array of Independent Disk，磁盘阵列）组合来实现存储。

事实上，前面提到的MySQL优化、数据库对象优化、数据库参数优化，以及应用程序优化等，都是通过减少或延缓磁盘的读写来减轻磁盘I/O的压力。另外，管理员可以通过增强磁盘I/O的吞吐量以及增强磁盘I/O本身的性能来提高数据库的整体性能。

下面针对磁盘I/O进行不同类型的优化。

1. 选择合适的RAID级别

RAID就是将N台硬盘通过RAID控制器结合成虚拟单台大容量硬盘。根据数据冗余和分布方式，RAID分为不同的级别，其中常见的几种RAID级别如表15.1所示。

表 15.1　常见的 RAID 级别

RAID 级别	特　性	优　点	缺　点
RAID 0	也叫条带化（Stripe），按一定的条带大小将数据依次分布到各个磁盘，没有数据冗余	数据并发读写速度快，无额外磁盘空间开销	数据无冗余保护，可靠性差
RAID 1	也叫磁盘镜像（Mirror），两个磁盘一组，所有数据都同时写入两个磁盘，但读时从任一磁盘读	可靠性高，并发读写性能优良	如果容量一定，需要2倍的磁盘，投资大
RAID 10	也叫ARID1+0,是RAID 0与RAID 1的组合体，集成了RAID 0的快速和RAID 1的安全	可靠性高，并发读写性能优良	RAID10会造成50%的磁盘的浪费。投资比较大
RAID 3	RAID3是把数据分成多个块，按照一定的容错算法，存放在 N+1 个硬盘上,事实上空间是 N 个磁盘的空间总和，第 N+1 个磁盘存储的信息是校验容错信息	当第 N+1 个硬盘中出现故障时，从其他 N 个硬盘中恢复原始数据	RAID 3 校验磁盘很容易成为整个系统的瓶颈。校验的负载如果很大，会导致整个 RAID 系统性能的下降

(续表)

RAID 级别	特性	优点	缺点
RAID 4	跟 RAID0 一样对磁盘组条带化，不同的是：需要额外增加一个磁盘来写各校验纠错数据	RAID 中一个磁盘损坏，其他数据可以通过校验纠错数据计算出来，读数据速度快	在失败恢复的时候，难度要比 RAID3 大，控制器的设计难度也要大许多，而且访问数据的效率不高
RAID 5	RAID5 是一种存储性能、数据安全和存储成本兼顾的存储解决方案，将每个条带的校验纠错数据块也分布到各个磁盘，而不是写到一个磁盘	跟 RAID 4 相似，只是其写性能和数据保护能力要更强一点	容错能力不及 RAID 1，在磁盘出现损坏时候，读写能力会下降

了解了RAID级别的特性后，就可以根据数据读写的特点、可靠性要求以及投资额度来选择合适的RAID级别了，比如：

（1）数据读写都比较频繁，最好选择RAID 10。
（2）数据读很频繁，写相对少一点，对可靠性有一定的要求，可以选择RAID 5。
（3）数据读写比较频繁，但是对可靠性要求不高，可以选择RAID10。

2. 使用Symbolic Links分布I/O

在不使用RAID或者逻辑卷的情况下，要想达到分布式磁盘I/O的目的，可以使用操作系统的Symbolic Links（符号链接）将不同的数据库或表、索引执行不同的物理磁盘上面。

可以通过SHOW VARIABLES LIKE 'have_symlink'语句检查系统是否支持符号链接。

```
mysql> SHOW VARIABLES LIKE 'have_symlink';
+---------------+-------+
| Variable_name | Value |
+---------------+-------+
| have_symlink  | YES   |
+---------------+-------+
```

（1）将一个数据库指向其他物理磁盘，首先在目标磁盘上创建目录，然后创建从MySQL数据目录到目标目录的符号链接，命令如下：

```
[root@localhost ~]# mkdir /otherdisk/databases/test
[root@localhost ~]# ln -s /otherdisk/databases/test /path/to/datadir
```

（2）对于新建的MyISAM表，可以使用DATA DIRECTORY和INDEX DIRECORY来将表的数据文件或索引文件指向其他物理磁盘，命令如下：

```
Create table test(id int primary key,name varchar(20))
Type = myisam
DATA DIRECOTORY = '/disk2/data'
INDEX DIRECTORY= '/disk3/index'
```

注意，对于已经存在的MyISAM表，可以将.myd或者.myi文件移到目标磁盘，然后建立符号链接。

15.5 MySQL并发设置优化

MySQL数据库并发性能是影响数据整体性能的一个重要的参数,可以通过提高并发连接数量来提升并发效率。下面来了解一下如何优化数据库连接的最大数。

首先查看一下并发连接的最大数和目前连接的最大数,命令如下:

```
mysql> SHOW VARIABLES LIKE 'max_connections';
+-----------------+-------+
| Variable_name   | Value |
+-----------------+-------+
| max_connections | 1000  |
+-----------------+-------+

mysql> SHOW GLOBAL STATUS LIKE 'max_used_connections';
+----------------------+-------+
| Variable_name        | Value |
+----------------------+-------+
| Max_used_connections | 125   |
+----------------------+-------+
```

参数max_connections表示允许同时连接MySQL数据库客户端的最大数量,max_used_connections表示同时使用的最大连接数。max_used_connections占max_connections的80%左右比较合适。一般来讲,在MySQL主机性能允许的范围内,max_used_connections参数设置为500~800比较合适,而参数max_used_connections是用户正在连接的最大数量,可以设置得大一些。

参数net_buffer_length设置的只是消息缓冲区的初始化大小,系统默认值是16KB,当然如果系统内存比较紧张,可以考虑将该参数降低到8KB左右。查看命令如下:

```
mysql> SHOW VARIABLES LIKE 'net%';
+-------------------+-------+
| Variable_name     | Value |
+-------------------+-------+
| net_buffer_length | 16384 |
| net_read_timeout  | 30    |
| net_retry_count   | 10    |
| net_write_timeout | 60    |
+-------------------+-------+
4 rows in set (0.00 sec)
```

参数max_allowed_packet表示在网络传输中,一次传递数据量的最大值。必须将该参数设置为1024的倍数,单位为字节。查看命令如下:

```
mysql> SHOW VARIABLES LIKE 'max_allowed_packet%';
+--------------------+---------+
| Variable_name      | Value   |
+--------------------+---------+
```

```
| max_allowed_packet | 67108864|
+--------------------+---------+
1 row in set (0.00 sec)
```

参数back_log表示在MySQL的连接请求等待队列中允许存放的最大连接请求数。系统默认的值是151。查看命令如下：

```
mysql> SHOW VARIABLES LIKE 'back_log%';
+---------------+-------+
| Variable_name | Value |
+---------------+-------+
| back_log      | 151   |
+---------------+-------+
```

15.6 服务器语句超时处理

在MySQL中可以设置服务器语句超时的限制，单位可以达到毫秒级别。

当中断的执行语句超过设置的毫秒数后，服务器将终止查询影响不大的事务或连接，然后将错误报给客户端。

设置服务器语句超时的限制，可以通过设置系统变量max_execution_time来实现。例如：

```
SET GLOBAL MAX_EXECUTION_TIME=2000;
```

默认情况下，MAX_EXECUTION_TIME的值为0，代表没有时间限制。通过上述设置后，如果SELECT语句执行时间超过2000毫秒，语句将会被终止。

15.7 线程和临时表的优化

除了前面几节介绍的优化方法以外，还有一些因素也能影响MySQL的整体性能，例如线程管理和临时表管理。

15.7.1 线程的优化

线程的优化主要是优化MySQL中与连接线程相关的系统参数以及状态变量。在MySQL中实现了一个线程池，将空闲的连接线程存放在线程池中，当有新的请求时，MySQL会检查线程池中是否有空闲的连接线程，如果没有空闲的线程，就创建新的线程；如果有空闲的线程，就直接从线程池中取出空闲的线程。

查看与连接线程相关的系统参数，SQL语句如下：

```
mysql> SHOW VARIABLES LIKE 'thread%';
+-------------------+---------------------------+
| Variable_name     | Value                     |
```

```
+-------------------+--------------------------+
| thread_cache_size | 9                        |
| thread_handling   | one-thread-per-connection|
| thread_stack      | 1048576                  |
+-------------------+--------------------------+
```

参数thread_cache_size表示数据库线程池中应该存放的连接线程数。thread_cache_size的默认值为9。建议根据物理内存设置，1GB的内存可以设置为9，2GB的内存可以设置为18，3GB内存可以设置为27，大于3GB的内存可以设置成64。

参数thread_stack 表示每个连接线程被创建的时候，MySQL分配内存的大小。系统默认值是1024KB，MySQL的 max_connections * thread_stack 应小于可用内存。

15.7.2 临时表的优化

临时文件就是为了各种不同的目的而产生的中间文件，使用完毕后会被及时回收和清理。临时表也是如此，它是MySQL在进行一些内部操作的时候生成的数据库表。这些操作主要包括GROUP BY、DISTINCT、ORDER BY、UNION，以及一些FROM语句中的子查询等。

可以使用EXPLAIN来分析查询语句，看看是否会用到临时表。EXPLAIN输出中的EXTRA列会指明是否"Using temporary"。事实上，大多数用户都不会去关注临时表的产生、使用与消亡的过程，因为它对用户是透明的。但是对于数据库管理员或者其他关心性能的人员而言，就不得不重视临时表，因为如果设置不当或者程序使用不当，就可能会产生大量的磁盘临时表，这对系统性能会产生很大的影响。

什么是磁盘临时表呢？除了直接产生的磁盘临时表外，大量磁盘临时表是由内存临时表转化来的。内存临时表存在于内存中，由MEMORY引擎进行处理，速度较快。而磁盘临时表则是在磁盘上创建、使用、销毁的。由于磁盘是慢速访问设备，因此磁盘临时表的操作效率要比内存临时表的操作效率差了几个数量级，具有较差的性能。因此，需要尽量避免磁盘临时表的产生。

查看临时文件、内存临时表和磁盘临时表的命令如下：

```
mysql> SHOW GLOBAL STATUS LIKE 'created_tmp%';
+-------------------------+-------+
| Variable_name           | Value |
+-------------------------+-------+
| Created_tmp_disk_tables | 0     |
| Created_tmp_files       | 5     |
| Created_tmp_tables      | 13    |
+-------------------------+-------+
```

每次数据库系统创建临时表时，Created_tmp_tables就会增加；如果是在磁盘上创建临时表，Created_tmp_disk_tables也会增加；Created_tmp_files表示MySQL数据库创建的临时文件的数量。

下面看一下MySQL数据库对临时表的配置，如下所示。

```
mysql> SHOW VARIABLES WHERE
Variable_name in('max_heap_table_size','tmp_table_size');
+---------------------+----------+
| Variable_name       | Value    |
+---------------------+----------+
| max_heap_table_size | 16777216 |
| tmp_table_size      | 16777216 |
+---------------------+----------+
```

max_heap_table_size参数表示heap数据表的最大长度，默认设置是16MB。值得注意的是，有text或者blog字段时，不能用内存临时表而要用磁盘临时表。

tmp_table_size参数规定了内存临时表的最大值（推荐该参数设置为200MB），如果内存临时表超过了限制，数据库会将数据存储在磁盘临时表中，存储的目录默认是/tmp/，如下所示。

```
mysql> SHOW VARIABLES LIKE 'tmpdir';
+---------------+-------+
| Variable_name | Value |
+---------------+-------+
| tmpdir        | /tmp  |
+---------------+-------+
```

15.8 增加资源组

MySQL中有一个资源组功能，用于调控线程优先级以及绑定CPU。MySQL用户需要有RESOURCE_GROUP_ADMIN权限才能创建、修改、删除资源组。

注意，在Linux环境下，MySQL进程需要有CAP_SYS_NICE 权限才能使用资源组的完整功能。

```
[root@localhost~]# sudo setcap cap_sys_nice+ep /usr/local/mysql 9.0/bin/mysqld
[root@localhost~]# getcap /usr/local/mysql 9.0/bin/mysqld
/usr/local/mysql 9.0/bin/mysqld = cap_sys_nice+ep
```

MySQL默认提供两个资源组，分别是USR_default和SYS_default。下面来讲述资源组的常用操作。

创建名称为"my_resouce_group"的资源组，SQL语句如下：

```
mysql> CREATE RESOURCE GROUP my_resouce_group type=USER vcpu=0, 1
thread_priority=5;
```

将当前线程加入资源组，SQL语句如下：

```
mysql> SET RESOURCE GROUP my_resouce_group;
```

查看资源组my_resouce_group中包含的线程，SQL语句如下：

```
mysql> SELECT * FROM Performance_schema.threads WHERE
RESOURCE_GROUP='my_resouce_group';
+-----------+--------------------------+------------+-----------------+-------
-----------+------------------+----------------+-------------
```

```
+-----------+---------------------------+------------+----------------+
-----------------+------------------+----------------+--------------------+---
--------------+----------------+----------------+
| THREAD_ID | NAME                      | TYPE       | PROCESSLIST_ID |
PROCESSLIST_USER | PROCESSLIST_HOST | PROCESSLIST_DB | PROCESSLIST_COMMAND |
PROCESSLIST_TIME | PROCESSLIST_STATE | PROCESSLIST_INFO                |
PARENT_THREAD_ID | ROLE | INSTRUMENTED | HISTORY | CONNECTION_TYPE | THREAD_OS_ID |
RESOURCE_GROUP   |
+-----------+---------------------------+------------+----------------+
-----------------+------------------+----------------+--------------------+---
--------------+----------------+----------------+
|        48 | thread/sql/one_connection | FOREGROUND |              9 | root
| localhost        | mysql            | Query          |                    0 | Sending data
| SELECT * FROM Performance_Schema.threads WHERE RESOURCE_GROUP='my_resouce_group' |
NULL | NULL | YES          | YES     | SSL/TLS         |         1352 | my_resouce_group |
+-----------+---------------------------+------------+----------------+
-----------------+------------------+----------------+--------------------+---
--------------+----------------+----------------+
```

资源组里有线程时，删除资源组会报错：

```
mysql> DROP RESOURCE GROUP my_resouce_group;
ERROR 3656 (HY000): Resource group test_resouce_group is busy.
```

修改资源组的SQL语句如下：

```
mysql> ALTER RESOURCE GROUP my_resouce_group vcpu = 2,3 THREAD_PRIORITY = 8;
Query OK, 0 rows affected (0.10 sec)
```

把资源组里的线程移出到默认资源组USR_default中，SQL语句如下：

```
mysql> SET RESOURCE GROUP USR_default FOR 48;
Query OK, 0 rows affected (0.00 sec)
```

删除资源组的SQL语句如下：

```
mysql> DROP RESOURCE group my_resouce_group;
Query OK, 0 rows affected (0.08 sec)
```

如果需要指定线程的ID，可以使用以下语句：

```
SET RESOURCE GROUP my_resouce_group FOR thread_id;
```

这里的thread_id为线程的ID。

查询线程ID的方法如下：

```
SELECT thread_id FROM Performance_Schema.threads;
```

例如，将ID为20的线程加入资源组my_resouce_group中，SQL语句如下：

```
SET RESOURCE GROUP test_resouce_group FOR 20;
```

第 16 章
MySQL性能监控

在MySQL数据库中，通常使用三方软件来监控数据库性能的健康状况。本章将重点介绍监控利器Nagios工具。使用Nagios监控工具不仅可以监控MySQL数据库性能，还可以对数据库集群、数据复制等操作进行监控。因为数据库性能的瓶颈不仅和数据库本身的优化情况有关，还和操作系统本身的情况以及网络负载的状况有关。因此，本章还将介绍如何监控Linux操作系统的I/O和CPU的运行状况。通过本章的学习，读者可以更好地对MySQL数据库以及服务器的性能进行全方位的监控。

16.1 监控系统的基本方法

在Linux操作系统中，用户可以使用一些分析系统性能的命令去分析数据库服务器的性能。本节主要介绍这些命令的使用方法。

16.1.1 ps 命令

ps命令主要用来获取某个进程的一些信息，其常用参数如下：

（1）-A：显示所有进程。
（2）-a：显示一个终端的所有进程，除了会话引线。
（3）-N：忽略选择。
（4）-d：显示所有进程，但省略所有的会话引线。
（5）-e：列出程序时，显示每个程序所使用的环境变量。
（6）-x：显示没有控制终端的进程，同时显示各个命令的具体路径。d和x不可合用。
（7）-p pid：进程使用CPU的时间。
（8）-u uid or username：选择有效的用户ID或者用户名。
（9）-g gid or groupname：显示组的所有进程。

(10)-f：全部列出，通常和其他选项联用。

(11)-l：长格式（有F、wchan、C等字段）。

(12)-j：作业格式。

(13)-e：命令之后显示环境。

例如，运行ps -aux命令显示所有进程的信息，如下所示。

```
[root@localhost ~]# ps -aux
Warning: bad syntax, perhaps a bogus '-'? See /usr/share/doc/procps-3.2.7/FAQ
USER       PID %CPU %MEM    VSZ   RSS TTY      STAT START   TIME COMMAND
root         1  0.0  0.1   1948   740 ?        Ss   Nov19   0:03 /sbin/init
root         2  0.0  0.0      0     0 ?        S<   Nov19   0:00 [kthreadd]
...
nagios   20727  0.0  0.1  12952  1012 ?        Ssl  11:12   0:03
/var/www/nagios/bin/nagios -d /
nagios   21037  0.0  0.1   4960   908 ?        Ss   11:23   0:00 ./nrpe -c
/var/www/nagios/etc/n
```

可以使用ps -ef命令查询所有进程及其环境变量信息，如下所示。

```
[root@localhost ~]# ps -ef
UID        PID  PPID  C STIME TTY          TIME CMD
root         1     0  0 Nov19 ?        00:00:03 /sbin/init
root         2     0  0 Nov19 ?        00:00:00 [kthreadd]
...
root     17334     1  0 Nov20 ?        00:00:00 /bin/sh
/usr/local/mysql/bin/mysqld_safe --datadir
nagios   21037     1  0 11:23 ?        00:00:00 ./nrpe -c
/var/www/nagios/etc/nrpe.cfg -d
nagios   23516 23515  0 14:52 ?        00:00:00 /var/www/nagios/libexec/check_ssh
127.0.0.1,192.16
root     23524 20118  3 14:52 pts/3    00:00:00 ps -ef
```

通常查看了进程信息后，如果需要终止某个进程，可以使用kill命令，如下所示。

```
[root@localhost ~]#kill -KILL [pid]
```

如果需要强行终止某个进程，可以使用kill -9 [pid]命令，如下所示。

```
[root@localhost ~]#kill -9 [pid]
```

16.1.2 top 命令

top命令常用于Linux操作系统下的性能分析工作，能够实时地显示操作系统中各个进程消耗资源的情况，例如显示CUP使用情况、内存使用情况和执行时间。下面执行top命令查看一下系统执行的情况：

```
[root@localhost ~]# top
top - 15:12:58 up 3 days,  4:13,  4 users,  load average: 0.29, 0.27, 0.28
Tasks: 138 total,   3 running, 133 sleeping,   1 stopped,   1 zombie
Cpu(s): 24.4%us,  9.2%sy,  0.0%ni, 66.4%id,  0.0%wa,  0.0%hi,  0.0%si,  0.0%st
```

```
Mem:    673164k total,    650156k used,    23008k free,    90248k buffers
Swap:   524280k total,       52k used,   524228k free,   185692k cached

  PID USER      PR  NI  VIRT  RES  SHR S %CPU %MEM    TIME+  COMMAND
 2464 root      20   0 47108  20m 7888 R 15.9  3.1  39:50.58 Xorg
20115 root      20   0 90728  18m  11m S  4.0  2.8   0:07.86 gnome-terminal
...
   62 root      15  -5     0    0    0 S  0.0  0.0   0:00.00 kacpi_notify
  120 root      15  -5     0    0    0 S  0.0  0.0   0:00.00 cqueue
  122 root      15  -5     0    0    0 S  0.0  0.0   0:00.04 ksuspend_usbd
  127 root      15  -5     0    0    0 S  0.0  0.0   0:01.09 khubd
```

下面分析一下top命令的统计信息的含义。首先分析top命令第一行信息：

```
top - 15:12:58 up 3 days,  4:13,  4 users,  load average: 0.29, 0.27, 0.28
```

该行信息的具体含义如下：

（1）15:12:58：表示系统运行的当前时间。

（2）up 3 days：表示系统运行时间。

（3）4 users：表示登录用户的数量。

（4）load average: 0.29, 0.27, 0.28：表示系统负载，即任务队列的平均长度，3个数值分别为1分钟、5分钟、15分钟到现在的平均值。

然后分析top命令的第二行和第三行的具体含义，这两行分别表示进程和CPU的性能的一些信息。

```
Tasks: 138 total,   3 running, 133 sleeping,   1 stopped,   1 zombie
Cpu(s): 24.4%us,  9.2%sy,  0.0%ni, 66.4%id,  0.0%wa,  0.0%hi,  0.0%si,  0.0%st
```

各个统计信息的具体含义如下：

（1）138 total：表示进程的总数。

（2）3 running：表示正在运行的进程数量。

（3）133 sleeping：表示睡眠的进程数量。

（4）1 stopped：表示停止的进程数量。

（5）1 zombie：表示僵尸进程数量。

（6）24.4%us：表示用户空间占用CPU的百分比。

（7）9.2%sy：表示内核空间占用CPU的百分比。

（8）0.0%ni：表示用户进程空间内改变过优先级的进程占用CPU的百分比。

（9）66.4%id：表示空闲CPU的百分比。

（10）0.0%wa：表示等待输入/输出的CPU的百分比。

接着分析top命令的第四行和第五行的具体含义，这两行分别表示内存的性能分析信息。

```
Mem:    673164k total,    650156k used,    23008k free,    90248k buffers
Swap:   524280k total,       52k used,   524228k free,   185692k cached
```

Mem各个统计信息的含义如下：

（1）673164k total：表示物理内存的总的大小。
（2）650156k used：表示使用的物理内存的大小。
（3）23008k free：表示空闲的物理内存的大小。
（4）90248k buffers：表示内核缓存内存空间的大小。

Swap各个统计信息的含义如下：

（1）524280k total：表示交换区的总量。
（2）52k used：表示使用的交换区总量。
（3）524228k free：表示空闲的交换区总量。
（4）185692k cached：表示缓冲的交换区总量。

内存中的内容被换出到交换区，而后又被换入内存，但使用过的交换区尚未被覆盖，该数值即为这些内容已存在于内存中的交换区的大小。相应的内存内容再次被换出时，可不必再对交换区进行写入。

16.1.3 vmstat 命令

vmstat命令可以显示Linux性能指标，该命令分别输出进程、内存、交互区、I/O、系统和CPU的情况。下面直接执行该命令，结果如下：

```
[root@localhost ~]# vmstat
procs -----------memory---------- ---swap-- -----io---- --system-- -----cpu------
 r  b   swpd   free   buff  cache   si   so    bi    bo   in   cs us sy id wa st
 2  0    404  12264 101280 164840    0    0     3    33    6  113  4  2 94  0  0
```

procs（进程）中有r、b两个字段，各字段说明如下：

（1）r：表示可运行进程的数量。
（2）b：表示阻塞进程的数量。

memory（内存）中有4个报告虚拟内存如何使用的字段，具体意义如下：

（1）swpd：表示已经使用的交换空间的数量。
（2）free：表示自由RAM的数量。
（3）buff：表示缓存使用的RAM的数量。
（4）cache：表示文件系统缓存使用的RAM数量。

swap（交换）字段的详细说明如下：

（1）si：表示从磁盘分页到内存的数量。
（2）so：表示从内存分页到磁盘的数量。

io字段的详细说明如下：

（1）bi：表示从磁盘读入的块。
（2）bo：表示写入磁盘的块。

system字段和CPU字段说明如下（CPU状态使用总CPU时间的百分比来表示）：

（1）in：表示系统中断。

（2）cs：表示进程上下文开关。

（3）us：表示用户模式。

（4）sy：表示内核模式。

（5）wa：表示等待I/O。

（6）id：表示空闲状态。

16.1.4　mytop 命令

mytop是一个类似于Linux下的top命令的MySQL监控工具，可以监控当前用户正在执行的命令。

首先，使用源码安装mytop-1.6.tar.gz，命令如下：

```
[root@localhost ~]# tar -zxvf mytop-1.6.tar.gz
[root@localhost ~]# cd mytop-1.6
[root@localhost mytop-1.6]# perl Makefile.PL
[root@localhost mytop-1.6]# make
[root@localhost mytop-1.6]# make test
[root@localhost mytop-1.6]# make install
[root@localhost mytop-1.6]#
```

直接输入mytop命令，结果如下：

```
[root@localhost mytop-1.6]# mytop
Can't locate DBI.pm in @INC (@INC contains:
/usr/lib/perl5/5.10.0/i386-linux-thread-multi /usr/lib/perl5/5.10.0
/usr/lib/perl5/site_perl/5.10.0/i386-linux-thread-multi
/usr/lib/perl5/site_perl/5.10.0 /usr/lib/perl5/site_perl/5.8.8
/usr/lib/perl5/site_perl/5.8.7 /usr/lib/perl5/site_perl/5.8.6
/usr/lib/perl5/site_perl/5.8.5 /usr/lib/perl5/site_perl
/usr/lib/perl5/vendor_perl/5.10.0/i386-linux-thread-multi
/usr/lib/perl5/vendor_perl/5.10.0 /usr/lib/perl5/vendor_perl/5.8.8
/usr/lib/perl5/vendor_perl/5.8.7 /usr/lib/perl5/vendor_perl/5.8.6
/usr/lib/perl5/vendor_perl/5.8.5 /usr/lib/perl5/vendor_perl .) at /usr/bin/mytop line
22.
    BEGIN failed--compilation aborted at /usr/bin/mytop line 22.
```

发现报错了，出现"Can't locate DBI.pm in @INC"这种问题的主要原因是系统没有安装DBI组件。要解决这个问题，可以安装DBI、Data-ShowTable、DBD-mysql（假设已安装了Perl和MySQL数据库）。

下面可以直接到 ftp://ftp.funet.fi/pub/languages/perl/CPAN/modules/by-module 下载 DBI-1.601.tar.gz、Data-ShowTable-3.3.tar.gz、DBD-mysql-3.0007_1.tar.gz文件，它们分别处于DBI、Data、DBD目录下。

首先，解压这3个压缩文件，命令如下：

```
[root@localhost ~]#tar zxvf DBI-1.601.tar.gz
[root@localhost ~]#tar zxvf Data-ShowTable-3.3.tar.gz
[root@localhost ~]#tar zxvf DBD-mysql-3.0007.tar.gz
```

然后安装DBI，命令如下：

```
[root@localhost ~]# cd DBI-1.601
[root@localhost DBI-1.601]# perl ./Makefile.PL
[root@localhost DBI-1.601]# make
[root@localhost DBI-1.601]# make test
[root@localhost DBI-1.601]# make install
```

再安装Data-ShowTable，命令如下：

```
[root@localhost ~]# cd Data-ShowTable-3.3
[root@localhost Data-ShowTable-3.3]# perl ./Makefile.PL
[root@localhost Data-ShowTable-3.3]# make
[root@localhost Data-ShowTable-3.3]# make install （注：无需make test）
```

最后安装DBD-mysql，命令如下：

```
[root@localhost src]# cd DBD-mysql-3.0007_1
[root@localhost DBD-mysql-3.0007_1]# perl ./Makefile.PL
--libs="-L/usr/local/mysql-6.0.9-alhpa/lib/mysql -lmysqlclient -lz -lrt -lcrypt -lnsl
-lm"  --cflags=" -I/usr/local/mysql-6.0.9-alpha/include/mysql -g -DUNIV_LINUX"
--testuser=root --testsocket=/home/cserken/mysql/tmp/mysql.sock
[root@localhost DBD-mysql-3.0007_1]# make
[root@localhost DBD-mysql-3.0007_1]# make test
[root@localhost DBD-mysql-3.0007_1]# make install
```

测试一下，结果发现问题还没有解决，因为没有安装TermReadKey。下面安装TermReadkey，命令如下：

```
[root@localhost ~]# tar -xzvf TermReadKey-2.30.tar.gz
[root@localhost ~]# cd TermReadKey-2.30
[root@localhost TermReadKey-2.30]# perl ./Makefile.PL
[root@localhost TermReadKey-2.30]# make
[root@localhost TermReadKey-2.30]# make install
```

再测试一下mytop命令，结果显示正常运行了，如下所示。

```
[root@localhost ~]# cd mytop-1.6
[root@localhost mytop-1.6]# mytop
MySQL on localhost (5.5.27-ndb-7.2.8-cluster-gpl-log)    up 4+11:10:38 [19:18:03]
 Queries: 3.0    qps:    0 Slow:     0.0       Se/In/Up/De(%):     00/00/00/00
[root@localhost ~]#
 Key Efficiency: 75.0%  Bps in/out:   0.0/  0.0

     Id      User         Host/IP          DB      Time   Cmd Query or State
     --      ----         -------          --      ----   --- --------
   3835      root         localhost        test       0   Query show full process
```

mytop命令的各个参数的含义如下：

（1）-u / --user <USERNAME>：指定用户名，预设是root。

（2）-p / --pass / --password <PASSWORD>：指定密码，预设是none。

（3）-h / --host <HOSTNAME[:PORT]>：指定MySQL服务器的主机名，预设是localhost。

（4）-P / --port <PORT>：指定连接MySQL服务器的端口，预设是3306。

（5）-s / --delay <SECONDS>：更新的秒数，预设是5秒。

（6）-d / --db / --database <DATABASE>：指定连接的资料库，预设是test。

（7）-b / --batch / --batchmode：指定为batch mode，每次更新不会清除旧的显示结果，会将更新资料显示在最上方，预设是unset。

（8）-S / --socket <PATH_TO_SOCKET>：指定使用MySQL socket直接连线，而不使用TCP/IP连线，预设是none（当mytop和MySQL在同一台机器上时才能使用）。

（9）--header or -noheader：是否要显示表头，预设是header。

（10）--color or --nocolor：是否要使用颜色，预设是color。

（11）-i / -idle or -noidle：idle的thread是否要出现在清单上，预设是idle。

【例16.1】mytop命令的使用。

先在MySQL数据库中创建xfimti用户，SQL语句如下：

```
mysql> grant all privileges on *.* to 'xfimti'@'localhost' identified by '123';
Query OK, 0 rows affected (0.01 sec)

mysql> flush privileges;
Query OK, 0 rows affected (0.02 sec)
```

再输入mytop命令，监控信息显示如下：

```
[root@localhost ~]# cd mytop-1.6
[root@localhost mytop-1.6]# ./mytop -u xfimti -p123 -d test -s /temp/mysql.socket

MySQL on localhost (5.5.27-ndb-7.2.8-cluster-gpl-log)     up 4+11:39:16 [19:46:41]
 Queries: 3.0     qps:    0 Slow:      0.0         Se/In/Up/De(%):    00/00/00/00

 Key Efficiency: 83.3%  Bps in/out:   0.0/  0.0

      Id      User         Host/IP          DB      Time   Cmd Query or State
      --      ----         -------          --      ----   --- ----------
    3875     xfimti       localhost        test        0   Query show full process
```

16.1.5　sysstat工具

sysstat工具可用于检测系统性能及效率。例如CPU的使用率、硬盘和网络吞吐数据，这些数据的收集和分析，有利于判断系统是否正常运行。如果要通过源码包安装sysstat，请到https://github.com/sysstat/sysstat下载源码包。

安装sysstat工具包的命令如下：

```
[root@localhost ~]# tar zxvf sysstat-10.0.0.tar.gz
[root@localhost sysstat-10.0.0]# ./configure
```

```
[root@localhost sysstat-10.0.0]# make
[root@localhost sysstat-10.0.0]# make install
```

sysstat工具包集成了如下工具：

（1）iostat工具：提供CPU使用率及硬盘吞吐效率的数据。
（2）mpstat工具：提供单个处理器或多个处理器的相关数据。
（3）sar工具：负责收集、报告并存储系统活跃的信息。
（4）sa1工具：负责收集并存储每天的系统动态信息到一个二进制文件中。
（5）sa2工具：负责把每天的系统活跃信息写入总结性的报告中。
（6）sadc工具：系统动态数据收集工具，收集的数据被写入一个二进制文件中。
（7）sadf工具：显示被sar工具通过多种格式收集的数据。

下面详细介绍iostat、mpstat和sar工具。

1. iostat工具

Iostat工具用于输出与CPU和磁盘I/O相关的统计信息，具体的语法格式如下：

```
iostat [-c|-d] [-k] [-t] [-V] [-x] [-p device|ALL] [间隔描述] [检测次数]
```

各参数的含义如下：

（1）-c：表示仅显示CPU的状态。
（2）-d：仅显示存储设备的状态，不可以和-c一起使用。
（3）-k：默认显示的是读入/读出的块信息。
（4）-t：显示搜集数据的时间。
（5）-V：显示版本号和帮助信息。
（6）-x：显示扩展信息。
（7）-p device|ALL：device为某个设备或者某个分区，如果使用ALL，就表示要显示所有分区和设备的信息。

直接运行iostat命令，结果如下：

```
[root@localhost ~]# iostat
Linux 2.6.25-14.fc9.i686 (localhost.localdomain)    11/23/2012    _i686_    (1 CPU)

avg-cpu:  %user   %nice %system %iowait  %steal   %idle
           3.60    0.04    1.53    0.49    0.00   94.34

Device:            tps    kB_read/s    kB_wrtn/s    kB_read    kB_wrtn
sda               1.66         2.47        32.28     800397   10474295
dm-0              8.34         2.46        32.28     797729   10473872
dm-1              0.00         0.00         0.00        592        404
sdb               0.00         0.01         0.00       4433          1
```

结果中关于CPU性能的参数的含义如下：

（1）%user：在用户级别运行所使用的CPU的百分比。

(2) %nice：nice操作所使用的CPU的百分比。

(3) %system：在系统级别（kernel）运行所使用CPU的百分比。

(4) %iowait：CPU等待硬件I/O时，所占用的CPU百分比。

(5) %idle：表示CPU空闲时间所占比例。

结果中关于磁盘I/O性能的参数的含义如下：

(1) Device：表示设备块的名字。

(2) tps：表示每秒发送的I/O请求数。

(3) kB_read/s：表示从该设备每秒读取的数据块数量。

(4) kB_wrtn/s：表示从该设备每秒写入的数据块数量。

(5) kB_read：表示从该设备读取的数据块总数。

(6) kB_wrtn：表示从该设备写入的数据块总数。

使用-x参数可以获得更多的统计信息，如下所示。

```
[root@localhost init.d]# iostat -d -x -k 1 10
Linux 2.6.25-14.fc9.i686 (localhost.localdomain)  11/23/2012   _i686_   (1 CPU)

Device:         rrqm/s   wrqm/s     r/s     w/s    rkB/s    wkB/s avgrq-sz avgqu-sz
await r_await w_await svctm  %util
  sda             0.03     6.65    0.25    1.42     2.47    32.28    41.81     0.16
98.24  28.60  110.40   6.41   1.07
  dm-0            0.00     0.00    0.27    8.07     2.46    32.28     8.33     1.35
161.51  34.13  165.77   1.28   1.06
  dm-1            0.00     0.00    0.00    0.00     0.00     0.00     8.00     0.00
53.71  52.81   55.04   7.02   0.00
  sdb             0.02     0.00    0.00    0.00     0.01     0.00    37.26     0.00
16.76  15.00  225.50   9.53   0.00
```

结果中各个的参数的含义如下：

(1) rrqm/s：表示每秒有多少与这个设备相关的读取请求被合并。

(2) wrqm/s：表示每秒有多少与这个设备相关的写入请求被合并。

(3) r/s：表示每秒请求读该设备的数量。

(4) w/s：表示每秒请求写该设备的数量。

(5) await：表示每一个I/O请求的平均处理时间。

(6) %util：表示在统计时间内所有I/O处理时间除以总共统计时间。

此外，还可以通过如下命令查询CPU的部分信息，如下所示。

```
[root@localhost ~]# iostat -c 1 10
avg-cpu:  %user   %nice    %sys %iowait   %idle
           1.98    0.00    0.35   11.45   86.22
avg-cpu:  %user   %nice    %sys %iowait   %idle
           1.62    0.00    0.25   34.46   63.67
```

也可以通过以下命令查询某个具体的设备块的信息，如下所示。

```
[root@localhost ~]# iostat -d -k 1 |grep sda10
Device:            tps    kB_read/s    kB_wrtn/s    kB_read    kB_wrtn
sda10            60.72        18.95        71.53  395637647 1493241908
sda10           299.02      4266.67       129.41       4352        132
sda10           483.84      4589.90      4117.17       4544       4076
sda10           218.00      3360.00       100.00       3360        100
sda10           546.00      8784.00       124.00       8784        124
sda10           827.00     13232.00       136.00      13232        136
```

2. mpstat工具

mpstat是系统实时监控工具，主要报告CPU的一些信息，该命令的语法如下：

```
mpstat [-P {|ALL}] [internal [count]]
```

各参数的含义如下：

（1）-P {|ALL}：表示需要监控哪个CPU。
（2）Internal：表示相邻的两次采样的时间间隔。
（3）Count：表示采样的次数。

下面通过案例来介绍一下该命令的具体用法。

【例16.2】mpstat命令的使用。

（1）显示所有CPU的信息，每秒执行一次。

```
[root@localhost ~]# mpstat -P ALL 1
Linux 2.6.25-14.fc9.i686 (localhost.localdomain)     11/23/2012    _i686_  (1 CPU)

07:49:21 AM
CPU    %usr   %nice    %sys %iowait    %irq   %soft  %steal  %guest   %idle
07:49:22 AM  all    0.99    0.00    0.99    0.00    0.00    0.00    0.00    0.00
98.02
07:49:22 AM    0    0.99    0.00    0.99    0.00    0.00    0.00    0.00    0.00
98.02

07:49:22 AM
CPU    %usr   %nice    %sys %iowait    %irq   %soft  %steal  %guest   %idle
07:49:23 AM  all    3.03    0.00    3.03    0.00    0.00    1.01    0.00    0.00
92.93
07:49:23 AM    0    3.03    0.00    3.03    0.00    0.00    1.01    0.00    0.00
92.93
```

（2）显示ID为0的CPU的信息，每秒执行一次。

```
[root@localhost ~]# mpstat -P 0 1
Linux 2.6.25-14.fc9.i686 (localhost.localdomain) 11/23/2012    _i686_    (1 CPU)

10:42:38 PM  CPU   %user   %nice %system %iowait    %irq   %soft   %idle    intr/s
10:42:43 PM  all    6.89    0.00   44.76    0.10    0.10    0.10   48.05   1121.60
10:42:43 PM    0    9.20    0.00   49.00    0.00    0.00    0.20   41.60    413.00
10:42:43 PM    1    4.60    0.00   40.60    0.00    0.20    0.20   54.60    708.40
```

```
10:42:43 PM   CPU    %user    %nice   %system  %iowait    %irq    %soft    %idle    intr/s
10:42:48 PM   all     7.60     0.00    45.30     0.30     0.00     0.10    46.70   1195.01
10:42:48 PM     0     4.19     0.00     2.20     0.40     0.00     0.00    93.21   1034.53
10:42:48 PM     1    10.78     0.00    88.22     0.40     0.00     0.00     0.20    160.48
Average:      CPU    %user    %nice   %system  %iowait    %irq    %soft    %idle    intr/s
Average:      all     7.25     0.00    45.03     0.20     0.05     0.10    47.38   1158.34
Average:        0     6.69     0.00    25.57     0.20     0.00     0.10    67.43    724.08
Average:        1     7.69     0.00    64.44     0.20     0.10     0.10    27.37    434.17
```

结果中各参数的含义如下：

（1）%user：表示处理用户进程所使用的CPU的百分比。

（2）%nice：表示使用nice命令对进程进行降级时所使用的CPU的百分比。

（3）%sys：表示内核进程使用的CPU百分比。

（4）%iowait：表示等待进行I/O所使用的CPU时间百分比。

（5）%irq：表示用于处理系统中断的CPU百分比。

（6）%soft：表示用于软件中断的CPU百分比。

（7）%idle：显示CPU的空闲时间。

（8）%intr/s：显示CPU每秒接收的中断总数。

3．sar工具

sar是目前Linux最为全面的系统性能分析工具之一，可以从多方面对系统的活动进行报告，包括文件的读写情况、系统调用的使用情况、磁盘I/O、CPU效率、内存使用状况、进程活动及IPC有关的活动等。

sar的语法格式如下：

```
sar [options] [-A] [-o file] t [n]
```

其中，t为采样间隔，n为采样次数，默认值是1；-o file表示将命令结果以二进制格式存放在文件中，file是文件名；options为命令行选项，选项的含义如下：

（1）-A：所有报告的总和。

（2）-u：输出CPU使用情况的统计信息。

（3）-v：输出inode、文件和其他内核表的统计信息。

（4）-d：输出每一个块设备的活动信息。

（5）-r：输出内存和交换空间的统计信息。

（6）-b：显示I/O和传送速率的统计信息。

（7）-a：文件读写情况。

（8）-c：输出进程统计信息，每秒创建的进程数。

（9）-R：输出内存页面的统计信息。

（10）-y：终端设备活动情况。

（11）-w：输出系统交换活动信息。

下面通过一个案例来理解sar工具的使用。

【例16.3】 sar工具的使用。

```
[root@localhost ~]# sar -u 1 5
Linux 2.6.25-14.fc9.i686 (localhost.localdomain)    11/24/2012    _i686_  (1 CPU)

02:33:56 AM       CPU     %user     %nice   %system   %iowait    %steal     %idle
02:33:57 AM       all      2.06      0.00      5.15      0.00      0.00     92.78
02:33:58 AM       all      3.00      0.00      6.00      0.00      0.00     91.00
02:33:59 AM       all     41.41      0.00      9.09      0.00      0.00     49.49
02:34:00 AM       all     96.77      0.00      3.23      0.00      0.00      0.00
```

结果中各参数的含义如下：

（1）CPU：all表示统计信息为所有CPU的平均值。

（2）%user：显示在用户级别（application）运行所占用的CPU总时间的百分比。

（3）%nice：显示在用户级别，用于nice操作所占用的CPU总时间的百分比。

（4）%system：在核心级别（kernel）运行所占用的CPU总时间的百分比。

（5）%iowait：显示用于等待I/O操作所占用的CPU总时间的百分比。

（6）%steal：管理程序（hypervisor）为另一个虚拟进程提供服务而等待虚拟CPU的百分比。

（7）%idle：显示CPU空闲时间占用CPU总时间的百分比。

> **提示** 如果%iowait的值过高，则表示硬盘存在I/O瓶颈；如果%idle的值高但系统响应慢，有可能是CPU在等待分配内存，此时应加大内存容量；如果%idle 的值持续低于10，则表示系统的CPU处理能力相对较低，表明系统中最需要解决的资源是CPU。

16.2 开源监控利器Nagios实战

Nagios是一个用来监控主机、服务和网络的开源软件。在实际工作中，需要监控的对象主要是主机资源监控和网络服务监控。主机资源监控包括系统负载、当前IP链接数、磁盘空间使用情况、当前进程数以及自定义的资源监控等；网络服务监控包括主机存活检查、Web服务监控、FTP服务监控、数据库服务监控以及自定义服务器监控等。

16.2.1 安装 Nagios 之前的准备工作

在安装Nagios之前，需要做以下必要的工作：

（1）使用Fedora操作系统，并且拥有root使用权限。

（2）在root用户下面安装Apache服务器。Apache服务器的安装步骤如下：

步骤01 下载httpd-2.2.20.tar.gz安装包，解压并复制到/user/src/目录下面。

```
[root@localhost ~]# mv httpd-2.2.0.tar.gz /usr/src
[root@localhost ~]# gunzip httpd-2.2.0.tar.gz
[root@localhost ~]# tar -xvf httpd-2.2.0.tar
[root@localhost ~]# cd /usr/src
[root@localhost src]# ls
httpd-2.2.0   httpd-2.2.0.tar   redhat
[root@localhost src]# cd httpd-2.2.0
[root@localhost httpd-2.2.0]# ls
ABOUT_APACHE         configure          LAYOUT            README
acinclude.m4         configure.in       libhttpd.dsp      README.platforms
Apache.dsw           docs               LICENSE           ROADMAP
apachenw.mcp.zip     emacs-style        Makefile.in       server
build                httpd.dsp          Makefile.win      srclib
BuildBin.dsp         httpd.spec         modules           support
buildconf            include            NOTICE            test
CHANGES              INSTALL            NWGNUmakefile     VERSIONING
config.layout        InstallBin.dsp     os

[root@localhost httpd-2.2.0]# ./configure --prefix=/usr/local/apache
[root@localhost httpd-2.2.0]# make
[root@localhost httpd-2.2.0]# make install
```

步骤02 执行/usr/local/apache/bin/apachetl -t命令,检测Apache的配置文件语法是否正确:

```
[root@localhost ~]# cd /usr/local/apache/bin/
[root@localhost bin]# ls
ab              apu-1-config    dbmmanage       htcacheclean    htpasswd        logresolve
apachectl       apxs            envvars         htdbm           httpd           rotatelogs
apr-1-config    checkgid        envvars-std     htdigest        httxt2dbm
[root@localhost bin]# ./apachectl -t
Syntax OK
```

步骤03 在/etc/profile文件中修改PATH路径,直接把/usr/local/apache/bin的路径加入环境变量PATH中,以方便执行Apache的命令。

```
[root@localhost ~]# vi /etc/profile
```

步骤04 在该文件的最后直接添加下面两句设置环境变量的语句:

```
PATH=$PATH:/usr/local/apache/bin
export PATH
```

这样可以直接运行apachetl start命令,从而启动apache服务,如下所示。

```
[root@localhost ~]#apachetl start
```

还可以在浏览器中输入"http://ip地址",测试Apache服务是否成功启动。

步骤05 将Apache服务设定为系统启动服务,命令如下:

```
[root@localhost ~]#chkconfig httpd on
```

16.2.2 安装 Nagios 主程序

用户可以在http://www.nagios.org页面中下载Nagios，然后进行安装。下面以安装nagios-3.4.1.tar.gz为例进行讲解，操作步骤如下：

步骤 01 解压nagios-3.4.1.tar.gz，命令如下：

```
[root@localhost ~]# useradd nagios
[root@localhost ~]# mkdir /usr/local/nagios
[root@localhost ~]# chown nagios.nagios /usr/local/nagios
[root@localhost ~]# ls
anaconda-ks.cfg                      MySQL Cluster.txt
Desktop                              nagios
Documents                            nagios-3.4.1.tar.gz
Download                             network.txt
install.log                          node.txt
install.log.syslog                   Pictures
libaio-0.3.96-3.i386.rpm             Public
Music                                Templates
my.cnf.bak                           Unsaved Document 1
Mysql5.5                             Videos
mysql-cluster-gpl-7.2.8-linux2.6-i686.tar
[root@localhost ~]# tar zxvf nagios-3.4.1.tar.gz
```

步骤 02 开始进行编译，命令如下：

```
[root@localhost ~]# cd nagios
[root@localhost nagios]# ./configure --prefix=/usr/local/nagios
[root@localhost nagios]# make all
```

步骤 03 如果在编译过程中没有发生错误，接下来就进行安装，如下所示。

```
[root@localhost nagios]# make install
[root@localhost nagios]# make install-config
[root@localhost nagios]# make install-init
[root@localhost nagios]# make install-commandmode
[root@localhost nagios]# ls /usr/local/nagios
bin  etc  libexec  sbin  share  var
```

Nagios服务器安装完成后，在安装目录/usr/local/nagios下生成如表16.1所示的目录。

表 16.1 Nagios 服务器的目录

目录	作用
bin	Nagios 执行程序所在的目录，这个目录只有一个文件 nagios
etc	Nagios 配置文件位置，初始安装完后，只有几个*.cfg-sample 文件
sbin	Nagios Cgi 文件所在的目录，也就是执行外部命令所需文件所在的目录
share	Nagios 网页文件所在的目录
var	Nagios 日志文件、spid 等文件所在的目录

步骤 04 将nagios的启动脚本添加到系统服务当中，命令如下：

```
[root@localhost ~]# vi /etc/profile
```

步骤 05 在/etc/profile文件的最后直接添加下面两句设置环境变量的语句：

```
PATH=$PATH:/usr/local/nagios/bin
export PATH
[root@localhost ~]# chkconfig –add nagios
```

步骤 06 将nagios添加到系统服务中，命令如下：

```
[root@localhost ~]# chkconfig nagios on
```

16.2.3 整合 Nagios 到 Apache 服务

Nagios安装完成后，即可添加到Apache服务中，具体操作步骤如下：

步骤 01 将Apache服务宿主用户加入Nagios组里面，这样做是为了让Apache有适当的权限通过CGI命令对Nagios进行调用，例如通过浏览器界面关闭Nagios服务。

```
[root@localhost ~]# cd /usr/local/nagios/
[root@localhost nagios]# usermode -G nagios apache
[root@localhost nagios]# usermod -G nagios apache
```

步骤 02 Apache的运行用户和运行组默认是daemon，因此打开/etc/httpd/conf/httpd.conf文件，把Apache的运行用户和运行组改成Nagios。

```
User nagios
Group nagios
```

步骤 03 将Nagios的信息加入Apache中，打开/etc/httpd/conf/httpd.conf文件，在文件最后添加如下代码：

```
#setting for nagios
ScriptAlias /nagios/cig-bin /usr/local/nagios/sbin
<Directory "/usr/local/nagios/sbin">
    AuthType Basic
    Options ExecCGI
    AllowOverride None
    Order allow,deny
    Allow from all
    AuthName "Nagios Access"
    AuthUserFile /usr/local/nagios/etc/htpasswd
    Require valid-user
</Directory>

Alias /nagios /usr/local/nagios/share
<Directory "/usr/local/nagios/share">
    AuthType Basic
    Options None
    AllowOverride None
    Order allow,deny
```

```
        Allow from all
        AuthName "nagios Access"
        AuthUserFile /usr/local/nagios/etc/htpasswd
        Require valid-usr
</Directory>
```

上述文本的作用主要是对Nagios目录进行用户验证，只有合法的用户才能访问Nagios的页面。配置完成之后，可以执行apachctl -t检测Apache配置文件的语法是否有错误。

```
[root@localhost ~]# apachctl -t
Syntax OK
```

步骤 04 创建Web接口，命令如下：

```
[root@localhost ~]# cd nagios
[root@localhost nagios]# make install-webconf
/usr/bin/install -c -m 644 sample-config/httpd.conf /etc/httpd/conf.d/nagios.conf

*** Nagios/Apache conf file installed ***
```

步骤 05 创建一个用于登录Nagios的用户名和密码，命令如下：

```
[root@localhost ~]# /usr/bin/htpasswd -c /usr/local/nagios/etc/htpasswd.users nagios
New password:
Re-type new password:
Adding password for user nagios
```

步骤 06 重启Apache服务，重启Nagios服务，命令如下：

```
[root@localhost ~]# httpd -k restart
[root@localhost ~]# service nagios restart
Running configuration check...done.
Stopping nagios: done.
Starting nagios: done.
```

步骤 07 在另外一台计算机上登录http://ip/nagios，此时会要求输入上面创建的用户名和密码，输入正确的用户名和密码之后，登录到Nagios服务器主页面，如图16.1所示。

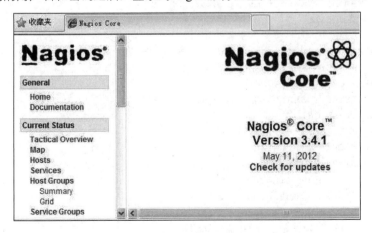

图 16.1　Nagios 服务器主页面

配置Nagios的时候，经常会遇到Internal Server Error错误：

```
Internal Server Error
The server encountered an internal error or misconfiguration and was unable to
complete your request.
    Please contact the server administrator, root@localhost and inform them of the time
the error occurred, and anything you might have done that may have caused the error.
    More information about this error may be available in the server error log.
```

此时查看Apache服务器的错误日志信息，如下所示。

```
[root@localhost ~]# vi /var/log/httpd/error_log
[Tue Nov 20 09:33:30 2012] [error] [client 192.168.0.208] (13)Permission denied:
exec of '/usr/local/nagios/sbin/status.cgi' failed, referer:
http://192.168.0.100/nagios/side.php
    [Tue Nov 20 09:33:30 2012] [error] [client 192.168.0.208] Premature end of script
headers: status.cgi, referer: http://192.168.0.100/nagios/side.php
    [Tue Nov 20 09:34:49 2012] [error] [client 192.168.0.208] PHP Warning: MagpieRSS:
Failed to fetch http://www.nagios.org/backend/feeds/frontpage/ and cache is off in
/usr/local/nagios/share/includes/rss/rss_fetch.inc on line 238, referer:
http://192.168.0.100/nagios/main.php
```

错误的含义是Apache开启了suexec的功能，对执行CGI的路径进行了限制。先检测一下计算机的Apache是否打开了suexec功能，如下所示。

```
[root@localhost sysconfig]# httpd -V
Server version: Apache/2.2.8 (Unix)
Server built:   Feb 25 2008 07:05:32
Server's Module Magic Number: 20051115:11
Server loaded:  APR 1.2.12, APR-Util 1.2.12
Compiled using: APR 1.2.12, APR-Util 1.2.12
Architecture:   32-bit
Server MPM:     Prefork
  threaded:     no
    forked:     yes (variable process count)
Server compiled with....
 -D APACHE_MPM_DIR="server/mpm/prefork"
 -D APR_HAS_SENDFILE
 -D APR_HAS_MMAP
 -D APR_HAVE_IPV6 (IPv4-mapped addresses enabled)
 -D APR_USE_SYSVSEM_SERIALIZE
 -D APR_USE_PTHREAD_SERIALIZE
 -D SINGLE_LISTEN_UNSERIALIZED_ACCEPT
 -D APR_HAS_OTHER_CHILD
 -D AP_HAVE_RELIABLE_PIPED_LOGS
 -D DYNAMIC_MODULE_LIMIT=128
 -D HTTPD_ROOT="/etc/httpd"
 -D SUEXEC_BIN="/usr/sbin/suexec"
 -D DEFAULT_PIDLOG="logs/httpd.pid"
 -D DEFAULT_SCOREBOARD="logs/apache_runtime_status"
 -D DEFAULT_LOCKFILE="logs/accept.lock"
 -D DEFAULT_ERRORLOG="logs/error_log"
```

```
-D AP_TYPES_CONFIG_FILE="conf/mime.types"
-D SERVER_CONFIG_FILE="conf/httpd.conf"
```

从结果中可以看出,Apache已经启用了suexec功能。执行下列命令,可以知道Apache如何限制CGI执行的路径:

```
[root@localhost sysconfig]# suexec -V
 -D AP_DOC_ROOT="/var/www"
 -D AP_GID_MIN=100
 -D AP_HTTPD_USER="apache"
 -D AP_LOG_EXEC="/var/log/httpd/suexec.log"
 -D AP_SAFE_PATH="/usr/local/bin:/usr/bin:/bin"
 -D AP_UID_MIN=500
 -D AP_USERDIR_SUFFIX="public_html"
```

CGI程序只有在/var/www/目录下才能正常运行,因为将Nagios安装在了/usr/local/nagios目录。最简单的解决方法就是将Nagios安装到/var/www/nagios。值得注意的是,编译的时候需要指定prefix=/var/www/nagios,其他编译安装选项不用改变。

如果不想重新编译,可以将/usr/local目录下面的nagios移动到/var/www目录下面,然后修改相关的配置文件。值得注意的是,nagios目录下面所有的cfg的路径都需要修改,因此建议读者重新编译安装,etc/profile环境变量也重新设置。笔者在这里重新安装一遍Nagios,上述问题就都解决了。

16.2.4 安装 Nagios 插件包

安装Nagios插件包的具体操作步骤如下:

步骤 01 下载Nagios插件包,然后解压缩,命令如下:

```
[root@localhost ~]# tar -zxvf nagios-plugins-1.4.16.tar.gz
[root@localhost ~]# cd nagios-plugins-1.4.16
```

步骤 02 编译并且开始安装Nagios插件,命令如下:

```
[root@localhost nagios-plugins-1.4.16]# ./configure --with-nagios-user=nagios
--with-nagios-group=nagios -prefix=/var/www/nagios
[root@localhost nagios-plugins-1.4.16]# make
[root@localhost nagios-plugins-1.4.16]# make install
```

步骤 03 验证Nagios的样例配置文件,命令如下:

```
[root@localhost nagios-plugins-1.4.16]# /var/www/nagios/bin/nagios -v
/var/www/nagios/etc/nagios.cfg

Nagios Core 3.4.1
Copyright (c) 2009-2011 Nagios Core Development Team and Community Contributors
Copyright (c) 1999-2009 Ethan Galstad
Last Modified: 05-11-2012
License: GPL

Website: http://www.nagios.org
Reading configuration data...
```

```
    Read main config file okay...
Processing object config file '/var/www/nagios/etc/objects/commands.cfg'...
Processing object config file '/var/www/nagios/etc/objects/contacts.cfg'...
Processing object config file '/var/www/nagios/etc/objects/timeperiods.cfg'...
Processing object config file '/var/www/nagios/etc/objects/templates.cfg'...
Processing object config file '/var/www/nagios/etc/objects/localhost.cfg'...
    Read object config files okay...

Running pre-flight check on configuration data...

Checking services...
    Checked 8 services.
Checking hosts...
    Checked 1 hosts.
Checking host groups...
    Checked 1 host groups.
Checking service groups...
    Checked 0 service groups.
Checking contacts...
    Checked 1 contacts.
Checking contact groups...
    Checked 1 contact groups.
Checking service escalations...
    Checked 0 service escalations.
Checking service dependencies...
    Checked 0 service dependencies.
Checking host escalations...
    Checked 0 host escalations.
Checking host dependencies...
    Checked 0 host dependencies.
Checking commands...
    Checked 24 commands.
Checking time periods...
    Checked 5 time periods.
Checking for circular paths between hosts...
Checking for circular host and service dependencies...
Checking global event handlers...
Checking obsessive compulsive processor commands...
Checking misc settings...

Total Warnings: 0
Total Errors:   0

Things look okay - No serious problems were detected during the pre-flight check
[root@localhost nagios-plugins-1.4.16]#
```

步骤 04 重新启动Nagios服务和Apache服务，命令如下：

```
[root@localhost nagios-plugins-1.4.16]# service nagios restart
Running configuration check...done.
Stopping nagios: done.
Starting nagios: done.
[root@localhost nagios-plugins-1.4.16]# service httpd restart
```

```
Stopping httpd:                                            [ OK ]
Starting httpd:                                            [ OK ]
```

16.2.5 监控服务器的 CPU、负载、磁盘 I/O 使用情况

在监控服务器之前，需要在服务器中安装NRPE（Nagios Remote Plugin Executor），配置Nagios以及使用htpasswd创建浏览器验证账号。具体操作步骤如下：

步骤 01 在MySQL服务器上安装NRPE，命令如下：

```
[root@localhost ~]# tar -zxvf nrpe-2.12.tar.gz
[root@localhost ~]# tar -zxvf nrpe-2.12.tar.gz
[root@localhost nrpe-2.12]# ./configure --prefix=/var/www/nagios
[root@localhost nrpe-2.12]# make all
[root@localhost nrpe-2.12]# make install-plugin
[root@localhost nrpe-2.12]# make install-daemon
[root@localhost nrpe-2.12]# make install-daemon-config
[root@localhost nrpe-2.12]# make install-xinetd
```

步骤 02 可以查看和修改NRPE的一些系统信息，比如可以通过以下操作修改端口为5666。

```
[root@localhost ~]# vi /etc/xinetd.d/nrpe
# description: NRPE (Nagios Remote Plugin Executor)
service nrpe
{
        flags           = REUSE
        socket_type     = stream
        port            = 5666
        wait            = no
        user            = nagios
        group           = nagios
        server          = /var/www/nagios/bin/nrpe
        server_args     = -c /var/www/nagios/etc/nrpe.cfg --inetd
        log_on_failure  += USERID
        disable         = no
        only_from       = 127.0.0.1
}
```

步骤 03 在/var/www/nagios/etc/objects/command.cfg命令定义文件中添加NRPE命令，如下所示：

```
[root@localhost ~]# vi /var/www/nagios/etc/objects/command.cfg
define command{
 command_name   check_nrpe
 command_line   $USER1$/check_nrpe -H $HOSTADDRESS$ -c $ARG1$
 }
```

这里有几点需要说明：

- 该命令的名字叫作check_nrpe。
- $USER1$/check_nrpe会通过应用resource.cfg来获得$USER1$变量的路径，从而获得/var/www/nagios/libexec/check_nrpe这个绝对路径。

- -H $HOSTADDRESS$ 用来获得指定被检测主机的IP地址。
- -c $ARG1$用来指定被检测主机上NRPE守护进程运行着的NRPE命令名。

步骤 04 修改客户端配置，命令如下：

```
vi /etc/nagios/nrpe.cfg
#下列配置表示允许127.0.0.1和192.168.121.55（Server）这两台机器访问当前机器的信息
allowed_hosts=127.0.0.1,192.168.121.55
```

步骤 05 测试客户端配置情况，如果没有启动NRPE服务，会提示"Connection refused by host"信息。

```
#启动NRPE
[root@localhost ~]# cd /var/www/nagios/bin
[root@localhost bin]# ./nrpe -c /var/www/nagios/etc/nrpe.cfg -d
[root@localhost bin]# netstat -an|grep 5666
tcp        0      0 0.0.0.0:5666            0.0.0.0:*               LISTEN
unix  3      [ ]         STREAM     CONNECTED     15666
/tmp/orbit-root/linc-aad-0-6c80032044a1a
```

步骤 06 接着测试是否可以监控客户端，命令如下：

```
[root@localhost ~]# cd /var/www/nagios/libexec
[root@localhost libexec]# ./check_nrpe -H localhost -c check_load
OK - load average: 0.06, 0.03, 0.12|load1=0.060;15.000;30.000;0;
load5=0.030;10.000;25.000;0; load15=0.120;5.000;20.000;0;
```

步骤 07 在Nagios监控服务器上配置服务端的信息，命令如下：

```
# 创建客户端配置（192.168.0.100为客户端IP）
[root@localhost ~]# vi  /var/www/nagios/etc/objects/192.168.0.100.cfg
define host{
        use linux-server
    host_name 127.0.0.1
alias client-1
        address 127.0.0.1
}

define service{
        use generic-service
host_name 127.0.0.1
    service_description check_disk2
    check_command check_nrpe!check_sda2
}

define service{
    use generic-service
    host_name 127.0.0.1
    service_description check_load
    check_command check_nrpe!check_load
}

define service{
    use generic-service
```

```
        host_name 127.0.0.1
    service_description check_swap
    check_command check_nrpe!check_swap
}
```

步骤 08 将上面配置添加到系统中，命令如下：

```
[root@localhost ~]# vi /var/www/nagios/etc/nagios.cfg
cfg_file=/etc/nagios/objects/192.168.0.100.cfg
```

步骤 09 重新启动Nagios服务和Apache服务，命令如下：

```
[root@localhost nagios-plugins-1.4.16]# service nagios restart
Running configuration check...done.
Stopping nagios: done.
Starting nagios: done.
[root@localhost nagios-plugins-1.4.16]# service httpd restart
Stopping httpd:                                            [ OK ]
Starting httpd:                                            [ OK ]
```

> **提示** 如果Nagios Web界面提示如下错误信息：
>
> ```
> It appears as though you do not have permission to view information for any
> of the services you requested...
> ```

可以打开cgi.cfg配置文件，里面有个参数：

```
use_authentication=1
```

为了保障系统的安全性，Nagios设置了这个参数，默认为1，将它改为0即可解决上面的问题。

打开Nagios页面左侧的【Services】，页面右侧中将会显示检测到的所有服务项，如图16.2所示。

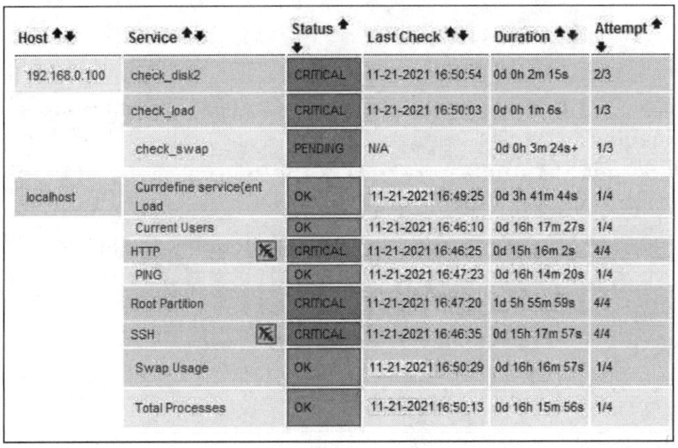

Host	Service	Status	Last Check	Duration	Attempt
192.168.0.100	check_disk2	CRITICAL	11-21-2021 16:50:54	0d 0h 2m 15s	2/3
	check_load	CRITICAL	11-21-2021 16:50:03	0d 0h 1m 6s	1/3
	check_swap	PENDING	N/A	0d 0h 3m 24s+	1/3
localhost	Currdefine service(ent Load	OK	11-21-2021 16:49:25	0d 3h 41m 44s	1/4
	Current Users	OK	11-21-2021 16:46:10	0d 16h 17m 27s	1/4
	HTTP	CRITICAL	11-21-2021 16:46:25	0d 15h 16m 2s	4/4
	PING	OK	11-21-2021 16:47:23	0d 16h 14m 20s	1/4
	Root Partition	CRITICAL	11-21-2021 16:47:20	1d 5h 55m 59s	4/4
	SSH	CRITICAL	11-21-2021 16:46:35	0d 15h 17m 57s	4/4
	Swap Usage	OK	11-21-2021 16:50:29	0d 16h 16m 57s	1/4
	Total Processes	OK	11-21-2021 16:50:13	0d 16h 15m 56s	1/4

图16.2　检测到的所有服务项

如果提示以下错误信息：

```
CHECK_NRPE: Error - Could not complete SSL handshake
```

可以使用下面的步骤解决。

步骤 01 检测NRPE机器上的防火墙有没有启动，如果启动，则要把5666端口打开。也可以直接关闭防火墙，命令如下：

```
[root@localhost ~]# service iptables stop
```

步骤 02 看看NRPE编译是否支持SSL（NRPE编译默认支持SSL）。另外，OpenSSL和openssl-devel也都要装上，命令如下：

```
[root@localhost ~]# tar -zxvf openssl-0.9.81.tar.gz
[root@localhost openssl-0.9.81]# ./config --prefix=/usr/local/ssl-0.9.81 shared zlib-dynamic enable-camellia
[root@localhost openssl-0.9.81]# make depend
[root@localhost openssl-0.9.81]# make
[root@localhost openssl-0.9.81]# make test
[root@localhost openssl-0.9.81]# make install
```

步骤 03 建立连接，命令如下：

```
[root@localhost openssl-0.9.81]# cd /usr/local/
[root@localhost local]# ln -s ssl-0.9.81 ssl
[root@localhost local]# vi /etc/ld.so.conf
include ld.so.conf.d/*.conf
/usr/local/ssl/lib

[root@localhost local]# ldconfig
```

步骤 04 配置环境变量，在/etc/profile文件的末尾添加以下信息：

```
[root@localhost local]# vi /etc/profile
PATH=$PATH:/var/local/ssl/bin
export PATH
```

步骤 05 测试SSL是否安装成功，命令如下：

```
[root@localhost ~]# cd /usr/local/
[root@localhost local]# ls
apache  doc  games    lib     man     mysql-cluster  share  ssl
bin     etc  include  libexec mysql   sbin           src    ssl-0.9.81
[root@localhost local]# ldd /usr/local/ssl/bin/openssl
        linux-gate.so.1 =>  (0x00110000)
        libssl.so.0.9.8 => /usr/local/ssl-0.9.81/lib/libssl.so.0.9.8 (0x00111000)
        libcrypto.so.0.9.8 => /usr/local/ssl-0.9.81/lib/libcrypto.so.0.9.8 (0x00157000)
        libdl.so.2 => /lib/libdl.so.2 (0x0090c000)
        libc.so.6 => /lib/libc.so.6 (0x00776000)
        /lib/ld-linux.so.2 (0x00756000)
[root@localhost local]# which openssl
/usr/bin/openssl
```

步骤 06 如果错误出现，有可能是配置文件有问题，需要修改nrpe.cfg文件，把allowed_hosts=127.0.0.1修改为allowed_hosts=127.0.0.1,192.168.0.100，然后重启其NRPE服务。

```
[root@localhost lib]# ps -ef|grep nrpe
nagios     6591     1  0 Nov20 ?        00:00:03 ./nrpe -c
/var/www/nagios/etc/nrpe.cfg -d
root      13152  1932  0 Nov21 pts/2    00:00:00 find / nrpe
root      20793 20118  0 11:14 pts/3    00:00:00 grep nrpe
[root@localhost lib]# kill -9 6591
[root@localhost lib]# cd
[root@localhost var]# cd /var/www/nagios/bin/
[root@localhost bin]# ./nrpe -c /var/www/nagios/etc/nrpe.cfg -d
```

步骤 07 如果还有错误，那么可以参照以下解决方法。

（1）确认check_nrpe和NRPE daemon的版本一致。

（2）确认check_nrpe和NRPE deamon同时启用或者禁用SSL支持。

（3）确认nrep.cfg可以被NRPE用户正常读取。

16.2.6 配置 Nagios 监控 MySQL 服务器

下面讲述通过配置Nagios实现监控MySQL服务器的方法。具体操作步骤如下：

步骤 01 首先建立nagios数据库，创建nagios数据库用户，授予nagios用户权限，命令如下：

```
[root@localhost ~]# mysql -uroot -p
Enter password:
Welcome to the MySQL monitor.  Commands end with ; or \g.
Your MySQL connection id is 3
Server version: 5.5.27-ndb-7.2.8-cluster-gpl-log MySQL Cluster Community Server
(GPL)

Copyright (c) 2000, 2011, Oracle and/or its affiliates. All rights reserved.

Oracle is a registered trademark of Oracle Corporation and/or its
affiliates. Other names may be trademarks of their respective
owners.

Type 'help;' or '\h' for help. Type '\c' to clear the current input statement.

mysql> create database nagios;
Query OK, 1 row affected (0.29 sec)

mysql> grant select on nagios.* to nagios@'%' identified by 'password';
Query OK, 0 rows affected (0.31 sec)
```

步骤 02 为了方便本地测试Nagios服务，授权给nagios@'localhost'，命令如下：

```
mysql> grant select on nagios.* to nagios@'localhost' identified by 'password';
Query OK, 0 rows affected (0.00 sec)

mysql> select User,Password,HOst from mysql.user;
+--------+-------------------------------------------+---------------------+
| User   | Password                                  | HOst                |
+--------+-------------------------------------------+---------------------+
| root   |                                           | localhost           |
| root   |                                           | localhost.localdomain |
```

```
|    root    |                                                      | 127.0.0.1             |
|    root    |                                                      | ::1                   |
|            |                                                      | localhost             |
|            |                                                      | localhost.localdomain |
| nagios     | *2470C0C06DEE42FD1618BB99005ADCA2EC9D1E19             | %                     |
| nagios     | *2470C0C06DEE42FD1618BB99005ADCA2EC9D1E19             | localhost             |
+------------+------------------------------------------------------+-----------------------+
8 rows in set (0.00 sec)
```

步骤 03 执行check_mysql命令,检测Nagios是否能够连接MySQL服务,命令如下:

```
[root@localhost ~]# cd /var/www/nagios/libexec/
[root@localhost libexec]# ./check_mysql -u nagios -d nagios -p password
./check_mysql: error while loading shared libraries: libmysqlclient.so.18: cannot open shared object file: No such file or directory
```

步骤 04 此时,发现加载libmysqlclient.so.18发生错误,需要建立一个链接,命令如下:

```
[root@localhost usr]# cd /usr/local/mysql/lib/
[root@localhost lib]# ls
libmysqlclient.a            libmysqlclient.so.18          libndbclient.so.6.0.0
libmysqlclient_r.a          libmysqlclient.so.18.0.0      libndbclient_static.a
libmysqlclient_r.so         libmysqld.a                   libtcmalloc_minimal.so
libmysqlclient_r.so.18      libmysqld-debug.a             ndb_engine.so
libmysqlclient_r.so.18.0.0  libmysqlservices.a            plugin
libmysqlclient.so           libndbclient.so
[root@localhost lib]# ln -s /usr/local/mysql/lib/libmysqlclient.so.18 /usr/lib/libmysqlclient.so.18
```

步骤 05 再次执行check_mysql命令,检测Nagios是否能够连接MySQL服务,命令如下:

```
[root@localhost ~]# cd /var/www/nagios/libexec/
[root@localhost libexec]# ./check_mysql -H 192.168.0.100 -u nagios -d nagios -p password
Uptime: 73815  Threads: 2  Questions: 16  Slow queries: 0  Opens: 33  Flush tables: 1  Open tables: 26  Queries per second avg: 0.000
```

步骤 06 打开Nagios配置文件,命令如下:

```
[root@localhost ~]# cd /var/www/nagios/objects/
[root@localhost objects]# vi commands.cfg
```

步骤 07 在commands.cfg文件的末尾添加如下代码:

```
#check_mysql
define command{
    command_name check_mysql
    command_line     $USER1$/check_mysql -H $HOSTADDRESS$ -u nagios -d nagios -p password
}
```

步骤 08 在192.168.0.100.cfg文件末尾追加check_mysql服务,命令如下:

```
[root@localhost ~]# vi  /var/www/nagios/etc/objects/192.168.0.100.cfg
define service{
```

```
    use generic-service
    host_name 192.168.0.100
    service_description MySQL
    normal_check_interval 1
    retry_check_interval 1
    check_command check_mysql
}
```

步骤09 打开Nagios页面左侧的【Services】，然后选择【MySQL】，可以看到如图16.3所示的信息。

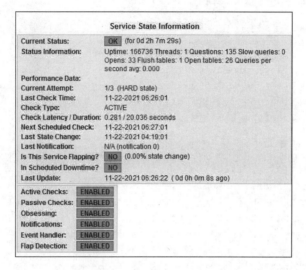

图 16.3　检测 MySQL 的状态信息

16.3　MySQL监控利器Cacti实战

本节讲述Cacti工具的使用方法。

16.3.1　Cacti 工具的安装

Cacti是一套基于PHP、MySQL、SNMP以及RRDTool开发的网络流量监测分析工具，它通过snmpget获得数据，使用RRDTool绘制图形。

在安装Cacti工具之前，需要安装MySQL服务器、Apache服务器和PHP。在安装PHP之前，还需要安装zlib、libpng、freetype、jpegsrc和Fontconfig，以便PHP能够支持GD库。GD库的下载地址是http://oss.oetiker.ch/rrdtool/pub/libs/。

具体操作步骤如下：

步骤01 安装zlib，命令如下：

```
[root@localhost ~]# tar zxvf zlib-1.2.3.tar.gz
[root@localhost ~]# cd zlib-1.2.3
```

步骤 02 安装libpng，命令如下：

```
[root@localhost zlib-1.2.3]# ./configure --prefix=/usr/local/zlib
[root@localhost zlib-1.2.3]# make
[root@localhost zlib-1.2.3]# make install
```

步骤 02 安装libpng，命令如下：

```
[root@localhost ~]# tar zxvf libpng-1.2.18.tar.gz
[root@localhost scripts]# mv makefile.linux ../makefile
[root@localhost scripts]# cd ..
[root@localhost libpng-1.2.18]# make
[root@localhost libpng-1.2.18]# make install
```

步骤 03 安装freetype，命令如下：

```
[root@localhost ~]# tar zxvf freetype-2.3.5.tar.gz
[root@localhost ~]# cd freetype-2.3.5
[root@localhost freetype-2.3.5]# ./configure --prefix=/usr/local/freetype
[root@localhost freetype-2.3.5]# make
[root@localhost freetype-2.3.5]# make install
```

步骤 04 安装jpegsrc，命令如下：

```
[root@localhost ~]# tar -zxvf jpegsrc.v6b.tar.gz
[root@localhost ~]# cd jpeg-6b/
[root@localhost jpeg-6b]# mkdir /usr/local/libjpeg
[root@localhost jpeg-6b]# mkdir /usr/local/libjpeg/include
[root@localhost jpeg-6b]# mkdir /usr/local/libjpeg/bin
[root@localhost jpeg-6b]# mkdir /usr/local/libjpeg/lib
[root@localhost jpeg-6b]# mkdir /usr/local/libjpeg/man
[root@localhost jpeg-6b]# mkdir /usr/local/libjpeg/man/man1
[root@localhost jpeg-6b]# ./configure --prefix=/usr/local/libjpeg --enable-shared --enable-static
[root@localhost jpeg-6b]# make
[root@localhost jpeg-6b]# make install
```

步骤 05 安装Fontconfig，命令如下：

```
[root@localhost ~]# tar -zxvf fontconfig-2.4.2.tar.gz
[root@localhost fontconfig-2.4.2]# ./configure --with-freetype-config=/usr/local/freetype
[root@localhost fontconfig-2.4.2]# make
[root@localhost fontconfig-2.4.2]# make install
```

步骤 06 安装GD，命令如下：

```
[root@localhost ~]#  tar -zxvf gd-2.0.35.tar.gz
[root@localhost ~]# cd gd-2.0.35
[root@localhost gd-2.0.35]# ./configure --prefix=/usr/local/libgd --with-png --with-freetype=/usr/local/freetype --with-jpeg=/usr/local/libjpeg
[root@localhost gd-2.0.35]# make
[root@localhost gd-2.0.35]# make install
```

步骤 07 安装PHP，命令如下：

```
[root@localhost ~]# tar -zxvf  php-5.2.3.tar.gz
```

```
[root@localhost ~]# cd php-5.2.3
[root@localhost php-5.2.3]# ./configure --prefix=/usr/local/php
--with-apxs2=/usr/local/apache/bin/apxs --with-mysql=/usr/local/mysql
--with-gd=/usr/local/libgd --enable-gd-native-ttf --with-ttf --enable-gd-jis-conv
--with-freetype-dir=/usr/local/freetype --with-jpeg-dir=/usr/local/libjpeg
--with-png-dir=/usr --with-zlib-dir=/usr/local/zlib --enable-xml --enable-mbstring
--enable-sockets
[root@localhost php-5.2.3]# make
[root@localhost php-5.2.3]# make install
[root@localhost php-5.2.3]# cp php.ini-recommended /usr/local/php/lib/php.ini
[root@localhost php-5.2.3]# ln -s /usr/local/php/bin/* /usr/local/bin/
```

步骤08 修改httpd.conf文件，命令如下：

```
[root@localhost ~]# vi /usr/local/apache/conf/httpd.conf
```

步骤09 查找AddType application/x-compress .Z，命令如下：

```
AddType application/x-gzip .gz .tgz
```

在其下加入AddType application/x-tar .tgz，命令如下：

```
AddType application/x-httpd-php .php
AddType image/x-icon .ico
```

步骤10 修改DirectoryIndex行，添加index.php，命令如下：

```
DirectoryIndex index.php index.html index.html.var
```

步骤11 创建test.php，测试是否成功安装PHP，命令如下：

```
[root@localhost ~]# vi /usr/local/apache/htdocs/test.php
```

在test.php文件末尾添加以下行，添加完成以后，按WQ键保存并退出。

```
<?php
phpinfo();
?>
```

步骤12 重启Apache服务器，命令如下：

```
[root@localhost ~]# /usr/local/apache/bin/httpd -k stop
[root@localhost ~]# /usr/local/apache/bin/httpd -k start
```

在浏览器中输入"http://192.168.0.100/test.php"进行PHP测试，确保PHP工作正常。

步骤13 下面开始安装RRDTool。安装rrdtool-1.2.23需要一些库文件的支持，需要将cgilib-0.5.tar.gz、zlib-1.2.3.tar.gz、libpng-1.2.18.tar.gz、freetype-2.3.5.tar.gz、libart_lgpl-2.3.17.tar.gz、rrdtool-1.2.23.tar.gz放到/root/rrdtool-1.2.23目录下，将脚本保存为"/root/rrdtool-1.2.23/rrdtoolinstall.sh"，并授予执行权限chmod u+x /root/rrdtool-1.2.23/rrdtoolinstall.sh。rrdtoolinstall.sh脚本信息如下：

```
#!/bin/sh
BUILD_DIR=/root/rrdtool-1.2.11
INSTALL_DIR=/usr/local/rrdtool
cd $BUILD_DIR
tar zxf cgilib-0.5.tar.gz
```

```
cd cgilib-0.5
make CC=gcc CFLAGS="-O3 -fPIC -I."
mkdir -p $BUILD_DIR/lb/include
cp *.h $BUILD_DIR/lb/include
mkdir -p $BUILD_DIR/lb/lib
cp libcgi* $BUILD_DIR/lb/lib
cd $BUILD_DIR
tar zxf zlib-1.2.2.tar.gz
cd zlib-1.2.2
env CFLAGS="-O3 -fPIC" ./configure --prefix=$BUILD_DIR/lb
make
make install
cd $BUILD_DIR
tar zxvf libpng-1.2.8-config.tar.gz
cd libpng-1.2.8-config
env CPPFLAGS="-I$BUILD_DIR/lb/include" LDFLAGS="-L$BUILD_DIR/lb/lib" CFLAGS="-O3 -fPIC" \
    ./configure --disable-shared --prefix=$BUILD_DIR/lb
make
make install
cd $BUILD_DIR
tar zxvf freetype-2.1.9.tar.gz
cd freetype-2.1.9
env CPPFLAGS="-I$BUILD_DIR/lb/include" LDFLAGS="-L$BUILD_DIR/lb/lib" CFLAGS="-O3 -fPIC" \
    ./configure --disable-shared --prefix=$BUILD_DIR/lb
make
make install

cd $BUILD_DIR
tar zxvf libart_lgpl-2.3.17.tar.gz
cd libart_lgpl-2.3.17
env CFLAGS="-O3 -fPIC" ./configure --disable-shared --prefix=$BUILD_DIR/lb
make
make install

IR=-I$BUILD_DIR/lb/include
CPPFLAGS="$IR $IR/libart-2.0 $IR/freetype2 $IR/libpng"
LDFLAGS="-L$BUILD_DIR/lb/lib"
CFLAGS=-O3
export CPPFLAGS LDFLAGS CFLAGS

cd $BUILD_DIR
tar zxf rrdtool-1.2.11.tar.gz
cd rrdtool-1.2.11
./configure --prefix=$INSTALL_DIR --disable-python --disable-tcl && make && make install
```

直接运行rrdtoolinstall.sh脚本即可，命令如下：

```
[root@localhost ~]# tar -zxvf rrdtool-1.2.11.tar.gz
[root@localhost ~]# cd rrdtool-1.2.11
[root@localhost rrdtool-1.2.11]# ls
```

```
cgilib-0.5.tar.gz          libpng-1.2.8-config.tar.gz   zlib-1.2.2.tar.gz
freetype-2.1.9.tar.gz      rrdtool-1.2.11.tar.gz
libart_lgpl-2.3.17.tar.gz  rrdtoolinstall.sh
[root@localhost rrdtool-1.2.11]# chmod u+x rrdtoolinstall.sh
[root@localhost rrdtool-1.2.11]# ./rrdtoolinstall.sh
```

步骤14 安装net-snmp，命令如下：

```
[root@localhost ~]# tar -zxvf net-snmp-5.2.4.tar.gz
[root@localhost ~]# cd net-snmp-5.2.4
[root@localhost net-snmp-5.2.4]# ./configure --prefix=/usr/local/net-snmp
--enable-developer
[root@localhost ~]# make
[root@localhost ~]# make install
```

步骤15 修改snmpd.conf，然后启动snmpd服务，命令如下：

```
[root@localhost net-snmp-5.2.4]# ln -s /usr/local/net-snmp/bin/* /usr/local/bin/
[root@localhost net-snmp-5.2.4]# cp EXAMPLE.conf
/usr/local/net-snmp/share/snmp/snmp.conf
[root@localhost net-snmp-5.2.4]# /usr/local/net-snmp/sbin/snmpd
```

步骤16 安装Cacti，命令如下：

```
[root@localhost ~]# tar -zxvf cacti-0.8.6j.tar.gz
[root@localhost ~]# mv -r cacti-0.8.6j /usr/local/apache/htdocs/cacti
[root@localhost ~]# vi /usr/local/apache/htdocs/cacti/include/config.php
$database_type = "mysql";
$database_default = "cacti";
$database_hostname = "localhost";
$database_username = "xfimti";
$database_password = "123";
```

步骤17 安装Cactid，命令如下：

```
[root@localhost ~]# tar -zxvf cacti-cactid-0.8.6i.tar.gz
[root@localhost ~]# cd cacti-cactid-0.8.6i
[root@localhost cacti-cactid-0.8.6i]# ./configure --with-mysql=/usr/local/mysql
--with-snmp=/usr/local/net-snmp
[root@localhost cacti-cactid-0.8.6i]# make
[root@localhost cacti-cactid-0.8.6i]# mkdir /usr/local/cactid
[root@localhost cacti-cactid-0.8.6i]# cp cactid cactid.conf /usr/local/cactid
[root@localhost ~]# vi  /usr/local/cactid/cactid.conf    //修改cactid配置文件
DB_Host         127.0.0.1
DB_Database     cacti
DB_User         xfimti
DB_Pass         123
```

步骤18 配置MySQL数据库，命令如下：

```
mysql> create database cacti;
Query OK, 1 row affected (0.06 sec)
[root@localhost cacti-0.8.6j]# vi cacti.sql
[root@localhost cacti-0.8.6j]# pwd
```

```
/usr/local/apache/htdocs/cacti/cacti-0.8.6j
[root@localhost cacti-0.8.6j]# mysql -u xfimti -p cacti < cacti.sql
Enter password:
```

> **提示** 直接将cacti.sql脚本导入数据库会发生错误，将建表语句TYPE=MyISAM改为engine=myisam即可解决问题。

至此Cacti安装完成，在浏览器地址栏中输入"http://192.168.0.100/cacti"，登录页面如图16.4所示，默认的用户名和密码是admin。

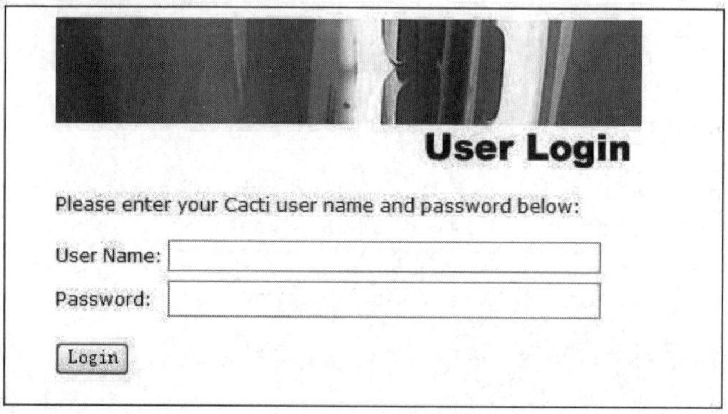

图 16.4 Cacti 登录页面

16.3.2 Cacti 监控 MySQL 服务器

登录Cacti后，主页面如图16.5所示，可以看到两个选项卡：console和graphs。console表示控制台，在控制台中可以进行所有的配置操作；graphs用来查看所有的服务器性能。

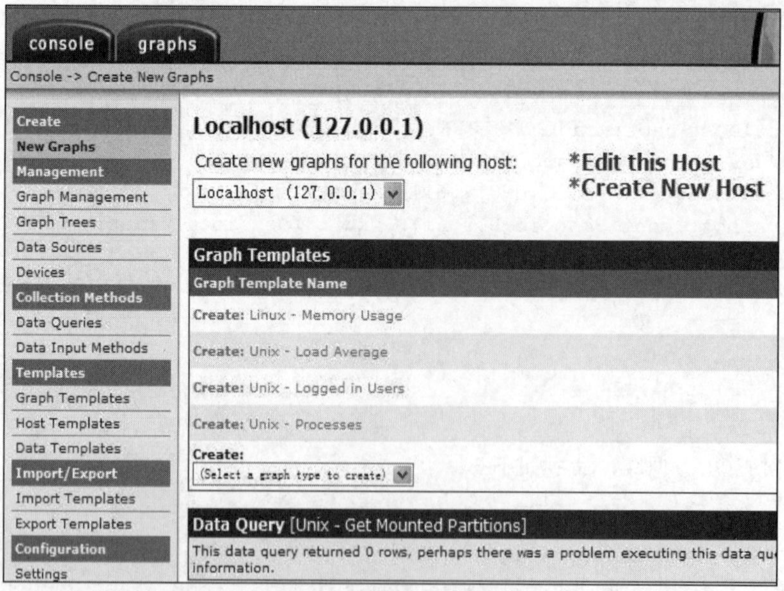

图 16.5 Cacti 登录后的页面

console中各菜单项的具体意义如下：

（1）New Graphs：创建新图像的快捷方式。

（2）Graph Management：可以删除和复制图像，Cacti会自动创建图像。

（3）Graph Trees：在[grahps]界面里，图像或设备是以树形结构显示的，可以在此设置树的结构。

（4）Data Sources：用来管理rrd文件。一般无须修改，Cacti会自己创建rrd文件。

（5）Devices：可以在此创建新的设备或修改其名称等信息。

（6）Data Queries和Data Input Methods：表示采集数据的方式。

（7）Graph Templates、Host Templates和Data Templates：分别表示图像模板、主机类型模板和数据模板。

（8）Import Templates和Export Templates：是模板（图像模板、主机类型模板和数据模板）的导入和导出操作。

（9）Setting：用来配置Cacti的主要配置菜单。

（10）System Utilities：用来显示Cacti系统的一些缓存和日志信息。

（11）User Management：用来对用户进行管理。可以在此添加或删除用户，并对每个用户设置详细的权限。

要实现使用Cacti监控MySQL服务器，首先需要下载监控MySQL的模板，可以从下面的地址下载：

```
http://code.google.com/p/mysql-cacti-templates/downloads/detail?name=better-ca
cti-templates-1.1.8.tar.gz
```

该模板主要是用来监控MySQL服务器的性能，包括InnoDB相关监控、MyISAM相关监控、Nginx状态监控。下面开始安装该模板，操作步骤如下：

步骤01 安装mysql-cacti-templates.tar.gz文件，命令如下：

```
[root@localhost ~]# tar zxf better-cacti-templates-1.1.8.tar.gz
[root@localhost ~]# cd better-cacti-templates-1.1.8
[root@localhost better-cacti-templates-1.1.8]# cd scripts/
[root@localhost scripts]# ls
ss_get_by_ssh.php  ss_get_mysql_stats.php
[root@localhost scripts]# cp ss_get_mysql_stats.php
/usr/local/apache/htdocs/cacti/scripts/
[root@localhost scripts]# cd  /usr/local/apache/htdocs/cacti/scripts/
[root@localhost scripts]# vi ss_get_mysql_stats.php
```

步骤02 编辑ss_get_mysql_stats.php文件，注意以下3行：

```
$mysql_user = 'xfimti';
$mysql_pass = 'xfimti';
$cache_dir  = '/tmp';
```

步骤03 在MySQL中创建具有SUPER和PROCESS权限的用户，命令如下：

```
mysql> grant super,process,select on *.* to xfimti@'%';
```

```
Query OK, 0 rows affected (0.00 sec)
```

步骤04 导入模板。单击管理界面左侧的【Import Templates】（见图16.5）链接，然后将模板文件cacti_host_template_x_mysql_server_ht_0.8.6i-server1.1.8.xml导入进去，如图16.6所示。

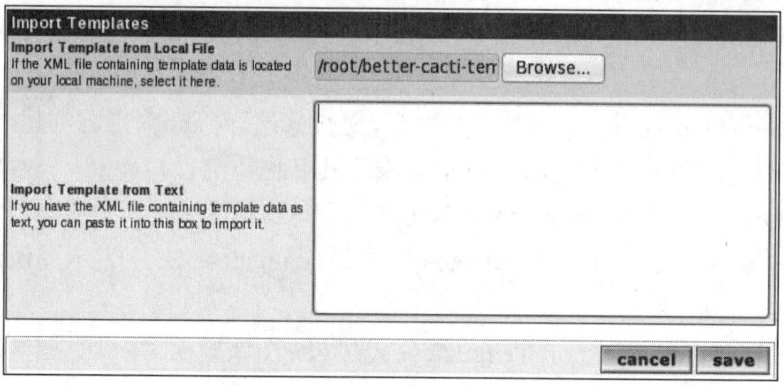

图16.6　导入监控 MySQL 的模板

步骤05 添加设备。在主页面中单击【Devices】链接，然后新建一个Devices，或者选择已有的Devices，将它设置为需要监控的主机，如图16.7所示。

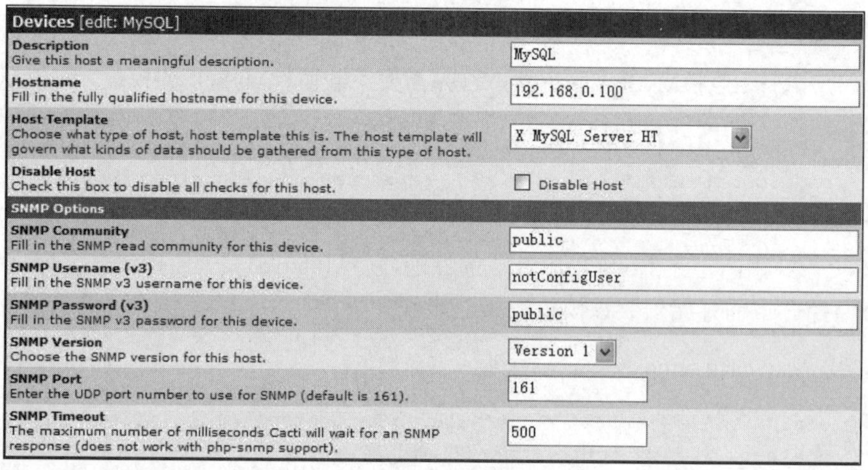

图16.7　创建一个新的 Devices

步骤06 保存设置，过一段时间就可以看见生成的图像了，如图16.8所示是部分监控的对象。

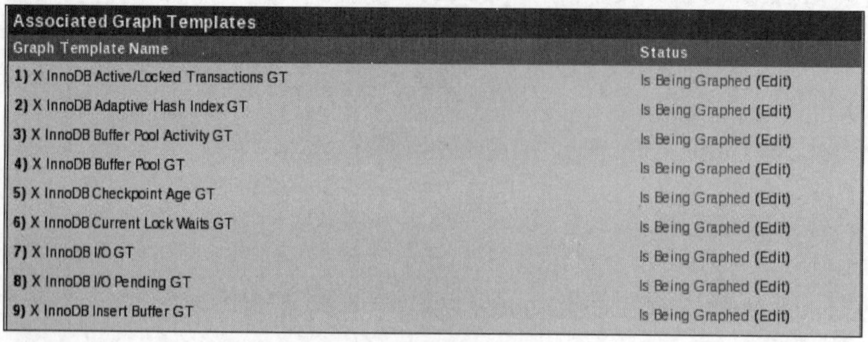

图16.8　部分监控的对象

创建设备之后，如果发现一个SNMP error的问题，如图16.9所示，可以通过如下步骤来解决：

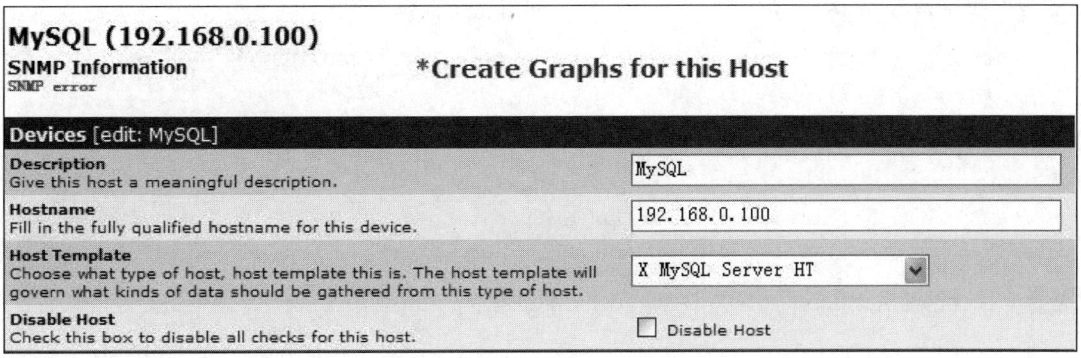

图16.9　SMNP error 出错界面

步骤01　确认被检测机器中snmp服务已启动，如果没有启动，可以执行service snmpd start启动该服务：

```
[root@localhost ~]# service snmpd status
snmpd (pid 29856) is running...
```

步骤02　snmp服务启动成功后，在检测机器上执行如下命令，检测能否返回数据：

```
[root@localhost ~]# snmpwalk -v 2c -c public 192.168.0.100 if
Timeout: No Response from 192.168.0.100
```

从结果中可以看出，被监控机器的snmp服务没有给监控机器授权。

步骤03　下面需要修改/etc/snmp/snmpd.conf文件，如下所示。

```
####
# First, map the community name "public" into a "security name"

#       sec.name  source              community
com2sec notConfigUser 192.168.0.100   public
```

步骤04　找到如下代码部分，将systemview改为all。

```
#      group          context sec.model sec.level prefix read    write    notif
access notConfigGroup ""      any       noauth    exact  systemview none none
```

修改之后的代码如下：

```
#      group          context sec.model sec.level prefix read  write notif
access notConfigGroup ""      any       noauth    exact  all   none  none
```

步骤05　然后找到如下代码部分，将"#"去掉。

```
##              incl/excl subtree           mask
#view all       included  .1                80
```

步骤06　最后关闭snmpd服务，再重新启动，命令如下：

```
[root@localhost ~]# kill -9 29856
```

```
[root@localhost ~]# service snmpd start
Starting snmpd:                                              [  OK  ]
```

步骤 07 重新在检测机器上执行如下命令，检测能否返回数据。

```
[root@localhost ~]# snmpwalk -v 2c -c public 192.168.0.100 if
...
IF-MIB::ifOutQLen.2 = Gauge32: 0
IF-MIB::ifOutQLen.3 = Gauge32: 0
IF-MIB::ifSpecific.1 = OID: SNMPv2-SMI::zeroDotZero
IF-MIB::ifSpecific.2 = OID: SNMPv2-SMI::zeroDotZero
IF-MIB::ifSpecific.3 = OID: SNMPv2-SMI::zeroDotZero
```

步骤 08 执行/usr/local/apache/htdocs/cacti/poller.php脚本，命令如下：

```
[root@localhost bin]# pwd
/usr/local/php/bin
[root@localhost bin]# ls
pear  peardev  pecl  php  php-config  phpize
[root@localhost bin]# ./php /usr/local/apache/htdocs/cacti/poller.php
sh: /usr/local/bin/rrdtool: No such file or directory
12/05/2012 12:46:02 AM - POLLER: Poller[0] Maximum runtime of 292 seconds exceeded. Exiting.
12/05/2012 12:46:02 AM - SYSTEM STATS: Time:292.5174 Method:cmd.php Processes:1 Threads:N/A Hosts:3 HostsPerProcess:3 DataSources:0 RRDsProcessed:0
 PHP Warning:  pclose(): 48 is not a valid stream resource in /usr/local/apache/htdocs/cacti/lib/rrd.php on line 47
```

步骤 09 在/usr/local/bin目录下建立链接rrdtool指向usr/local/rrdtool/bin/rrdtool，命令如下：

```
[root@localhost bin]# ./php /usr/local/apache/htdocs/cacti/poller.php
12/05/2012 01:43:35 AM - POLLER: Poller[0] Maximum runtime of 292 seconds exceeded. Exiting.
12/05/2012 01:43:35 AM - SYSTEM STATS: Time:293.0108 Method:cmd.php Processes:1 Threads:N/A Hosts:3 HostsPerProcess:3 DataSources:0 RRDsProcessed:0
 PHP Warning:  pclose(): 48 is not a valid stream resource in /usr/local/apache/htdocs/cacti/lib/rrd.php on line 47
```

步骤 10 重新单击【Devices】右侧窗体中的【MySQL】链接，如图16.10所示。

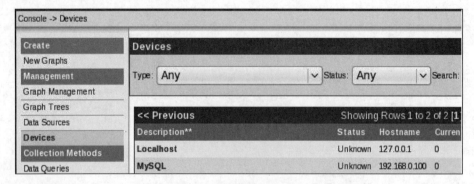

图 16.10 Devices 右侧的 MySQL 链接界面

SNMP error错误问题被解决了，如图16.11所示。

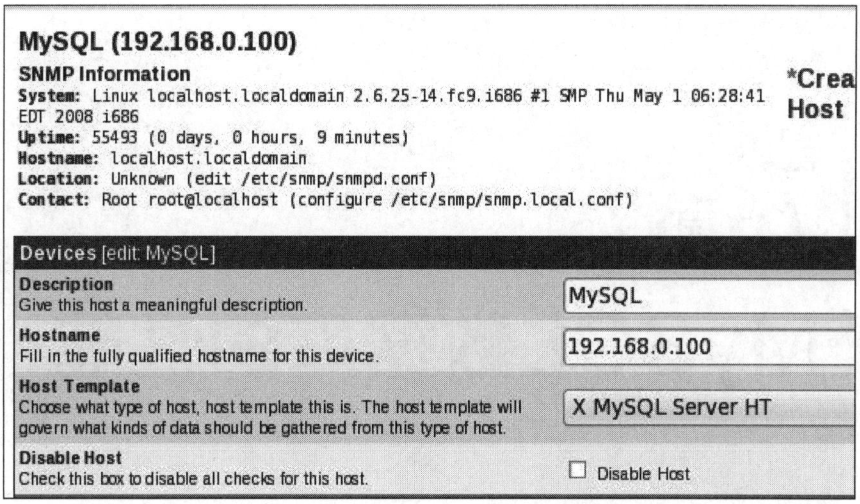

```
MySQL (192.168.0.100)
SNMP Information                                          *Crea
System: Linux localhost.localdomain 2.6.25-14.fc9.i686 #1 SMP Thu May 1 06:28:41
EDT 2008 i686                                             Host
Uptime: 55493 (0 days, 0 hours, 9 minutes)
Hostname: localhost.localdomain
Location: Unknown (edit /etc/snmp/snmpd.conf)
Contact: Root root@localhost (configure /etc/snmp/snmp.local.conf)

Devices [edit: MySQL]
Description                                               MySQL
Give this host a meaningful description.
Hostname                                                  192.168.0.100
Fill in the fully qualified hostname for this device.
Host Template                                             X MySQL Server HT
Choose what type of host, host template this is. The host template will
govern what kinds of data should be gathered from this type of host.
Disable Host                                              ☐ Disable Host
Check this box to disable all checks for this host.
```

图 16.11　SNMP error 问题被成功解决

第 17 章 提升MySQL数据库的性能

MySQL的特性包括使用新的字符集（utf8mb4）、自增变量持久化、实现了统计直方图、日志分类更详细、支持JSON类型等。充分利用这些特性可以编写更高效的查询语句和处理更复杂的数据操作，从而提升MySQL数据库的性能。

17.1 默认字符集改为utf8mb4

在MySQL 5.7.8之前的版本中，默认字符集为latin1。网站开发人员在设计数据库的时候，往往会将编码修改为utf8字符集，如果忘记修改默认编码，就会出现乱码的问题。而在MySQL 8和MySQL 9中，数据库的默认编码改为utf8mb4，从而避免了乱码的问题。

下面通过案例来对比不同的版本中默认字符集的变化。

【例17.1】对比不同版本中默认字符集的变化。

在MySQL 5.7版本中，查看数据库的默认编码，结果如下：

```
mysql> SHOW VARIABLES LIKE 'character_set_database';
+------------------------+---------+
| Variable_name          | Value   |
+------------------------+---------+
| character_set_database | latin1  |
+------------------------+---------+
```

在MySQL 5.7版本中，查看数据表的默认编码，结果如下：

```
mysql> SHOW CREATE TABLE tb_emp1\G
*************************** 1. row ***************************
       Table: tb_emp1
Create Table: CREATE TABLE `tb_emp1` (
  `id` int(11) DEFAULT NULL,
  `name` varchar(25) DEFAULT NULL,
```

```
  `deptId` int(11) DEFAULT NULL,
  `salary` float DEFAULT NULL
) ENGINE=MyISAM DEFAULT CHARSET=latin1
```

在MySQL 9.0版本中,查看数据库的默认编码,结果如下:

```
mysql> SHOW VARIABLES LIKE 'character_set_database';
+------------------------+---------+
| Variable_name          | Value   |
+------------------------+---------+
| character_set_database | utf8mb4 |
+------------------------+---------+
```

在MySQL 9.0版本中,查看数据表的默认编码,结果如下:

```
mysql> SHOW CREATE TABLE tb_emp1\G
*** 1. row ***
       Table: tb_emp1
Create Table: CREATE TABLE `tb_emp1` (
  `id` int(11) DEFAULT NULL,
  `name` varchar(25) DEFAULT NULL,
  `deptId` int(11) DEFAULT NULL,
  `salary` float DEFAULT NULL
) ENGINE=InnoDB DEFAULT CHARSET=utf8mb4 COLLATE=utf8mb4_0900_ai_ci
```

对比结果,MySQL 9.0已经将默认编码改为了utf8mb4。

17.2 自增变量的持久化

在MySQL 8.0之前,自增主键AUTO_INCREMENT的值如果大于max(primary key)+1,在MySQL重启后,会重置AUTO_INCREMENT=max(primary key)+1,这种现象在某些情况下会导致业务主键冲突或者其他难以发现的问题。在MySQL 9.0中,对自增主键AUTO_INCREMENT进行了持久化。

下面通过案例来对比不同版本中自增变量是否持久化。

【例17.2】对比不同版本中自增变量是否持久化。

在MySQL 5.7版本中,测试步骤如下:

步骤01 创建数据表,其中包含自增主键的id字段,SQL语句如下:

```
mysql> CREATE TABLE test1(id int auto_increment primary key);
Query OK, 0 rows affected (0.08 sec)
```

步骤02 插入4个空值,SQL语句如下:

```
mysql> INSERT INTO test1 values(0),(0),(0),(0);
Query OK, 4 rows affected (0.08 sec)
Records: 4  Duplicates: 0  Warnings: 0
```

步骤 03 查询数据表test1中的数据,结果如下:

```
mysql> SELECT * FROM test1;
+----+
| id |
+----+
|  1 |
|  2 |
|  3 |
|  4 |
+----+
```

步骤 04 删除id为4的记录,SQL语句如下:

```
mysql> DELETE FROM test1 where id=4;
Query OK, 1 row affected (0.04 sec)
```

步骤 05 再次插入一个空值,SQL语句如下:

```
mysql> INSERT INTO test1 values(0);
Query OK, 1 row affected (0.03 sec)
```

步骤 06 查询此时数据表test1中的数据,结果如下:

```
mysql> SELECT * FROM test1;
+----+
| id |
+----+
|  1 |
|  2 |
|  3 |
|  5 |
+----+
```

从结果中可以看出,虽然删除了id为4的记录,但是再次插入空值时,并没有重用被删除的4,而是分配了5。

步骤 07 删除id为5的记录,结果如下:

```
mysql> DELETE FROM test1 where id=5;
Query OK, 1 row affected (0.04 sec)
```

步骤 08 重启数据库,重新插入一个空值,SQL语句如下:

```
mysql> INSERT INTO test1 values(0);
Query OK, 1 row affected (0.02 sec)
```

步骤 09 再次查询数据表test1中的数据,结果如下:

```
mysql> SELECT * FROM test1;
+----+
| id |
+----+
|  1 |
|  2 |
```

```
| 3 |
| 4 |
+----+
```

从结果中可以看出，新插入的0值分配的是4，但是按照重启前的操作逻辑，此处应该分配6。出现上述结果的主要原因是自增主键没有持久化。

在MySQL 5.7中，对于自增主键的分配规则是由InnoDB数据字典内部的一个计数器来决定的，而该计数器只在内存中维护，并不会持久化到磁盘中。当数据库重启时，该计数器会通过下面这种方式初始化。

```
SELECT MAX(ai_col) FROM table_name FOR UPDATE;
```

在MySQL 9.0版本中，执行与MySQL 5.7版本中相同的步骤，步骤09的结果如下：

```
mysql> SELECT * FROM test1;
+----+
| id |
+----+
| 1 |
| 2 |
| 3 |
| 6 |
+----+
```

从结果中可以看出，自增变量已经持久化了。

MySQL 9.0将自增主键的计数器持久化到重做日志中。每次计数器发生改变，都会将其写入重做日志中。如果数据库重启，InnoDB会根据重做日志中的信息来初始化计数器的内存值。为了尽量减小对系统性能的影响，计数器写入重做日志时，并不会马上刷新数据库系统。

17.3 GROUP BY不再隐式排序

在MySQL 9.0版本中，MySQL对GROUP BY 字段不再隐式排序。如果确实需要排序，必须加上ORDER BY子句。

下面通过案例来对比不同版本中GROUP BY字段的排序情况。

【例17.3】对比不同版本中GROUP BY字段的排序情况。

分别在MySQL 5.7版本和MySQL 9.0版本中创建数据表、插入数据和查询数据，SQL语句如下：

```
mysql> CREATE TABLE bs1
(
bsid            INT(11) NOT NULL,
bscount         INT(11) DEFAULT '0'
);

mysql> INSERT INTO bs1 (bsid,bscount)
```

```
    VALUES(101,5),
    (102,5),
    (103,5),
    (105,0),
    (106,0),
    (107,8),
    (108,8),
    (109,8);

mysql> SELECT * FROM bs1;
+------+---------+
| bsid | bscount |
+------+---------+
| 101  |    5    |
| 102  |    5    |
| 103  |    5    |
| 105  |    0    |
| 106  |    0    |
| 107  |    8    |
| 108  |    8    |
| 109  |    8    |
+------+---------+
```

在MySQL 5.7中查看数据表bs1的结构，结果如下：

```
mysql> SHOW CREATE TABLE bs1\G
*************************** 1. row ***************************
       Table: bs1
Create Table: CREATE TABLE `bs1` (
  `bsid` int(11) NOT NULL,
  `bscount` int(11) DEFAULT '0'
) ENGINE=MyISAM DEFAULT CHARSET=latin1
```

在MySQL 9.0中查看数据表bs1的结构，结果如下：

```
mysql>SHOW CREATE table bs1\G
*************************** 1. row ***************************
       Table: bs1
Create Table: CREATE TABLE `bs1` (
  `bsid` int(11) NOT NULL,
  `bscount` int(11) DEFAULT '0'
) ENGINE=InnoDB DEFAULT CHARSET=utf8mb4 COLLATE=utf8mb4_0900_ai_ci
```

在MySQL 5.7中分组查询，结果如下：

```
mysql>SELECT count(bsid), bscount FROM bs1 GROUP BY bscount;
+-------------+---------+
| count(bsid) | bscount |
+-------------+---------+
|      2      |    0    |
|      3      |    5    |
|      3      |    8    |
+-------------+---------+
```

从结果中可以看出，字段bscount按升序自动排列。

在MySQL 9.0中分组查询，结果如下：

```
mysql>SELECT count(bsid), bscount FROM bs1 GROUP BY bscount;
+-------------+---------+
| count(bsid) | bscount |
+-------------+---------+
|           3 |       5 |
|           2 |       0 |
|           3 |       8 |
+-------------+---------+
```

从结果中可以看出，字段bscount没有按升序排列。

在MySQL 9.0中添加ORDER BY子句实现排序效果，结果如下：

```
mysql>SELECT COUNT(bsid), bscount FROM bs1 GROUP BY bscount ORDER BY bscount;
+-------------+---------+
| count(bsid) | bscount |
+-------------+---------+
|           2 |       0 |
|           3 |       5 |
|           3 |       8 |
+-------------+---------+
```

17.4 统计直方图

MySQL 9.0支持统计直方图。利用直方图，用户可以对一张表中的一列做数据分布的统计，特别是没有索引的字段。这可以帮助查询优化器找到更优的执行计划。

17.4.1 直方图的优点

在数据库中，查询优化器负责将SQL转换成最有效的执行计划。但有时，查询优化器找不到最优的执行计划，导致花费了更多不必要的时间。造成这种情况的主要原因是，查询优化器无法准确地知道以下几个问题的答案：

- 每张表有多少行？
- 每一列有多少不同的值？
- 每一列的数据分布情况如何？

例如，销售表production中包括id、tm、count三个字段，分别表示编号、销售时间和销售数量。

对比以下两个查询语句：

```
SELECT * FROM production WHERE tm BETWEEN "22:00:00" AND "23:59:00"
SELECT * FROM production WHERE tm BETWEEN "08:00:00" AND "12:00:00"
```

如果销售时间大部分集中在上午8点到12点，在查询销售情况时，第一个查询语句耗费的时间会远远大于第二个查询语句。因为没有统计数据，优化器会假设tm的值是均匀分配的。

如何才能使查询优化器知道数据的分布情况呢？一个解决方法就是在列上建立统计直方图。

直方图能近似获得一列数据的分布情况，从而让数据库知道它含有哪些数据。直方图有多种形式，MySQL支持两种：等宽直方图（singleton）和等高直方图（equi-height）。

- 等宽直方图：指每个bucket的宽度相等。可以用来了解数据的范围和分布。
- 等高直方图：指每个bucket的高度相等。可以用来了解数据的数量分布。

MySQL会自动将数据划分到不同的bucket中，也会根据数据情况自动决定创建哪种类型的直方图。当需要关注数据的范围分布时，可以使用等宽直方图；当需要关注数据的数量分布时，可以使用等高直方图。

17.4.2　直方图的基本操作

创建直方图的语法格式如下：

```
ANALYZE TABLE table_name [UPDATE HISTOGRAM on col_name with N BUCKETS |DROP HISTOGRAM ON clo_name]
```

buckets的值必须指定，可以设置为1～1024，默认值是100。统计直方图的信息存储在数据字典表column_statistcs中，可以通过视图information_schema.COLUMN_STATISTICS访问。直方图以灵活的JSON格式存储。ANALYZE TABLE会基于表的大小自动判断是否要进行取样操作，也会基于表中列的数据分布情况以及bucket的数量，来决定是建立等宽直方图还是建立等高直方图。

【例17.4】 创建直方图。

创建用于测试的数据表production，SQL语句如下：

```
mysql> CREATE TABLE production (id int,tm TIME,count int);
```

在数据表production的字段tm上创建直方图，SQL语句如下：

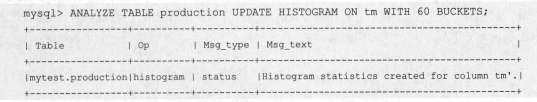

设置buckets值时，可以先设置得低一些，如果没有满足需求，再增大。

对于不同的数据集合，buckets的值取决于以下几个因素：

- 这列有多少不同的值。
- 数据的分布情况。
- 需要多高的准确性。

在数据表production的字段tm和字段count上创建直方图，SQL语句如下：

```
mysql>ANALYZE TABLE production UPDATE HISTOGRAM ON tm,count WITH 60 BUCKETS;
+------------------+---------+----------+------------------------------------------------+
| Table            | Op      | Msg_type | Msg_text                                       |
+------------------+---------+----------+------------------------------------------------+
|mytest.production|histogram|status    |Histogram statistics created for column 'count'.|
|mytest.production|histogram|status    |Histogram statistics created for column 'tm'.   |
+------------------+---------+----------+------------------------------------------------+
```

再次创建直方图时，将会将上一个直方图重写。

如果需要删除已经创建的直方图，用DROP HISTOGRAM即可实现：

```
mysql>ANALYZE TABLE production DROP HISTOGRAM ON tm,count;
+------------------+---------+----------+------------------------------------------------+
| Table            | Op      | Msg_type | Msg_text                                       |
+------------------+---------+----------+------------------------------------------------+
|mytest.production|histogram|status    |Histogram statistics removed for column 'count'.|
|mytest.production|histogram|status    |Histogram statistics removed for column 'tm'.   |
+------------------+---------+----------+------------------------------------------------+
```

直方图统计了表中某些字段的数据分布情况，为选择高效的执行计划提供了参考。直方图与索引有着本质的区别：维护一个索引有代价，每一次的插入、更新、删除都需要更新索引，这会对性能有一定的影响；而直方图一次创建无须更新，除非明确去更新它，所以不会影响插入、更新、删除的性能。

建立直方图的时候，MySQL服务器会将所有数据读到内存中，然后在内存中进行操作，包括排序。如果对一张很大的表建立直方图，可能需要将几百兆字节的数据都读到内存中。为了规避这种风险，MySQL会根据给定的histogram_generation_max_mem_size的值，计算应该将多少行数据读到内存中。

设置histogram_generation_max_mem_size值的方法如下：

```
mysql>SET histogram_generation_max_mem_size = 10000;
```

17.5　日志分类更详细

在MySQL 9.0版本中，日志分类更加详细。例如在错误信息中添加了错误信息编号[MY-010311]和错误所属子系统[Server]。

下面通过一个案例来对比不同版本中日志分类的情况。

【例17.5】 对比不同版本中日志分类的情况。

在MySQL 5.7版本中，部分错误日志如下：

```
2018-06-08T09:07:20.114585+08:00 0 [Warning] 'proxies_priv' entry '@
root@localhost' ignored in --skip-name-resolve mode.
```

```
    2018-06-08T09:07:20.117848+08:00 0 [Warning] 'tables_priv' entry 'user
mysql.session@localhost' ignored in --skip-name-resolve mode.
    2018-06-08T09:07:20.117868+08:00 0 [Warning] 'tables_priv' entry 'sys_config
mysql.sys@localhost' ignored in --skip-name-resolve mode.
```

在MySQL 9.0版本中,部分错误日志如下:

```
# MySQL 9.0
    2024-06-21T17:53:13.040295+08:00 28 [Warning] [MY-010311] [Server] 'proxies_priv'
entry '@ root@localhost' ignored in --skip-name-resolve mode.
    2024-06-21T17:53:13.040520+08:00 28 [Warning] [MY-010330] [Server] 'tables_priv'
entry 'user mysql.session@localhost' ignored in --skip-name-resolve mode.
    2024-06-21T17:53:13.040542+08:00 28 [Warning] [MY-010330] [Server] 'tables_priv'
entry 'sys_config mysql.sys@localhost' ignored in --skip-name-resolve mode.
```

由对比结果可知,MySQL 9.0版本中的错误日志分类更加详细。

17.6 支持不可见索引

不可见索引的特性对于性能调试非常有用。在MySQL 9.0中,索引可以被"隐藏"和"显示"。当一个索引被隐藏时,它不会被查询优化器使用。也就是说,管理员可以隐藏一个索引,然后观察该索引对数据库的影响:如果数据库性能有所下降,说明这个索引是有用的,于是将其"恢复显示";如果数据库性能看不出变化,说明这个索引是多余的,可以将它删除。

下面通过一个案例来了解如何隐藏和显示索引。

【例17.6】隐藏和显示索引。

创建不可见索引,SQL语句如下:

```
mysql> CREATE TABLE test1(a int,b int,index idx_a_b(a,b desc) invisible );
Query OK, 0 rows affected (0.18 sec)
```

查看索引idx_a_b的属性Visible的值,SQL语句如下:

```
mysql>SHOW INDEX FROM test1 \G
*************************** 1. row ***************************
        Table: test1
   Non_unique: 1
     Key_name: idx_a_b
 Seq_in_index: 2
  Column_name: b
    Collation: D
  Cardinality: 0
     Sub_part: NULL
       Packed: NULL
         Null: YES
   Index_type: BTREE
      Comment:
Index_comment:
```

```
        Visible: NO
     Expression: NULL
```

从结果中可以看出,Visible的属性值为NO。

显示不可见索引,SQL语句如下:

```
mysql>ALTER TABLE test1 ALTER index idx_a_b visible;
Query OK, 0 rows affected (0.12 sec)
Records: 0  Duplicates: 0  Warnings: 0
```

再次查看索引idx_a_b的属性Visible的值,SQL语句如下:

```
mysql>SHOW INDEX FROM test1 \G
*************************** 1. row ***************************
        Table: test1
   Non_unique: 1
     Key_name: idx_a_b
 Seq_in_index: 2
  Column_name: b
    Collation: D
  Cardinality: 0
     Sub_part: NULL
       Packed: NULL
         Null: YES
   Index_type: BTREE
      Comment:
Index_comment:
      Visible: YES
   Expression: NULL
```

从结果中可以看出,Visible的属性值变为了YES。

> **注意** 当索引被隐藏时,它的内容仍然和正常索引一样实时更新。如果一个索引需要长期隐藏,那么可以将它删除,因为索引的存在会影响插入、更新和删除的性能。

再次隐藏索引,SQL语句如下:

```
mysql>ALTER TABLE test1 ALTER index idx_a_b invisible;
Query OK, 0 rows affected (0.07 sec)
Records: 0  Duplicates: 0  Warnings: 0
```

> **注意** 数据表中的主键不能被设置为invisible。

17.7 支持JSON类型

MySQL是一个关系数据库,在MySQL 8.0之前,没有提供对非结构化数据的支持,但是如果用户有这样的需求,也可以通过MySQL的BLOB来存储非结构化的数据。

【例17.7】在MySQL 8.0及之前，通过BLOB来存储非结构化的数据。下面举例说明：

```
mysql> create table tp(json_data blob);
Query OK, 0 rows affected (0.01 sec)

mysql> insert into tp values('{"key1":"data1", "key2":2,
"key3":{"sub_key1":"sub_val1"}}');
Query OK, 1 row affected (0.01 sec)

SELECT * FROM tp;
+-------------------------------------------------------------+
| json_data                                                   |
+-------------------------------------------------------------+
| {"key1":"data1", "key2":2, "key3":{"sub_key1":"sub_val1"}}  |
+-------------------------------------------------------------+
1 row in set (0.01 sec)
```

在本例中，使用BLOB来存储JSON数据，使用这种方法，需要用户保证插入的数据是一个能够转换成JSON格式的字符串，MySQL并不保证任何正确性。在MySQL看来，这就是一个普通的字符串，并不会进行任何有效性检查。此外，提取JSON中的字段，也需要在用户的代码中完成。例如，在Python语言中提取JSON中的字段，代码如下：

```python
#!/usr/bin/python
import pymysql
import json

try:
    conn = pymysql.connect(host="127.0.0.1", db="test", user="root",
passwd="123456", port=3306)
    sql = "select * from tp"
    cur = conn.cursor()
    cur.execute(sql)
    rows = cur.fetchall()
    print json.dumps(json.loads(rows[0][0]), indent=4)
except:
    conn.close()
```

这种方式虽然也能够实现JSON的存储，但是有诸多缺点，最为显著的缺点有：

- 需要用户保证JSON的正确性，如果用户插入的数据并不是一个有效的JSON字符串，MySQL并不会报错。
- 所有对JSON的操作，都需要在用户的代码里进行处理，不够友好。
- 即使只提取JSON中的某一个字段，也需要读出整个BLOB，效率不高。
- 无法在JSON字段上创建索引。

在MySQL 9.0中，实现了对JSON类型的支持。MySQL本身已经是一个比较完备的数据库系统，其底层存储并不适合有太大的改动，那么MySQL是如何支持JSON格式的呢？

MySQL 9.0支持JSON的做法是在服务层提供一些便于操作JSON的函数，简单地将JSON编码成BLOB，然后交由存储引擎层进行处理。MySQL 9.0的JSON支持与存储引擎没有关系，MyISAM存储引擎也支持JSON格式。下面举例说明。

【例17.8】 InnoDB和MyISAM都支持JSON格式。

```
mysql> create table tb(data json)engine=innodb;
Query OK, 0 rows affected (0.18 sec)

mysql> insert into tb values('{"key":"val"}');
Query OK, 1 row affected (0.03 sec)

mysql> create table ts(data json)engine=myisam;
Query OK, 0 rows affected (0.02 sec)

mysql> insert into ts values('{"key":"val"}');
Query OK, 1 row affected (0.00 sec)
```

MySQL 9.0中的一些JSON函数包括：

- JSON_EXTRACT()：从JSON数据中提取值。
- JSON_UNQUOTE()：从JSON数据中去除引号。
- JSON_OBJECT()：创建一个JSON对象。
- JSON_ARRAY()：创建一个JSON数组。
- JSON_MERGE()：合并两个JSON数据。
- JSON_SEPARATE()：将JSON 据分解为多个部分。

这些函数使用户能够更轻松地处理JSON数据，提高数据库操作的性能。

当将JSON数据存储在MySQL中的BLOB对象中时，数据会被组织成以下格式：

- 首先存放JSON元素的个数。
- 然后存放转换成BLOB以后的字节数。
- 接下来存放key pointers和value pointers。

为了加快查找速度，MySQL内部会对key进行排序，以提高处理速度。这样，在查询JSON数据时，可以更快地定位到相应的key和value。

在MySQL 9.0中，key的长度只用2个字节（65535）保存，如果超过这个长度，MySQL将报错，如下所示：

```
mysql> insert into tb values(JSON_OBJECT(repeat('a', 65535), 'val'));
Query OK, 1 row affected (0.37 sec)

mysql> insert into tb values(JSON_OBJECT(repeat('a', 65536), 'val'));
ERROR 3151 (22032): The JSON object contains a key name that is too long.
```

在MySQL的源码中，与JSON相关的文件有：

- json_binary.cc。
- json_binary.h。

- json_dom.cc。
- json_dom.h。
- json_path.cc。
- json_path.h。

其中，json_binary.cc处理JSON的编码、解码，json_dom.cc是JSON的内存表示，json_path.cc用以将字符串解析成 JSON。

对于JSON的编码，入口是json_binary.cc文件中的serialize函数；对于JSON的解码，即将BLOB解析成JSON对象，入口是json_binary.cc文件中的parse_binary函数。只要搞清楚了JSON的存储格式，这两个函数就很好理解了。

17.8　全文索引的加强

MySQL 9.0支持更灵活和更加优化的全文检索。例如，全文索引支持外部的分析器，就像MyISAM；插件可以替代内置分析器，也可以作为一个前端来使用。另外，MySQL 9.0实现了标记优化器，这个优化器可以将查询结果传递到InnoDB，因此InnoDB可以跳过全文检索部分。

在InnoDB上实现了支持CJK（中文、日文和韩文）的全文检索。MySQL 9.0为CJK提供了一个默认的全文分析器（n-gram分析器）。

在全文索引中，n-gram就是一段文字里面连续的n个字的序列。例如，用n-gram来对"春花秋月"进行分词，得到的结果如表17.1所示。其中n由参数ngram_token_size控制，表示分词的大小，默认是2。

表 17.1　"春花秋月"的分词

n 的值	分词结果
1	n=1 : '春', '花', '秋', '月'
2	n=2 : '春花', '花秋', '秋月'
3	n=3 : '春花秋', '花秋月'
3	n=4 : '春花秋月'

下面通过举例来说明全文检索功能。

【例17.9】MySQL 9.0的全文检索。

创建数据表，并设置全文检索，SQL语句如下：

```
mysql>CREATE TABLE tft (
      id int(11) DEFAULT NULL,
      name varchar(512) DEFAULT NULL,
      content text,
      FULLTEXT KEY idx_name(name),
      FULLTEXT KEY idx_content(content) WITH PARSER ngram)
```

插入演示数据，SQL语句如下：

```
mysql> INSERT INTO tft (id,name,content) VALUES (1,'春花秋月','经典古诗');
Query OK, 1 row affected (0.00 sec)
```

普通检索必须使用整个词才能检索到，SQL语句如下：

```
mysql> SELECT * FROM tft WHERE MATCH (name) AGAINST ('春花秋月');
+------+----------+----------+
| id   | name     | content  |
+------+----------+----------+
|    1 | 春花秋月 | 经典古诗 |
+------+----------+----------+
```

只有部分词是不能检索出信息的，SQL语句如下：

```
mysql>SELECT * FROM tft WHERE MATCH (name) AGAINST ('秋月');
Empty set (0.00 sec)
```

新的全文检索功能可以检索任意两个组合的记录，SQL语句如下：

```
mysql> SELECT * FROM tft WHERE MATCH (content) AGAINST ('古诗');
+------+----------+----------+
| id   | name     | content  |
+------+----------+----------+
|    1 | 春花秋月 | 经典古诗 |
+------+----------+----------+
```

再次使用全文检索功能检索任意两个组合的记录，SQL语句如下：

```
mysql> SELECT * FROM tft WHERE MATCH (content) AGAINST ('典古');
+------+----------+----------+
| id   | name     | content  |
+------+----------+----------+
|    1 | 春花秋月 | 经典古诗 |
+------+----------+----------+
```

MySQL 9.0的全文检索功能为用户提供了高效的文本搜索能力，特别适合需要对大规模文本数据进行快速查询的场景。通过创建全文索引并使用MATCH ... AGAINST语法，用户可以实现自然语言处理、布尔搜索以及查询扩展等多种搜索模式，从而满足不同的搜索需求。

17.9　动态修改InnoDB缓冲池的大小

从MySQL 5.7.5版本开始，支持在不重启系统的情况下动态调整innodb_buffer_pool_size。调整大小的过程是以innodb_buffer_pool_chunk_size为单位迁移pages到新的内存空间的过程，迁移进度可以通过innodb_buffer_pool_resize_status查看。当在线修改缓冲池大小时，以chunk为单位进行增长或收缩。

缓冲池大小是innodb_buffer_pool_chunk_size × innodb_buffer_pool_instances的倍数（128MB的倍数），如果不是，系统将会适当调大innodb_buffer_pool_size，以满足要求。因此，可能会出现缓冲池大小的实际分配比配置文件中指定的size要大的情况。

下面举例说明如何在线调整缓冲池的大小。

【例17.10】 在线调整缓冲池的大小。

查看当前缓冲池的大小，SQL语句如下：

```
mysql> show variables like '%innodb_buffer_pool_size%';
+-------------------------+----------+
| Variable_name           | Value    |
+-------------------------+----------+
| innodb_buffer_pool_size | 16777216 |
+-------------------------+----------+
```

查看缓冲池中实例的个数，SQL语句如下：

```
mysql> show variables like 'innodb_buffer_pool_instances';
+------------------------------+-------+
| Variable_name                | Value |
+------------------------------+-------+
| innodb_buffer_pool_instances | 1     |
+------------------------------+-------+
```

动态修改缓冲池的大小为1000MB，SQL语句如下：

```
mysql> set global innodb_buffer_pool_size=1048576000;
Query OK, 0 rows affected, 1 warning (0.00 sec)
```

查看警告信息，SQL语句如下：

```
mysql> show warnings;
+---------+------+-------------------------------------------------------------------------+
| Level   | Code | Message                                                                 |
+---------+------+-------------------------------------------------------------------------+
| Warning | 1292 | Truncated incorrect innodb_buffer_pool_size value: '1048576000'         |
+---------+------+-------------------------------------------------------------------------+
1 row in set (0.00 sec)
```

出现上述警告信息的原因是，1000MB不是innodb_buffer_pool_chunk_size × innodb_buffer_pool_instances的倍数，即不是128MB的倍数。

查看缓冲池的设置进度，SQL语句如下：

```
mysql> show global status like 'Innodb_buffer_pool_resize_status';
+----------------------------------+-------------------------------------------------+
| Variable_name                    | Value                                           |
+----------------------------------+-------------------------------------------------+
| Innodb_buffer_pool_resize_status | Completed resizing buffer pool at 180816 17:00:31.|
+----------------------------------+-------------------------------------------------+
1 row in set (0.01 sec)
```

查看当前缓冲池的大小，SQL语句如下：

```
mysql> show variables like '%innodb_buffer_pool_size%';
+-------------------------+------------+
```

```
| Variable_name          | Value      |
+------------------------+------------+
| innodb_buffer_pool_size | 1056964608 |
+------------------------+------------+
```

从结果可以看出，缓冲池的大小被设置成了1056964608字节（约1024MB），因为1024是128的整数倍，缓冲池大小比配置文件里指定的size要大。

17.10 表空间数据加密

在MySQL 9.0中，InnoDB Tablespace Encryption支持对独享表空间的InnoDB数据文件加密，其依赖keyring plugin来进行密钥的管理，开启加密功能需要启动参数--early-plugin-load。

在MySQL的配置文件my.ini中开启--early-plugin-load参数：

```
[mysqld]
early-plugin-load=keyring_file.so
```

启动参数后，查看服务器是否支持加密功能：

```
mysql> SELECT PLUGIN_NAME, PLUGIN_STATUS FROM INFORMATION_SCHEMA.PLUGINS where PLUGIN_NAME LIKE 'keyring%';
+--------------------------+----------------------+
| PLUGIN_NAME              | PLUGIN_STATUS        |
+--------------------------+----------------------+
| keyring_file             | ACTIVE               |
+--------------------------+----------------------+
```

创建加密表空间：

```
mysql> create table t(id int,name varchar(10))engine = innodb default charset utf8mb4 encryption='y';
Query OK, 0 rows affected (0.17 sec)
```

查看表t的定义：

```
mysql> SHOW CREATE table t\G
*************************** 1. row ***************************
       Table: t
Create Table: CREATE TABLE `t` (
  `id` int(11) DEFAULT NULL,
  `name` varchar(10) DEFAULT NULL
) ENGINE=InnoDB DEFAULT CHARSET=utf8mb4 ENCRYPTION='y'
```

ENCRYPTION='y'表示使用加密功能加密所有存储在该表中的数据。

17.11 跳过锁等待

在MySQL 5.7版本中，SELECT...FOR UPDATE语句在执行的时候，如果获取不到锁，会一直等待，直到innodb_lock_wait_timeout超时。

在MySQL 9.0版本中，通过添加NOWAIT和SKIP LOCKED语句，能够立即返回。如果查询的行已经加锁，那么NOWAIT会立即报错返回；SKIP LOCKED也会立即返回，只是返回的结果中不包含被锁定的行。

下面通过案例来理解在MySQL 9.0版本中如何跳过锁等待，如表17.2所示。

表 17.2 跳过锁等待的例子

Session1	Session2

创建表bbs1，然后添加数据，命令如下：
 	
``
mysql> CREATE TABLE bbs1(
    -> id int,
    -> name varchar(20));
mysql>INSERT INTO bbs1 (id,name)
    -> VALUES(101, 'lili'),
    -> (102, 'zhangfeng'),
    -> (103, 'wangming'),
    -> (104, 'wangxiao');
``
	添加 NOWAIT 语法：
	
	``
	mysql> SELECT * FROM bbs1 WHERE id = 102 FOR UPDATE NOWAIT;
	ERROR 3572 (HY000): Statement aborted because lock(s) could not be acquired immediately and NOWAIT is set.
	``

向数据表bbs1中添加排它锁，命令如下：
 	
``
mysql> BEGIN;
mysql> SELECT * FROM bbs1 WHERE id = 102 FOR UPDATE;
+------+-----------+
| id
	name |
	+------+-----------+
	| 102
		zhangfeng |
		+------+-----------+
		``
			添加 SKIP LOCKED 语法：
			
			``
			mysql> SELECT * FROM bbs1 WHERE id = 102 FOR UPDATE SKIP LOCKED;
			Empty set (0.00 sec)
			``

17.12 MySQL 9.0新特性1——支持将JSON输出保存到用户变量

从MySQL 9.0版本开始支持将EXPLAIN FORMAT的JSON输出保存到用户变量。下面通过一个案例来介绍该新特性。

步骤01 创建演示数据表tb，SQL语句如下：

```
CREATE TABLE tb
(
id      INT,
name    VARCHAR(25)
);
```

步骤 02 插入演示数据，SQL语句如下：

```
INSERT INTO tb (id ,name) VALUES (1,'电视机') ,(2,'空调');
```

步骤 03 将EXPLAIN FORMAT的JSON输出保存到用户变量@myvariable中，SQL语句如下：

```
EXPLAIN FORMAT =JSON INTO @myvariable
UPDATE tb SET name = "洗衣机" WHERE id =2;
```

步骤 04 查看用户变量@myvariable，执行结果如下：

```
mysql> SELECT @myvariable\G
*************************** 1. row ***************************
@myvariable: {
  "query_block": {
    "select_id": 1,
    "table": {
      "update": true,
      "table_name": "tb",
      "access_type": "ALL",
      "rows_examined_per_scan": 2,
      "filtered": "100.00",
      "attached_condition": "(`test_db`.`tb`.`id` = 2)"
    }
  }
}
```

17.13　MySQL 9.0新特性2——支持准备语句

MySQL 9.0支持准备语句（Prepared Statements），这是一种预编译的SQL语句，可以包含占位符，这些占位符在执行时会被实际的值替换。准备语句的优势如下：

（1）准备语句可以提高性能，因为SQL解析和编译只发生一次，然后可以被多次执行，每次都可以使用不同的参数。

（2）准备语句能增强应用程序的安全性，因为它们有助于防止SQL注入攻击。

下面通过案例来理解准备语句的使用方法。

步骤 01 使用PREPARE语句准备一个SQL语句模板，该模板可以包含一个或多个占位符（?）。

```
mysql> SET @s = 'SELECT SQRT(POW(?,2) + POW(?,2)) AS hypotenuse';
mysql> PREPARE stmt1 FROM @s;
```

步骤 02 使用SET语句为占位符设置具体的值。

```
mysql> SET @a = 6;
mysql> SET @b = 8;
```

步骤 03 使用EXECUTE语句执行准备好的SQL语句，并用之前设置的参数替换占位符。

```
mysql> EXECUTE stmt1 USING @a, @b;
+------------+
| hypotenuse |
+------------+
|         10 |
+------------+
```

步骤 04 执行完成后，使用DEALLOCATE PREPARE语句释放准备好的语句。

```
mysql> DEALLOCATE PREPARE stmt1;
```

准备语句还可以用于动态选择需要查询的表，它将表名作为用户变量，并在执行时替换到SQL语句中。例如，首先设置用户变量@table的值为表名，然后构造一个包含该表名的SQL查询字符串，最后使用PREPARE和EXECUTE语句执行该查询。这种灵活性使得准备语句在处理动态SQL场景时非常有用。

17.14　MySQL 9.0新特性3——支持面向AI的向量存储

MySQL 9.0 增加了一个新的向量数据类型：VECTOR。它是一种可以存储N个数据项的数据结构（数组），语法格式如下：

```
VECTOR(N)
```

其中，每个数据项都是一个4字节的单精度浮点数。默认的数据项为2048个，最大值为16383。

向量类型的数据可以使用二进制字符串或者列表分隔的字符串表示，例如：

```
CREATE TABLE mytb1(id int, rgb vector(3));
INSERT INTO mytb1 VALUES (1, to_vector('[255,255,255]'));
INSERT INTO mytb1 VALUES (2, to_vector('[128,255,0]'));
INSERT INTO mytb1 VALUES (3, to_vector('[0,65,225]'));
```

MySQL 9.0同时还增加了一些用于操作VECTOR数据的向量函数。

1. STRING_TO_VECTOR函数

STRING_TO_VECTOR 函数用于将字符串形式的向量数据转换为二进制。STRING_TO_VECTOR函数的参数是一个字符串，包含一组由逗号分隔的浮点数，并且使用方括号（[]）进行引用。例如：

```
mysql> SELECT HEX(STRING_TO_VECTOR("[1.08, -18.8, 88]"));
+--------------------------------------------+
| HEX(STRING_TO_VECTOR("[1.08, -18.8, 88]")) |
+--------------------------------------------+
| 713D8A3F666696C10000B042                   |
+--------------------------------------------+
```

2. VECTOR_TO_STRING函数

VECTOR_TO_STRING函数用于将向量数据转换为字符串。例如：

```
mysql> SELECT VECTOR_TO_STRING(STRING_TO_VECTOR("[1.08, -18.8, 88]"));
+--------------------------------------------------------+
| VECTOR_TO_STRING(STRING_TO_VECTOR("[1.08, -18.8, 88]")) |
+--------------------------------------------------------+
| [1.08000e+00,-1.88000e+01,8.80000e+01]                 |
+--------------------------------------------------------+

mysql> SELECT VECTOR_TO_STRING(0x00000040000040444000A0400000E400);
+------------------------------------------------------+
| VECTOR_TO_STRING(0x00000040000040444000A0400000E400) |
+------------------------------------------------------+
| [2.00000e+00,7.68000e+02,5.00003e+00,2.09385e-38]    |
+------------------------------------------------------+
```

输出结果中的浮点数使用科学记数法表示。

3. VECTOR_DIM函数

VECTOR_DIM函数用于返回向量数据的维度，也就是数据项的个数。例如：

```
mysql> SELECT VECTOR_DIM(rgb) FROM mytb1;
+----------------+
| VECTOR_DIM(rgb) |
+----------------+
|              3 |
|              3 |
|              3 |
+----------------+
```

第 18 章
MySQL终极优化实战

数据库存储引擎是数据库底层软件组件,数据库管理系统(DBMS)使用数据库存储引擎进行创建、查询、更新和删除数据的操作。选择合适的存储引擎可以很大程度地提高数据库性能,尤其是在高并发场景中。另外,通过使用分区表、优化数据表的锁和优化事务控制,也可以提升MySQL数据库的性能。

18.1 选择合适的存储引擎

MySQL提供了多个不同的存储引擎,包括处理事务安全表的引擎和处理非事务安全表的引擎。在MySQL中,不需要在整个服务器中使用同一种存储引擎,针对具体的要求,可以对每一张表使用不同的存储引擎。MySQL支持的存储引擎有InnoDB、MyISAM、Memory、NDB、Merge、Archive、Federated、CSV、BLACKHOLE等。

可以使用SHOW ENGINES语句查看系统所支持的存储引擎类型,结果如下。

```
mysql> SHOW ENGINES \G
*************************** 1. row ***************************
      Engine: MEMORY
     Support: YES
     Comment: Hash based, stored in memory, useful for temporary tables
Transactions: NO
          XA: NO
  Savepoints: NO
*************************** 2. row ***************************
      Engine: MRG_MYISAM
     Support: YES
     Comment: Collection of identical MyISAM tables
Transactions: NO
          XA: NO
  Savepoints: NO
```

```
*************************** 3. row ***************************
      Engine: CSV
     Support: YES
     Comment: CSV storage engine
Transactions: NO
          XA: NO
  Savepoints: NO
*************************** 4. row ***************************
      Engine: FEDERATED
     Support: NO
     Comment: Federated MySQL storage engine
Transactions: NULL
          XA: NULL
  Savepoints: NULL
*************************** 5. row ***************************
      Engine: PERFORMANCE_SCHEMA
     Support: YES
     Comment: Performance Schema
Transactions: NO
          XA: NO
  Savepoints: NO
*************************** 6. row ***************************
      Engine: MyISAM
     Support: YES
     Comment: MyISAM storage engine
Transactions: NO
          XA: NO
  Savepoints: NO
*************************** 7. row ***************************
      Engine: InnoDB
     Support: DEFAULT
     Comment: Supports transactions, row-level locking, and foreign keys
Transactions: YES
          XA: YES
  Savepoints: YES
*************************** 8. row ***************************
      Engine: ndbinfo
     Support: NO
     Comment: MySQL Cluster system information storage engine
Transactions: NULL
          XA: NULL
  Savepoints: NULL
*************************** 9. row ***************************
      Engine: BLACKHOLE
     Support: YES
     Comment: /dev/null storage engine (anything you write to it disappears)
Transactions: NO
          XA: NO
  Savepoints: NO
*************************** 10. row ***************************
      Engine: ARCHIVE
```

```
       Support: YES
       Comment: Archive storage engine
  Transactions: NO
            XA: NO
    Savepoints: NO
*************************** 11. row ***************************
        Engine: ndbcluster
       Support: NO
       Comment: Clustered, fault-tolerant tables
  Transactions: NULL
            XA: NULL
    Savepoints: NULL
```

Support列的值表示某种引擎是否能使用：YES表示可以使用，NO表示不能使用，DEFAULT表示该引擎为当前默认存储引擎。

查看当前默认的存储引擎，可以使用下面的语句：

```
mysql> SHOW VARIABLES LIKE '%storage_engine%';
+--------------------------------+-----------+
| Variable_name                  | Value     |
+--------------------------------+-----------+
| default_storage_engine         | InnoDB    |
| default_tmp_storage_engine     | InnoDB    |
| disabled_storage_engines       |           |
| internal_tmp_mem_storage_engine| TempTable |
+--------------------------------+-----------+
```

尽管MySQL数据库支持多种数据存储引擎，但是不管使用哪种存储引擎，所有的存储数据都被记录到.frm文件中，该文件还记录了表的一些属性值。值得注意的是，不管使用哪种数据存储引擎，都使用了高速缓存，数据库在读取.frm文件信息后会将表的信息缓存起来，以提高服务器下次读取数据的速度。

不同的存储引擎都有各自的特点，以适应不同的需求，常用存储引擎的对比情况如表18.1所示。

表 18.1 常用存储引擎的对比情况

特　　点	InnoDB	MyISAM	Memory	Merge	Archive
存储限制	64TB	有	有	没有	没有
事务安全	支持				
锁机制	行锁	表锁	表锁	表锁	表锁
B 数索引	支持	支持	支持	支持	支持
哈希索引			支持		
全文索引		支持			
集群索引	支持				
数据缓存	支持		支持		
索引缓存	支持	支持	支持	支持	支持
数据可压缩		支持			支持
空间使用	高	低	N/A	低	

（续表）

特　点	InnoDB	MyISAM	Memory	Merge	Archive
内存使用	高	低	中等	低	表锁
批量插入速度	低	高	高	高	支持
支持外键	支持				

如果要提供提交、回滚和崩溃恢复能力的事务安全（ACID兼容）能力，并要求实现并发控制，InnoDB是一个很好的选择；如果数据表主要用来插入和查询记录，则MyISAM存储引擎能提供较高的处理效率；如果只是临时存放数据，数据量不大，并且不需要较高的数据安全性，可以选择将数据保存在内存中的Memory引擎中，MySQL中使用该引擎作为临时表，存放查询的中间结果；如果只有INSERT和SELECT操作，可以选择Archive存储引擎，它支持高并发的插入操作，但是本身并不是事务安全的。Archive存储引擎非常适合存储归档数据，如记录日志信息可以使用Archive引擎。

使用哪一种引擎要根据需要灵活选择，一个数据库中的多张表可以使用不同引擎，以满足各种性能和实际需求。使用合适的存储引擎，将会提高整个数据库的性能。

MySQL数据库在创建表的时候，可以添加默认的存储引擎，示例如下：

```
create table books(
id int,
name varchar(20) not null
)engine=MyISAM default charset= utf8mb4;
```

更改表的存储引擎的语法格式如下：

```
ALTER TABLE <表名> ENGINE=<更改后的存储引擎名>;
```

【例18.1】将数据表books的存储引擎修改为InnoDB。

在修改存储引擎之前，先使用SHOW CREATE TABLE查看表tb_deptment3当前的存储引擎，结果如下：

```
mysql> SHOW CREATE TABLE books \G
*************************** 1. row ***************************
       Table: books
Create Table: CREATE TABLE `books` (
  `id` int DEFAULT NULL,
  `name` varchar(20) NOT NULL
) ENGINE=MyISAM DEFAULT CHARSET=utf8mb4 COLLATE=utf8mb4_0900_ai_ci
```

可以看到，表books当前的存储引擎为NGINE=MyISAM。

接下来修改存储引擎类型，SQL语句如下：

```
mysql> ALTER TABLE books ENGINE=InnoDB;
```

使用SHOW CREATE TABLE再次查看表books的存储引擎。

```
mysql> SHOW CREATE TABLE books \G
*************************** 1. row ***************************
       Table: books
Create Table: create table `books` (
```

```
    `id` int DEFAULT NULL,
    `name` varchar(20) NOT NULL
) ENGINE= InnoDB DEFAULT CHARSET=utf8mb4 COLLATE=utf8mb4_0900_ai_ci
```

表books的存储引擎变成了"InnoDB"。结果如下:

18.2 通过分区表提升MySQL执行效率

MySQL 5.1及高版本支持分区表(Partitioned Table),分区表的使用大大提升了MySQL执行效率。表分区是将一个大表根据特定条件分割成若干小表,每个分区包含一部分数据。这种方法特别适用于处理大数据量的场景,例如记录数超过600万的用户表,可以根据入库日期或所在地进行分区。

18.2.1 认识分区表

分区表是一种特殊的表结构,通过一些特殊的语句创建独立的存储空间,实际上每个分区都是有索引的独立表。这种设计使得分区看上去像一张单独的表,但在物理存储上却进行了优化。分区表的主要目的是通过物理数据库设计技术,减少特定查询操作的响应时间,同时对应用来说分区是完全透明的。

MySQL的分区主要有两种形式:水平分区(Horizontal Partitioning)和垂直分区(Vertical Partitioning)。

- 水平分区是根据表的行进行分割,这种形式的分区一定是将表的某个属性作为分割的条件。例如,将某张表里面的日期为2011年的数据和日期为2012年的数据分割开,就可以采用这种分区形式。
- 垂直分区是通过对表的垂直划分来减少目标表的宽度,是某些特定的列被划分到特定的分区。

下面介绍MySQL的各种分区表的常用操作案例。

18.2.2 RANGE 分区

RANGE分区使用values less than操作符来进行定义,把连续且不相互重叠的字段分配给分区。

【例18.2】创建RANGE分区,命令如下:

```
mysql> create table emp(
       empno varchar(20) not null,
       empname varchar(20),
       deptno int,
       birthdate date,
       salary int
```

```
        )
        partition by range(salary)
        (
        partition p1 values less than(1000),
        partition p2 values less than(2000),
        partition p3 values less than(3000)
        );
Query OK, 0 rows affected (0.01 sec)

mysql> insert into emp values(1000,'kobe',12,'1888-08-08',1500);
Query OK, 1 row affected (0.01 sec)

mysql> insert into emp values(1000,'kobe',12,'1888-08-08',3500);
ERROR 1526 (HY000): Table has no partition for value 3500
```

此时，按照工资级别（字段salary）进行表分区，partition by range的语法类似于switch...case的语法，如果salary小于1000，数据存储在p1分区；如果salary小于2000，数据存储在p2分区；如果salary小于3000，数据存储在p3分区。

上面插入的第二条数据的工资级别（字段salary）为3500，此时没有分区来存储该范围的数据，所以发生了错误。为了解决这种问题，加入"partition p4 values less than maxvalue"语句即可，命令如下：

```
mysql> drop table emp;
Query OK, 0 rows affected (0.00 sec)

mysql> create table emp(
        empno varchar(20) not null,
        empname varchar(20),
        deptno int,
        birthdate date,
        salary int
        )
        partition by range(salary)
        (
        partition p1 values less than(1000),
        partition p2 values less than(2000),
        partition p3 values less than(3000),
        partition p4 values less than maxvalue
        );
Query OK, 0 rows affected (0.01 sec)

mysql> insert into emp values(1000,'kobe',12,'1888-08-08',1000);
Query OK, 1 row affected (0.00 sec)

mysql> insert into emp values(1000,'durant',12,'1888-08-08',3500);
Query OK, 1 row affected (0.00 sec)
```

maxvalue表示可能的最大整数值。值得注意的是，values less than子句中可以只使用一个表达式，不过表达式结果不能为NULL。下面按照日期进行分区，命令如下：

```
mysql> drop table emp;
Query OK, 0 rows affected (0.00 sec)
mysql> create table emp(
      empno varchar(20) not null,
      empname varchar(20),
      deptno int,
      birthdate date,
      salary int
      )
      partition by range(year(birthdate))(
        partition p0 values less than(1980),
        partition p1 values less than(1990),
        partition p2 values less than(2000),
        partition p3 values less than maxvalue
      );
Query OK, 0 rows affected (0.01 sec)
```

该方案中，1980年以前出生的员工的信息存储在p0分区中，1990年以前出生的员工的信息存储在p1分区中，2000年以前出生的员工的信息存储在p2分区中，2000年以后出生的员工的信息存储在p3分区中。

RANGE分区是很有用的，常常用于以下几种情况：

（1）当要删除某个时间段的数据时，只需删除分区即可。例如，要删除1980年以前出生的员工的所有信息，此时会执行"alter table emp drop partition p0"要比执行"delete from emp where year(birthdate)<=1980"高效得多。

（2）如果使用包含日期或者时间的列，可以考虑使用RANGE分区。

（3）经常运行直接依赖于分割表的列的查询，比如，执行"select count(*) from emp where year(birthdate) = 1999 group by empno"，此时MySQL数据库可以很迅速地确定只有分区p2需要扫描，因为其他分区不符合查询条件。

18.2.3 LIST 分区

LIST分区类似RANGE分区，它们的区别主要在于：LIST分区中每个分区的定义和选择是基于某列的值从属于一个集合，而RANGE分区是从属于一个连续区间值的集合。

【例18.3】创建LIST分区，命令如下：

```
mysql> create table employees(
      empname varchar(20),
      deptno int,
      birthdate date not null,
      salary int
      )
      partition by list(deptno)
      (
        partition p1 values in (10,20),
        partition p2 values in (30),
        partition p3 values in (40)
```

```
    );
Query OK, 0 rows affected (0.01 sec)
```

以上示例以部门编号来划分分区，10号部门和20号部门的员工信息存储在p1分区，30号部门的员工信息存储在p2分区，40号部门的员工信息存储在p3分区。同RANG分区一样，如果插入数据的部门编号不在分区值列表中时，那么插入操作将失败并报错。

18.2.4 HASH 分区

HASH分区基于用户定义的表达式的返回值来选择分区，该表达式使用将要插入表中的这些行的列值进行计算。hash()函数中可以包含MySQL中有效的、产生非负整数值的任何表达式。

HASH分区主要用来确保数据在预先确定数目的分区中平均分布。在RANGE和LIST分区中，必须明确指定一个给定的列值或列值集合应该保存在哪个分区中；而在HASH分区中，MySQL自动完成这些工作，用户所要做的只是基于将要被哈希的列值指定一个列值或表达式，以及指定被分区的表将要被分割成的分区数量。

【例18.4】创建HASH分区，命令如下：

```
mysql> create table htable(
    id int,
    name varchar(20),
    birthdate date not null,
    salary int
    )
    partition by hash(year(birthdate))
    partitions 4;
Query OK, 0 rows affected (0.00 sec)
```

当使用"partition by hash"时，MySQL将基于用户函数结果的模数来确定使用哪个编号的分区。将要保存记录的分区编号为N = MOD(表达式, num)。例如，表htable中插入一条birthdate为"2010-09-23"的记录，可以通过如下方法计算该记录的分区：

```
mod(year('2010-09-23'),4)=mode(2010,4) =2
```

此时，该条记录的数据将会存储在分区编号为2的分区空间。

18.2.5 线性 HASH 分区

线性HASH分区和HASH分区的区别在于，线性哈希功能使用的是一个线性的2的幂运算法则，而HASH分区使用的是哈希函数的模数。

先看下面的例子：

```
mysql> create table htable2(
    id int not null,
    name varchar(20),
    hired date not null default '1999-09-09',
    deptno int
```

```
        )
        partition by linear hash(year(hired))
        partitions 4;
Query OK, 0 rows affected (0.03 sec)
```

如果表htable2中插入一条hired为"2010-09-23"的记录，记录将要保存到的分区是num个分区中的分区N，可以通过如下方法计算N：

（1）找到下一个大于num的2的幂次方数，把这个值称作V，即V = POWR(2,CEILING(LOG(2,num)))。假设，num的值是13，那么LOG(2,13)就是3.70043，CEILING(3.70043)就是4，则V = POWER(2,4)=16。

（2）计算N = F(column_list) & (V – 1)，当N≥num时，执行V = CEIL(V/2)，此时N = N & (V−1)。

【例18.5】演示通过线性哈希分区算法计算分区N的值。线性哈希分区表t1通过下面的语句创建。

```
mysql> create table th1(
    col1 int,
    col2 char(5),
    col3 date
    )
    partition by linear hash( year(col3) )
    partitions 6;
Query OK, 0 rows affected (0.59 sec)
```

现在假设要插入两条记录到表th1中，其中一条记录的col3列的值为"2003-04-14"，另一条记录的cols列值为"1998-10-19"。第一条记录要保存到的分区序号的计算过程如下：

记录将要保存到的num个分区中的分区N，假设num是7个分区，表t1使用线性HASH分区且有4个分区。

```
V = POWR(2,CEILING(LOG(2,num)))
 V = POWR(2,CEILING(LOG(2,7))) = 8
N = YEAR('2003-04-14') & (8 - 1)
 = 2003 & 7
 = 3
```

N的值是3，很显然3≥4不成立，所以附件条件不执行，第一条记录的信息将存储在3号分区中。

第二条记录将要保存到的分区序号计算如下：

```
V = POWR(2,CEILING(LOG(2,num)))
V = POWR(2,CEILING(LOG(2,7))) = 8
 = YEAR('1998-10-19') & (8 - 1)
 = 1998 & 7
 = 6
```

N的值是6，很显然6≥4成立，所以会继续执行。

```
V = CEIL(6/2) = 3
N = N & (V-1)
  = 6 & 2
  = 2
```

此时2≥4不成立，记录将被保存到2号分区中。线性哈希分区的优点在于增加、删除、合并和拆分分区更加快捷，有利于处理含有极大量（1000GB）数据的表。它的缺点在于，与使用常规HASH分区得到的数据分布相比，各个分区间数据的分布可能不大均衡。

18.2.6 KEY 分区

KEY分区类似于HASH分区，区别在于KEY分区只支持计算一列或多列，且MySQL服务器提供其自身的哈希函数，这些函数基于与PASSWORD()一样的运算法则。

【例18.6】创建KEY分区。

```
mysql> create table keytable(
       id int,
       name varchar(20) not null,
       deptno int,
       birthdate date not null,
       salary int
       )
       partition by key(birthdate)
       partitions 4;
Query OK, 0 rows affected (0.11 sec)
```

在KEY分区中使用关键字LINEAR和在HASH分区中使用具有同样的作用，分区的编号是通过2的幂运算得到的，而不是通过模数算法。

18.2.7 复合分区

复合分区是分区表中每个分区的再次分割，子分区既可以使用HASH分区，也可以使用KEY分区。

复合分区需要注意以下问题：

（1）如果一个分区中创建了复合分区，则其他分区也要有复合分区。
（2）如果创建了复合分区，则每个分区中的复合分区数必须相同。
（3）同一分区内的复合分区的名字不相同，不同分区内的复合分区的名字可以相同。

下面通过案例讲述不同的复合分区的创建方法。

1. RANGE - HASH复合分区

创建RANGE - HASH复合分区的命令如下：

```
mysql> create table rhtable(
    no varchar(20) not null,
```

```
      name varchar(20),
      deptno int,
      birthdate date not null,
      salary int
      )
      partition by range(salary)
      subpartition by hash(year(birthdate))
      subpartitions 3
      (
        partition p1 values less than (2000),
        partition p2 values less than maxvalue
      );
Query OK, 0 rows affected (0.23 sec)
```

2. RANGE - KEY复合分区

创建RANGE - KEY复合分区的命令如下:

```
mysql> create table rktable(
      no varchar(20) not null,
      name varchar(20),
      deptno int,
      birth date not null,
      salary int
      )
      partition by range(salary)
      subpartition by key(birth)
      subpartitions 3
      (
        partition p1 values less than (2000),
        partition p2 values less than maxvalue
      );
Query OK, 0 rows affected (0.07 sec)
```

3. LIST - HASH复合分区

创建LIST - HASH复合分区的命令如下:

```
mysql> create table lhtable(
      no varchar(20) not null,
      name varchar(20),
      deptno int,
      birth date not null,
      salary int
      )
      partition by list(deptno)
      subpartition by hash( year(birth) )
      subpartitions 3
      (
        partition p1 values in (10),
        partition p2 values in (20)
      );
Query OK, 0 rows affected (0.08 sec)
```

4. LIST - KEY复合分区

创建LIST - KEY复合分区的命令如下：

```
mysql> create table lktable(
    no varchar(20) not null,
    name varchar(20),
    deptno int,
    birthdate date not null,
    salary int
)
partition by list(deptno)
subpartition by key(birthdate)
subpartitions 3
(
    partition p1 values in (10),
    partition p2 values in (20)
);
Query OK, 0 rows affected (0.09 sec)
```

18.3 优化数据表的锁

MySQL与其他数据库在锁定机制方面最大的不同之处在于，对于不同的存储引擎支持不同的锁定机制。例如，InnoDB存储引擎支持行级锁和表级锁，默认情况下采用行级锁；MyISAM存储引擎采用的是表级锁。

18.3.1 MyISAM 表级锁优化建议

对于MyISAM存储引擎，虽然使用表级锁在实现过程方面比行级锁和页级锁所带来的附加成本要小，所消耗的资源也是最小的，不过MyISAM表级锁的颗粒比较大，在数据库并发处理过程中产生的数据资源争用的情况会比其他的锁定级别要多，从而在较大程度上降低了并发处理能力。MyISAM表级锁的优化建议如下：

（1）MyISAM表级锁的锁定级别是固定的，所以在考虑MyISAM表级锁优化时，重点考虑如何提高并发的效率。

（2）减少锁定的时间，让查询的时间尽可能短。要减少比较复杂的查询语句，可以考虑将复杂的查询分解成多个小的查询。尽可能建立足够高效的索引，让数据检索更迅速。尽量让MyISAM存储引擎的表控制字段类型。利用合理的机会优化MyISAM表数据文件。

（3）MyISAM表级锁可以分离能并行的操作。对于读锁互相阻塞的表级锁，可能会觉得存储引擎的表上只能是完全的串行化，没有办法再并行了，可是MyISAM的存储引擎还有一个非常有用的特性，就是Concurrent Insert特性。可以考虑设置Concurrent Insert的值为2，此时无论MyISAM存储引擎的数据文件的中间部分是否存在空洞（因为删除数据而留下的空闲空间），都允许在数据文件尾部进行插入操作。

（4）MyISAM的表级锁定对于读和写有不同优先级别设定，默认情况下写操作的优先级别高于读操作的优先级别，可以考虑根据应用的实际情况来设置读锁和写锁的优先级别，即通过设置系统参数low_priority_updates=1，设置写的优先级比读的优先级低。

18.3.2　InnoDB 行级锁优化建议

InnoDB存储引擎由于实现了行级锁，而行级锁的颗粒更小、实现更为复杂，因此带来的性能损耗比表级锁更高，但是InnoDB行级锁在并发性能上要远远高于表级锁。当系统并发量较高的时候，InnoDB的整体性能和MyISAM相比优势就比较明显了，所以说在选择使用哪种锁的时候，应该考虑应用是否有很大的并发量。想要合理使用InnoDB的行级锁，应该扬长避短，尽量做到以下几点：

（1）尽量控制事务的大小，减少锁定的资源量和锁定时间长度。

（2）尽可能让所有的数据检索都通过索引来完成，从而避免因为无法通过索引加锁而升级为表级锁定。

（3）尽可能减少基于范围的数据检索过滤条件，避免因为间隙锁带来的负面影响而锁定了不该锁定的记录。

（4）在业务环节允许的情况下，尽量使用较低级别的事务隔离，以减少因为事务隔离级别锁而带来的附加成本。

（5）合理使用索引，让InnoDB在索引上面加锁的时候更加准确。

（6）在应用中，尽可能按照相同的访问顺序来访问，防止产生死锁。

（7）在同一个事务中，尽可能做到一次锁定所需的所有资源，减少产生死锁的概率。

（8）对于容易产生死锁的业务，可以放弃使用InnoDB行级锁定，尝试使用表级锁定来减少死锁产生的概率。

（9）不要申请超过实际需要的锁级别。

18.4　优化事务控制

MySQL通过SET AUTOCOMMIT、START TRANSACTION、COMMIT和ROLLBACK等语句控制本地事务，具体语法如下：

```
START TRANSACTION | BEGIN [WORK];
COMMIT [WORK] [AND [NO] CHAIN] [[NO] RELEASE]
ROLLBACK [WORK] [AND [NO] CHAIN] [[NO] RELEASE]
SET AUTOCOMMIT = {0|1}
```

其中START TRANSACTION表示开启事务、COMMIT表示提交事务、ROLLBACK表示回滚事务、SET AUTOCOMMIT用于设置是否自动提交事务。

默认情况下，MySQL事务是自动提交的，如果需要通过明确的COMMIT和ROLLBACK来提交和回滚事务，那么需要通过明确的事务控制命令来开始事务，这是和Oracle的事务管理明显不同的地方。如果应用从Oracle数据库迁移到MySQL数据库，则需要确保应用中对事务进行了明确的管理。

MySQL的AUTOCOMMIT（自动提交）默认是开启，对MySQL的性能有一定影响。举个例子，如果用户插入了1000条数据，MySQL会提交事务1000次。这时可以把自动提交关闭，通过程序来提交事务一次就可以了。

关闭自动提交功能的命令如下：

```
mysql> set @@autocommit=0;
Query OK, 0 rows affected (0.00 sec)
```

查看自动提交功能是否被关闭，命令如下：

```
mysql> show variables like "autocommit";
+---------------+-------+
| Variable_name | Value |
+---------------+-------+
| autocommit    | ON    |
+---------------+-------+
1 row in set (0.02 sec)
```

下面通过两个Session（Session1和Session2）来理解事务控制的过程。

【例18.7】 事务的控制过程。

（1）在Session1中，打开自动提交事务功能，然后创建表ctable并插入两条记录，命令如下：

```
mysql> set @@autocommit=1;
Query OK, 0 rows affected (0.00 sec)

mysql> create table ctable (data INT);

mysql> insert into ctable values(1);
Query OK, 1 row affected (0.02 sec)

mysql> insert into ctable values(2);
Query OK, 1 row affected (0.00 sec)
```

（2）在Session2中，打开自动提交事务功能，然后查询表ctable，命令如下：

```
mysql> set @@autocommit=1;
Query OK, 0 rows affected (0.00 sec)

mysql> select * from ctable;
+------+
| data |
+------+
|    1 |
|    2 |
+------+
2 rows in set (0.00 sec)
```

（3）在Session1中，关闭自动提交事务功能，然后向表ctable中插入两条记录，命令如下：

```
mysql> set @@autocommit=0;
Query OK, 0 rows affected (0.00 sec)

mysql> insert into ctable values(3);
Query OK, 1 row affected (0.00 sec)

mysql> insert into ctable values(4);
Query OK, 1 row affected (0.00 sec)
```

（4）在Session2中，查询表ctable，命令如下：

```
mysql> select * from ctable;
+------+
| data |
+------+
|    1 |
|    2 |
+------+
2 rows in set (0.00 sec)
```

从结果中可以看出，在Session1中新插入的两条记录没有被查询出来。

（5）在Session1中，提交事务，命令如下：

```
mysql> commit;
Query OK, 0 rows affected (0.01 sec)
```

（6）在Session2中，查询表ctable，命令如下：

```
mysql> select * from ctable;
+------+
| data |
+------+
|    1 |
|    2 |
|    3 |
|    4 |
+------+
4 rows in set (0.00 sec)
```

此时，Session1中新插入的两条记录被查询出来了。

在表的锁定期间，如果使用start transaction命令开启一个新的事务，会造成一个隐含的unlock tables被执行，该操作存在一定的隐患。下面通过一个案例来理解。

【例18.8】在表的锁定期间，使用start transaction命令开启一个新的事务。

（1）首先创建数据表nbaplayer，命令如下：

```
create table nbaplayer(
        id int,
        name varchar(20),
        salary int);
```

（2）在Session1中，查询nbaplayer表，结果为空：

```
mysql> select * from nbaplayer;
Empty set (0.00 sec)
```

（3）在Session2中，查询nbaplayer表，结果为空：

```
mysql> select * from nbaplayer;
Empty set (0.00 sec)
```

（4）在Session1中，对表nbaplayer加写锁，命令如下：

```
mysql> lock table nbaplayer write;
Query OK, 0 rows affected (0.00 sec)
```

（5）在Session2中，向表nbaplayer中增加一条记录，命令如下：

```
mysql> insert into nbaplayer values(1,'kobe',10000);
等待
```

（6）在Session1中，插入一条记录，命令如下：

```
mysql> insert into nbaplayer values(2,'durant',40000);
Query OK, 1 row affected (0.02 sec)
```

（7）在Session1中，回滚刚才插入的记录，命令如下：

```
mysql> rollback;
Query OK, 0 rows affected (0.00 sec)
```

（8）在Session1中，开启一个新的事务，命令如下：

```
mysql> start transaction;
Query OK, 0 rows affected (0.00 sec)
```

（9）在Session2中，由于表锁被释放，此时成功增加该条记录，结果如下：

```
mysql> insert into nbaplayer values(1,'kobe',10000);
Query OK, 1 row affected (2 min 32.99 sec)
```

（10）在Session2中，查询nbaplayer，命令如下：

```
mysql> select * from nbaplayer;
+------+--------+--------+
| id   | name   | salary |
+------+--------+--------+
|    2 | durant |  40000 |
|    1 | kobe   |  10000 |
+------+--------+--------+
2 rows in set (0.00 sec)
```

从结果中可以看出，Session1的回滚操作并没有被成功执行。

MySQL提供的LOCK IN SHARE MODE锁，可以保证停止任何对它要读的数据行的更新或者删除操作。下面通过一个例子来理解。

【例18.9】 MySQL的LOCK IN SHARE MODE锁。

（1）在Session1中，开启一个新的事务，然后查询数据表nbaplayer的salary列的最大值，命令如下：

```
mysql> begin;
Query OK, 0 rows affected (0.00 sec)
mysql> select max(salary) from nbaplayer lock in share mode;
+-------------+
| max(salary) |
+-------------+
|       40000 |
+-------------+
1 row in set (0.00 sec)
```

（2）在Session2中，尝试做更新操作，命令如下：

```
mysql> update nbaplayer set salary = 90000 where id = 1;
等待
```

（3）在Session1中，提交事务，命令如下：

```
mysql> commit;
Query OK, 0 rows affected (0.00 sec)
```

（4）在Session2中，等Session1的事务提交之后，更新操作才成功执行，结果如下：

```
mysql> update nbaplayer set salary = 90000 where id = 1;
Query OK, 1 row affected (16.25 sec)
Rows matched: 1  Changed: 1  Warnings: 0
```

由结果可以看出，在Session1中对salary列加了LOCK IN SHARE MODE锁，在Session1未提交前，不能在Session2中对salary列进行更新操作。

第 19 章
企业人事管理系统数据库设计

本章将设计一个企业人事管理系统数据库。通过本系统的讲述，使读者掌握数据库设计的流程及MySQL在实际项目开发中涉及的重要技术。

19.1 需求分析

需求调查是任何一个软件项目的第一项工作，人事管理系统也不例外。软件首先从登录界面开始，验证用户名和密码之后，根据登录用户的不同权限打开软件，展示不同的功能模块。软件的主要功能模块是人事管理、备忘录、员工生日提醒、数据库的维护等。

通过需求调查，总结出如下需求信息：

（1）由于该系统的使用对象较多，要有较好的权限管理，每个用户可以具备对不同功能模块操作的权限。

（2）对员工的基础信息进行初始化。

（3）记录公司内部员工基本档案信息，提供便捷的查询功能。

（4）在查询员工信息时，可以对当前员工的家庭情况、培训情况进行添加、修改、删除操作。

（5）按照指定的条件对员工进行统计。

（6）可以将员工信息以表格的形式导出到Word文档中，以便进行打印。

（7）具备灵活的数据备份、还原及清空功能。

19.2 系统功能结构

企业人事管理系统以操作简单方便、界面简洁美观、系统运行稳定、安全可靠为开发原则，以功能需求为开发目标。

根据具体需求分析，设计企业人事管理系统的功能结构，如图19.1所示。

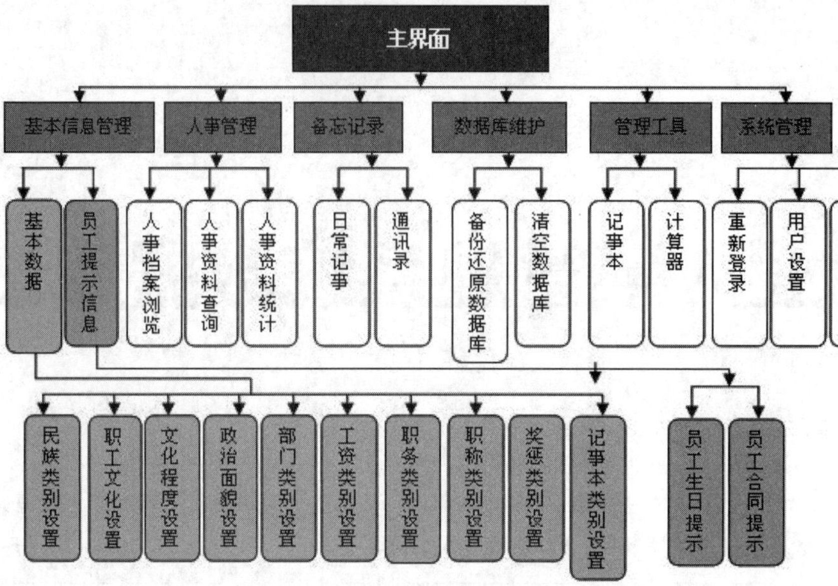

图 19.1　企业人事管理系统功能结构

19.3　数据库设计

数据库设计的好坏，直接影响着软件开发效率的高低、维护的方便与否，以及以后能否对功能进行扩充。因此，数据库设计非常重要，良好的数据库结构可以让我们事半功倍。

19.3.1　数据库实体 E-R 图

公司人事管理系统主要侧重于员工的基本信息及工作简历、家庭成员、奖惩记录等，数据量的多少由公司员工的多少来决定。MySQL数据库系统在安全性、准确性和运行速度方面有绝对的优势。

系统开发过程中，数据库占据重要的地位，数据库的设置依据需求分析而定，通过上述需要分析及系统功能的确定，规划出系统中使用的数据库实体对象，它们的E-R图如图19.2～图19.8所示。

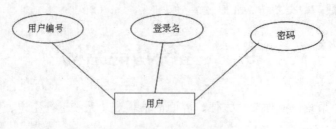

图 19.2　用户实体 E-R 图

第19章 企业人事管理系统数据库设计

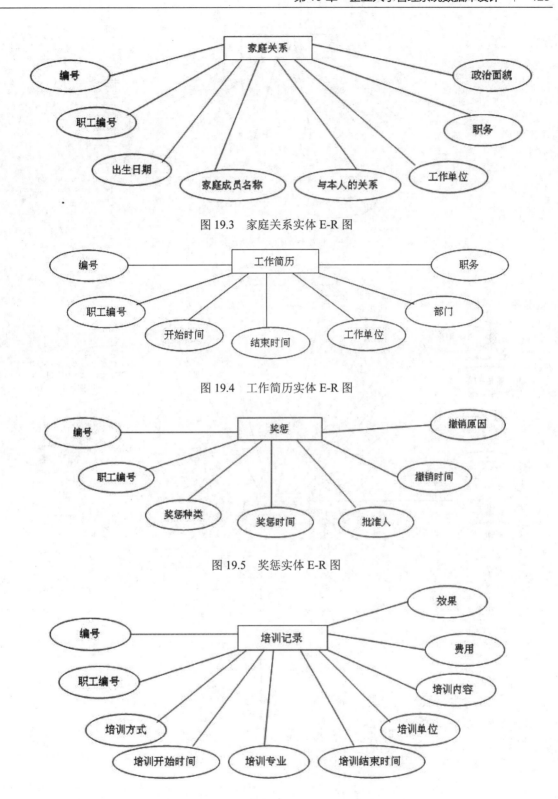

图 19.3 家庭关系实体 E-R 图

图 19.4 工作简历实体 E-R 图

图 19.5 奖惩实体 E-R 图

图 19.6 培训记录实体 E-R 图

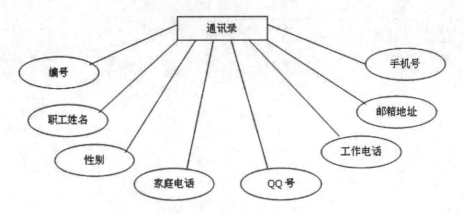

图 19.7 通讯录实体 E-R 图

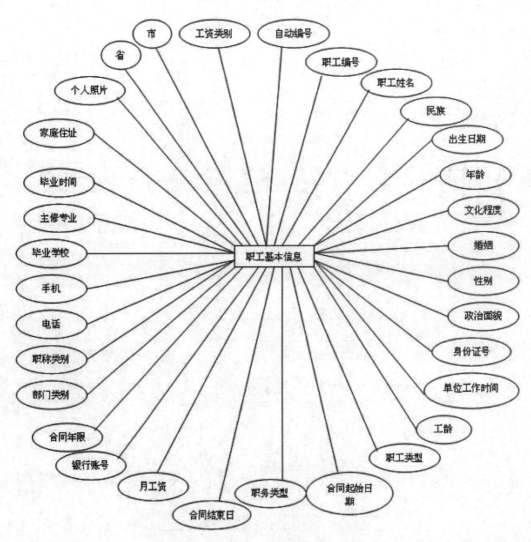

图 19.8 职工基本信息实体 E-R 图

19.3.2 数据库表的设计

E-R图设计完之后，将根据实体E-R图设计数据表结构，下面列出主要数据表。

（1）tb_login（用户登录表），用来记录操作者的用户名和密码，如表19.1所示。

表 19.1　tb_login（用户登录表）

列　名	描　述	数据类型	空/非空	约束条件
id	用户编号	INT	非空	主键，自动增长
uid	用户登录名	VARCHAR（50）	非空	
pwd	密码	VARCHAR（50）	非空	

（2）tb_family（家庭关系表），如表19.2所示。

表 19.2　tb_family（家庭关系表）

列　名	描　述	数据类型	空/非空	约束条件
id	编号	INT	非空	主键，自动增长
sut_id	职工编号	VARCHAR（50）	非空	外键
leaguername	家庭成员名称	VARCHAR（20）		
nexus	与本人的关系	VARCHAR（20）		
birthdate	出生日期	DATE		
workunit	工作单位	VARCHAR（50）		
business	职务	VARCHAR（20）		
visage	政治面貌	VARCHAR（20）		

（3）tb_workResume（工作简历表），如表19.3所示。

表 19.3　tb_workresume（工作简历表）

列　名	描　述	数据类型	空/非空	约束条件
id	编号	INT	非空	主键，自动增长
sut_id	职工编号	VARCHAR（50）	非空	外键
begindate	开始时间	DATE		
enddate	结束时间	DATE		
workunit	工作单位	VARCHAR（50）		
branch	部门	VARCHAR（20）		
business	职务	VARCHAR（20）		

（4）tb_randp（奖惩表），如表19.4所示。

表 19.4　tb_randp（奖惩表）

列　　名	描　　述	数据类型	空/非空	约束条件
id	编号	INT	非空	主键，自动增长
sut_id	职工编号	VARCHAR（50）	非空	外键
rpkind	奖惩种类	VARCHAR（20）		
rpdate	奖惩时间	DATE		
sealman	批准人	VARCHAR（20）		
quashdate	撤销时间	DATE		
quashwhys	撤销原因	VARCHAR（100）		

（5）tb_trainNote（培训记录表），如表19.5所示。

表 19.5　tb_trainNote（培训记录表）

列　　名	描　　述	数据类型	空/非空	约束条件
id	编号	INT	非空	主键，自动增长
sut_id	职工编号	VARCHAR（50）	非空	外键
trainfashion	培训方式	VARCHAR（20）		
begindate	培训开始时间	DATE		
enddate	培训结束时间	DATE		
speciality	培训专业	VARCHAR（20）		
trainunit	培训单位	VARCHAR（50）		
kulturmemo	培训内容	VARCHAR（50）		
charger	费用	FLOAT		
effects	效果	VARCHAR（20）		

（6）tb_addressBook（通讯录表），如表19.6所示。

表 19.6　tb_addressbook（通讯录表）

列　　名	描　　述	数据类型	空/非空	约束条件
id	编号	INT	非空	主键，自动增长
sutname	职工姓名	VARCHAR（20）	非空	
sex	性别	VARCHAR（4）		
phone	家庭电话	VARCHAR（18）		
qq	QQ 号	VARCHAR（15）		
workphone	工作电话	VARCHAR（18）		
e-mail	邮箱地址	VARCHAR（100）		
handset	手机号	VARCHAR（12）		

（7）tb_stuffbusic（职工基本信息表），如表19.7所示。

表 19.7 tb_stuffbusic（职工基本信息表）

列　　名	描　　述	数据类型	空/非空	约束条件
id	编号	INT	非空	自动增长/主键
stu_id	职工编号	VARCHAR（50）	非空	唯一
stuffname	职工姓名	VARCHAR（20）		
folk	民族	VARCHAR（20）		
birthday	出生日期	DATE		
age	年龄	INT		
kultur	文化程度	VARCHAR（14）		
marriage	婚姻	VARCHAR（4）		
sex	性别	VARCHAR（4）		
visage	政治面貌	VARCHAR（20）		
idcard	身份证号	VARCHAR（20）		
workdate	单位工作时间	DATE		
worklength	工龄	INT		
employee	职工类型	VARCHAR（20）		
business	职务类型	VARCHAR（10）		
laborage	工资类别	VARCHAR（10）		
branch	部门类别	VARCHAR（20）		
duthcall	职称类别	VARCHAR（20）		
phone	电话	VARCHAR（14）		
handset	手机	VARCHAR（11）		
school	毕业学校	VARCHAR（50）		
speciality	主修专业	VARCHAR（20）		
graduatedate	毕业时间	DATE		
address	家庭住址	VARCHAR（50）		
photo	个人照片	BLOB		
beaware	省	VARCHAR（30）		
city	市	VARCHAR（30）		
m_pay	月工资	FLOAT		
bank	银行账号	VARCHAR（20）		
pact_b	合同起始日期	DATE		
pact_e	合同结束日期	DATE		
pact_y	合同年限	FLOAT		

19.4 使用MySQL Workbench创建数据表

在MySQL Workbench工作空间的SQL Development工作模式下，可以创建一个新的数据库连接，编辑并运行SQL语句；跟其他数据库管理软件一样，可以在图形化界面中管理数据库表的基本信息。本节中将介绍通过MySQL Workbench 在图形化界面中创建并管理数据库信息的方法。

19.4.1 创建数据库连接

在MySQL Workbench工作空间下，对数据库数据进行管理之前，需要创建数据库连接，具体操作步骤如下：

步骤 01 在MySQL Workbench的欢迎界面中，单击【New Connection】按钮⊕，如图19.9所示。

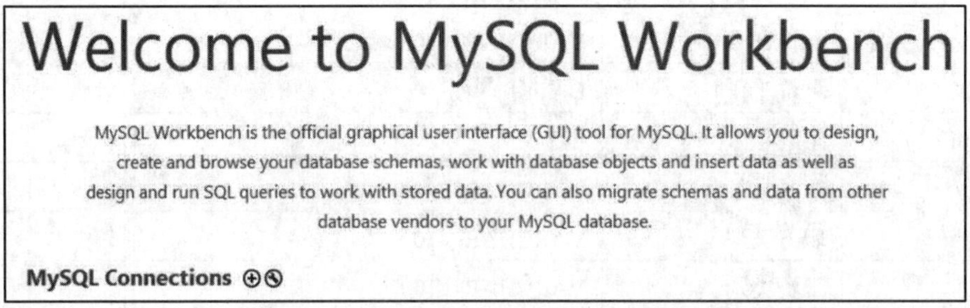

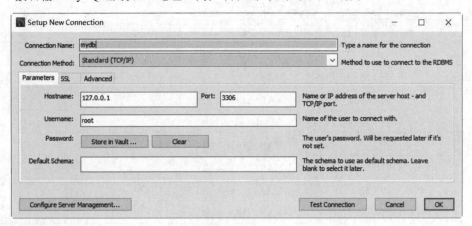

图 19.9　MySQL Workbench 的欢迎界面

步骤 02 打开【Setup New Connection】对话框，在【Connection Name】中输入数据库连接的名称，接着输入MySQL服务器IP地址、用户名和密码，如图19.10所示。

图 19.10　【Setup New Connection】对话框

步骤 03 单击【OK】按钮，连接MySQL数据后的界面如图19.11所示。Query1窗口用来执行SQL语句。

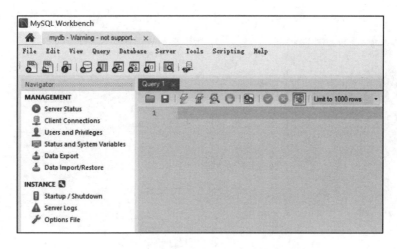

图 19.11　MySQL Workbench 连接数据库

19.4.2　创建新的数据库

成功创建数据库连接后,在左侧的SCHEMAS下面可以看到test数据库。用户可以创建新的数据库,具体操作步骤如下:

步骤01 单击工具栏上面的创建数据库的小图标,如图19.12所示。

图 19.12　单击创建数据库的小图标

步骤02 展现如图19.13所示的界面,此时需要输入新的数据库名字,在这里输入的是"crm",然后单击【Apply】按钮。

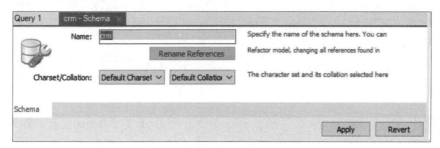

图 19.13　输入新的数据库名

步骤03 弹出一个新的界面,在该界面中看到创建数据库的语句,单击【Apply】按钮,如图19.14所示。

步骤04 弹出一个新的界面,在该界面中单击【Finish】按钮,完成创建数据库的操作,如图19.15所示。

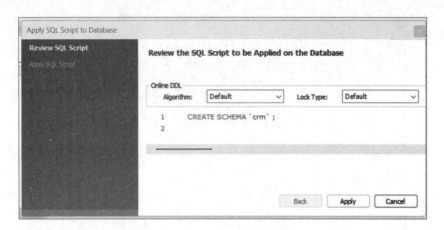

图 19.14 创建新的数据库的语句

图 19.15 完成数据库的创建

19.4.3 创建数据表

成功创建crm数据库后，即可创建和编辑数据表，具体操作步骤如下：

步骤 01 在左侧的SCHEMAS列表中展开crm节点，选择【Tables】选项，右击并在弹出的快捷菜单中选择【Create Table】菜单命令，如图19.16所示。

步骤 02 在弹出的窗口中可以添加表的信息，比如表的名称和表中各列的相关信息，如图19.17所示。

> **提示** 在创建表的时候，默认的数据存储引擎是InnoDB。Comments表示描述信息，要增加该表各字段信息，可以直接单击【Column Name】下面的列表，输入表的名称和数据类型。例如，需要创建主键id，直接勾选后面的PK(Primary Key)和AI(Auto Increment)复选框即可。

步骤 03 设置完数据表的基本信息后，单击【Apply】按钮。弹出一个确认数据表的对话框，在该对话框上有自动生成的SQL语句，如图19.18所示。

第 19 章 企业人事管理系统数据库设计 | 431

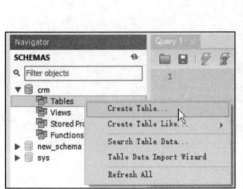

图 19.16 选择【Create Table】菜单命令

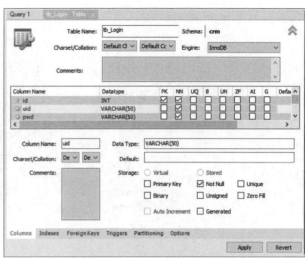

图 19.17 products-Table 窗口

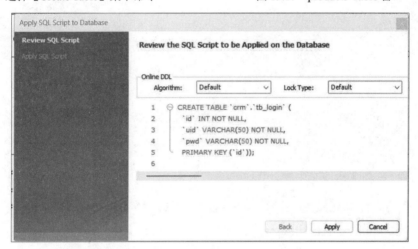

图 19.18 确认数据表的对话框

步骤 04 确认无误后单击【Apply】按钮，然后在弹出的对话框中单击【Finish】按钮，即可完成创建数据表的操作，如图19.19所示。

图 19.19 完成数据表的创建

步骤 05 表tb_login创建完之后，会在Tables节点下面展现出来，如图19.20所示。

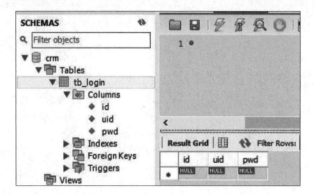

图 19.20　创建的新数据表

其他数据表的创建方法和上述操作类似，这里不再重复讲述。

第 20 章
在线购物系统数据库设计

> MySQL在互联网行业也被广泛应用。互联网的发展让各个产业突破了传统的发展领域，产业功能不断进化，实现同一内容的多领域共生，前所未有地扩大了传统产业链。目前，整个文化创意产业掀起跨界融合浪潮，不断释放出全新生产力，激发产业活力。当今无论是企业还是个人，都可以很方便地开发在线购物系统用于商品交易。本章通过在线购物系统数据库设计，进一步学习MySQL在互联网开发中的应用技能。

20.1 系统分析

本案例介绍一个在线购物系统，包含前台的分级搜索商品功能。客户可以浏览商品，进入前台界面购买商品，系统管理人员可以进入后台管理界面进行管理操作。

20.1.1 系统总体设计

在线购物系统案例在移动互联时代非常常见，是应用广泛的一个项目类型，我们日常生活中常用的京东网、当当网、天猫网都是在线购物网站。本示例系统从买家的角度去实现相关的管理功能，在线购物系统设计功能图如图20.1所示。

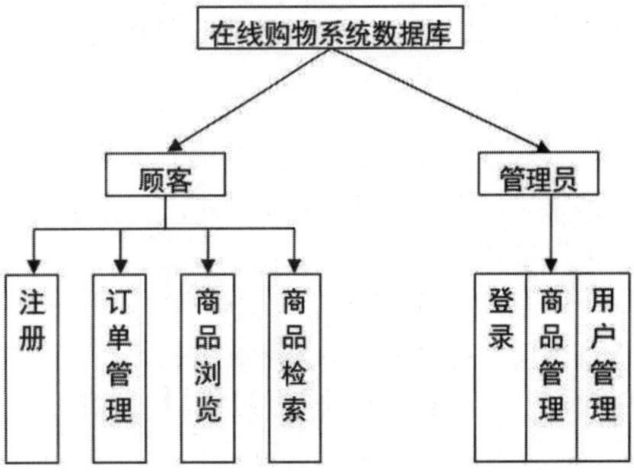

图 20.1 在线购物系统结构图

20.1.2 系统界面设计

对于在线购物系统的界面设计，考虑以下几点：

（1）使用单色调。使用单色调设计可以确保界面视觉上的统一性和一致性，避免视觉上的干扰。

（2）考虑用户使用习惯。

（3）不能对系统的使用产生影响。

基于以上考虑，在线购物系统的界面设计如图20.2所示。

图 20.2　系统界面设计

20.2　系统主要功能

在线购物系统的主要功能应包括商品管理、用户管理、商品检索、订单管理、购物车管理等，具体描述如下：

（1）商品管理：商品分类的管理，包括商品种类的添加、删除、类别名称的更改等功能；商品信息的管理，包括商品的添加、删除、商品信息（包括优惠商品、最新热销商品等信息）的变更等功能。

（2）用户管理：用户注册，用户注册为会员后，就可以使用在线购物的功能；用户信息管理，用户可以更改个人私有信息，如密码等。

（3）商品查询：商品速查，根据查询条件快速查询用户所需商品；商品分类浏览，按照商品的类别列出商品目录。

（4）订单管理：包括管理订单信息、浏览订单结算、订单维护。

（5）购物车管理：包括购物车中商品的增删、采购数量的改变，生成采购订单。

（6）后台管理：包括商品分类管理、商品基本信息管理、订单处理、会员信息管理。

20.3 数据库与数据表设计

在线购物系统是购物信息系统，数据库是其基础组成部分。系统的数据库根据系统的基本功能需求制定。

20.3.1 数据库实体 E-R 图

在系统开发过程中，数据库占据重要的地位，数据库的设置依据需求分析而定，通过前面的系统分析及系统功能的确定，规划出系统中使用的数据库实体对象，它们的E-R图，如图20.3～图20.8所示。

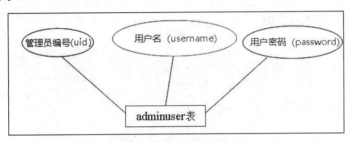

图 20.3 管理员实体 E-R 图

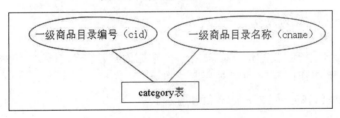

图 20.4 商品类别实体 E-R 图

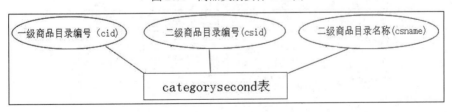

图 20.5 二级商品分类实体 E-R 图

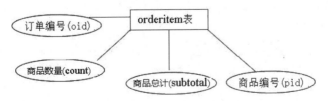

图 20.6 订单实体 E-R 图

图 20.7 商品明细实体 E-R 图

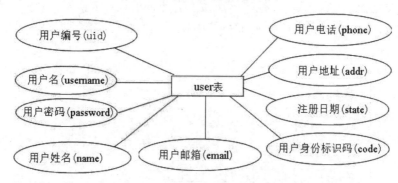

图 20.8 用户实体 E-R 图

20.3.2 数据库和数据表分析

根据本在线购物系统的实际情况,采用一个数据库,数据库的名称定为shopping。整个数据库包含了系统几大模块所需要的所有数据信息。

数据库shopping总共分6张表,如表20.1所示。

表 20.1 数据库中包含的数据表

表 名 称	说 明
adminuser	管理员表
category	商品类别表
categorysecond	二级分类表
orderitem	订单表
product	商品表
user	用户表

1)管理员表

管理员表用于存储后台管理用户信息,表名为adminuser,结构如表20.2所示。

表 20.2　adminuser 表

字段名称	字段类型	说　　明	备　　注
uid	INT	管理员编号	NOT NULL
username	VARCHAR（255）	用户名	NULL
password	VARCHAR（255）	用户密码	NULL

2）一级商品分类表

一级商品分类表用于存储商品大类信息，表名为category，结构如表20.3所示。

表 20.3　category 表

字段名称	字段类型	说　　明	备　　注
cid	INT	一级商品目录编号	NOT NULL
cname	VARCHAR（255）	一级商品目录名称	NULL

3）二级商品分类表

二级商品分类表用来存储商品大类下的小类信息，表名为categorysecond，结构如表20.4所示。

表 20.4　categorysecond 表

字段名称	字段类型	说　　明	备　　注
csid	INT	二级商品目录编号	NOT NULL
csname	VARCHAR（255）	二级商品目录名称	NULL
cid	INT	一级商品目录编号	NULL

4）订单表

订单表用来存储用户的订单信息，表名为orderitem，结构如表20.5所示。

表 20.5　orderitem 表

字段名称	字段类型	说　　明	备　　注
oid	INT	订单编号	NOT NULL
count	INT	商品数量	NULL
subtotal	INT	商品总计	NULL
pid	INT	商品编号	NULL

5）商品明细表

商品明细表用于存储出售的商品信息，表名为product，结构如表20.6所示。

表 20.6　product 表

字段名称	字段类型	说　　明	备　　注
pid	INT	商品编号	NOT NULL
pname	VARCHAR（255）	商品名称	NULL
market_price	FLOAT	商品单价	NULL

(续表)

字段名称	字段类型	说明	备注
shop_price	FLOAT	商品售价	NULL
image	VARCHAR（255）	订单编号	NULL
pdesc	VARCHAR（255）	商品描述	NULL
is_hot	INT	是否热卖商品	NULL
pdate	DATE	商品生产日期	NULL
csid	INT	二级商品分类目录	NULL

6）用户表

用户表存储买家个人信息，表名为user，结构如表20.7所示。

表 20.7 user 表

字段名称	字段类型	说明	备注
uid	INT	用户编号	NOT NULL
username	VARCHAR（255）	用户名	NULL
password	VARCHAR（255）	用户密码	NULL
name	VARCHAR（255）	用户姓名	NULL
email	VARCHAR（255）	用户邮箱	NULL
phone	VARCHAR（255）	用户电话	NULL
addr	VARCHAR（255）	用户地址	NULL
state	DATE	注册日期	NULL
code	VARCHAR（64）	用户身份标识码	NULL

20.4　使用MySQL Workbench数据建模

MySQL Workbench为MySQL提供了E-R图设计和数据库物理建模，用户可以通过它建立E-R图和生成数据库对象。本节将介绍如何在MySQL Workbench中建立物理模型，然后导出SQL语句。

20.4.1　建立 E-R 模型

在线购物系统中包括用户、商品和订单的物理模型。用户和订单之间是1个用户对应1个订单，是1:1的关系；而订单和商品之间是一对多（1:n）的关系。具体建模的操作步骤如下：

步骤01　在MySQL Workbench主界面中，选择【File】选项下的【New Model】菜单命令，如图20.9所示。

步骤02　打开【Model Overview】窗口，在该窗口下创建新的Schema。单击+（新增）按钮，此时会产生一个新的Schema，然后输入Schema的名字"shopping"，如图20.10所示。输入完成后，单击▣（关闭）按钮，最后单击【Add Diagram】按钮。

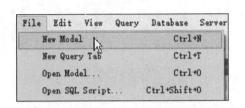

图 20.9　选择【New Model】菜单命令

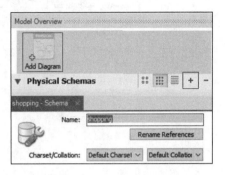

图 20.10　创建一个新的模型

步骤 03　打开Diagram窗口，单击▣（创建表）按钮，如图20.11所示。

步骤 04　在窗口的空白处单击，此时会创建一个table1，如图20.12所示。

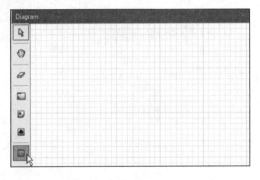

图 20.11　开始添加数据库物理模型

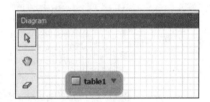

图 20.12　创建 table1

步骤 05　双击table1之后会出现编辑table1的窗口，修改Table Name为"user"，然后添加主键uid以及各个列的相关信息。添加列之后，可以选择数据类型和约束，如图20.13所示。

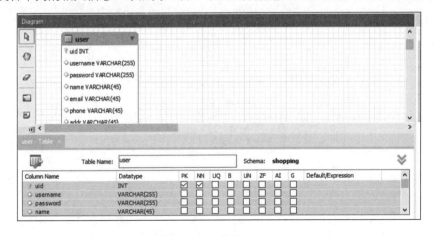

图 20.13　编辑 table1

步骤 06　使用相同的方法创建表orderitem和product，然后给表添加各个列的相关信息，如图20.14所示。

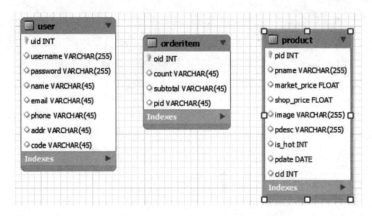

图 20.14 创建表 orderitem 和 product

步骤 07 单击【1:1】按钮，然后分别单击user表和orderitem表，使两张表建立1:1的关系，如图20.15所示。

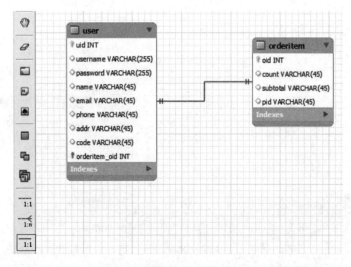

图 20.15 建立 1:1 关系

步骤 08 修改刚刚建立的一个外键的名称。双击表user，然后修改表中列的信息，如图20.16所示。

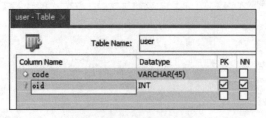

图 20.16 修改外键的名称

步骤 09 切换到【Foreign Key】选项，可以很清楚地看到user表的外键直接引用的是orderitem表，如图20.17所示。用户表和订单表是1:1的关系，已经设计完成。

步骤 10 单击【1:n】按钮，然后分别单击product表和orderitem表，此时会为两张表建立1:n的关系，如图20.18所示。

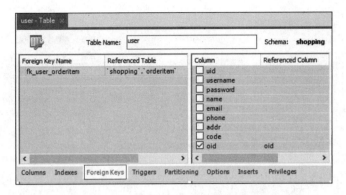

图 20.17　外键连接

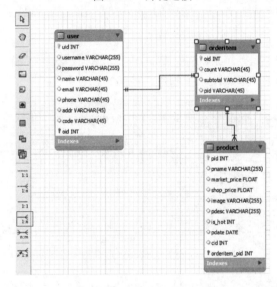

图 20.18　建立 1:n 的关系

步骤11 双击products表后选择【Foreign Key】选项，界面显示如图20.19所示，此时已经建立了一个fk_product_orderitem1的外键。

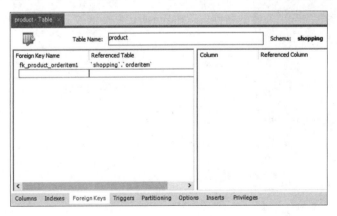

图 20.19　建立外键

参照上述操作步骤，创建数据表adminuser、category和categorysecond，如图20.20所示。

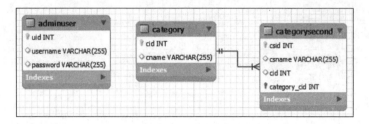

图 20.20　创建数据表 adminuser、category 和 categorysecond

20.4.2　导入 E-R 模型

上一小节中讲述了如何创建E-R模型，下面介绍如何将E-R模型直接转换成SQL脚本，具体操作步骤如下：

步骤 01 将创建的物理模型保存起来，选择【File】菜单下的【Export】菜单命令，之后选择【Forward Engineer SQL CREATE Script...】选项，如图20.21所示。

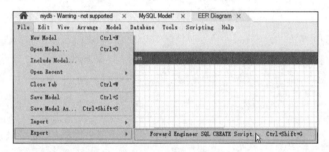

图 20.21　选择【Forward Engineer SQL CREATE Script...】选项

步骤 02 在打开的对话框中输入SQL脚本的导出的地址，如图20.22所示。然后单击【Next】按钮，即可依次按照提示导出脚本。

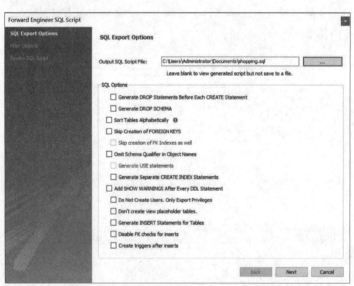

图 20.22　导出路径

步骤 03 在SQL Development工作模式下可以导入脚本，单击 （导入文件）按钮，如图20.23所示。然后根据提示打开上一步保存的SQL脚本文件即可。

步骤 04 单击 （执行）按钮，即可开始自动执行脚本，如图20.24所示。

图 20.23　导入 SQL 脚本

步骤 05 执行脚本后，单击左侧窗口的 （刷新）按钮，此时会出现shopping数据库，如图20.25所示。

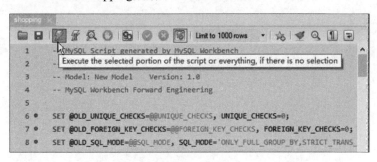

图 20.24　执行 SQL 脚本

图 20.25　执行 SQL 脚本的结果

至此，在线购物系统的数据库设计完成。